The Oregon Water Handbook

The Oregon Water Handbook

A Guide to Water and Water Management

◙ *Rick Bastasch*

Oregon State University Press
Corvallis

The paper in this book meets the guidelines for permanence and durability of the Committee on Production Guidelines for Book Longevity of the Council on Library Resources and the minimum requirements of the American National Standard for Permanence of Paper for Printed Library Materials Z39.48-1984.

Library of Congress Cataloging-in-Publication Data
Bastasch, Rick.
  The Oregon water handbook : a guide to water and water management / Rick Bastasch.
      p. cm.
  Rev. ed. of: Waters of Oregon. 1998.
  Includes bibliographical references and index.
  ISBN-13: 978-0-87071-181-7 (alk. paper)
  ISBN-10: 0-87071-181-4 (alk. paper)
  1. Water-supply—Oregon. 2. Water-supply—Oregon—Management.
I. Title.
  TD224.O7B37 2006
  363.6'109795--dc22

                                                    2006021298

Oregon State University Press
500 Kerr Administration
Corvallis OR 97331-2122
541-737-3166 • fax 541-737-3170
http://oregonstate.edu/dept/press

⊡ Contents

Preface

TO MAKE A LONG STORY SHORT, Oregon's out of easy water. The state has undergone about 150 years of development, all of it in some way water-dependent. We've taken full advantage, and then some, of the remarkable richness of our water resources. Oregon has developed a complex system to manage those resources over the last 100 years. The system began as a progressive effort to bring order to chaos through state management of a publicly owned resource. It succeeded by creating certainty and promoting development. But its successes have not kept pace with a changing Oregon. Now there is simply not enough water to get everyone what they want when they want it.

This is the second edition of a book that was originally titled *The Waters of Oregon*. In the ten years since it was published, Oregon's water world has continued to evolve. There is now more attention being paid to water supply issues generally, and the State has eliminated a water use permit backlog, is focusing more on water transfers, and is exploring how to mitigate impacts of past water use through water banking experiments.

But the biggest issue remains: water is an increasingly scarce resource as we enter the twenty-first century—and we're largely unprepared to deal with this reality. Global climate change will intensify scarcity already resulting from population growth, increased economic demands, and the needs of threatened species. In the past ten years, we have seen the eruption of environmental (I would argue, water) crises in the Columbia basin and the Klamath basin—and more could easily develop. As other sources are fully tapped, the Willamette River (once assumed to be too dirty to drink) is fast becoming one of Oregon's largest sources of drinking water. Oregonians may have effectively scrapped the state's land-use planning program, and a number of groundwater protections along with it, even as new groundwater limited areas have been added. It is hard to imagine a topic more vital to the future of Oregon than its water supply, but the State still has no plan (or if that's still a dirty word, call it a program, strategy, approach, or initiative) to deal with the problems we know of, let alone the contingencies that doubtless await us just a few years out.

So, this book is about water supply, not water quality or water pollution (at least not directly) because it does not need to be. In keeping with their environmental image, most Oregonians have at least some notion of water quality issues, perhaps fed by memories of Tom McCall's fabled Willamette River cleanup or by recent reminders not to dump used motor oil down storm drains. They know there are laws and somebody to call if they see water pollution.

Yet few Oregonians know it's illegal to pump any water out of a stream without a water right and perfectly legal to pump the last water out of a stream with one. There has been little reason for citizens to know where their water comes from or how it gets there. But the reasons have arrived. Urban growth, expanding agricultural markets, increasing recreational demand, and plummeting fish populations are all competing for a limited amount of water. It is time for citizens to understand and act upon the knowledge that together they own all the water in Oregon. In short, it's time for an owners' manual—a need this book attempts to meet.

Oregon's water supply and management system is a cranky contraption: big, complicated, contradictory, ever-changing, and increasingly unpredictable. It's been tinkered with for nearly a century, as each generation brings the machinery to bear on different objectives. Now rambling, cross-wired, and unwieldy, it is understood only by a few insiders. The public knows next to nothing about how its waters are managed. When the people are effectively cut off from the management of their water, decisions are left to the insiders and experts. Despite the best of their intentions, over time that isolation can lead to questionable, and sometimes surprising, results.

This book is intended to give interested citizens an overview of how water is managed in Oregon; to describe how pressing water issues have come about and what they mean; to call out a few interesting water facts; and generally be a resource to allow more, and more informed, participation in Oregon's water management decisions. More issues are raised than resolved. But resolutions can only come from an informed and respectful dialog among Oregonians. The more informed the debate, the better. If this book is a starting point for that kind of discussion, it has done its job.

As a water resources owners' manual for Oregonians, the book is primarily intended as a readable reference, organized to answer a sequence of hopefully logical questions:

- How much water is there?
- What is the process for sharing it?
- How do people use it once they've got it?
- How are land and water systems managed to keep water benefits coming?

As a reference work, the book includes frequent bibliographic citations. The references (especially those to Oregon Revised Statutes and Oregon Administrative Rules) are intended to allow the reader either to learn more about a topic and/or take issue with my assertions.

A word about maps: many map bases were commissioned from the Oregon Water Resources Department, but the final map products are the work of the author. Any errors, data misinterpretations, or cartographic offenses (beyond those, if any, lurking undiscovered in the base maps) are mine alone. This is also true of all the diagrams and charts, which were similarly author-generated.

Readers should employ a bit of caution vis-à-vis this book's purpose and content. First, it is intended as a thorough, but general, reference—not a source of legal advice. Should readers become seriously entangled with Oregon's water management system, they should hire the water professional of their choice: namely, attorneys, engineers, hydrologists, etc.

Second, those familiar with Oregon's system of government may detect that I have employed an administrative short-cut throughout the book. Oregon's natural resource agencies generally have an unpaid citizen board or commission overseeing the activities of a paid, professional department staff. Though statutory authority is usually vested in the commissions, in practice most of it is delegated to staff. Accordingly and for simplicity, I most often attribute actions or authorities to departments, whereas the statutes may explicitly reference commissions.

Third, laws and rules change. Although the system's foundation has remained pretty much unchanged through the decades, there is continual re-modeling of the superstructure. As the book dates, readers are advised to consult the most recent edition of the Oregon Revised Statutes (issued by Legislative Counsel) and Administrative Rules (maintained by the Secretary of State).

Lastly, I am grateful to a number of people who have contributed inspiration, information, and ideas during the writing of this book, both first and second editions. I remain indebted to the many people who helped me with the first edition, including: Danielle Clair for her invaluable assistance, from pre-draft expressions of confidence to her thoughtful review of early drafts; the staff of the Water Resources Department, generally—who labor mightily administering that antique and unwieldy collection of impossibilities known as water law—and in particular GIS specialists Jerry Sauter and Michael Ciscell for their work on various base maps; and Ray Sterner for permission to use his striking relief map of Oregon. I also benefited from the help of many in the writing of the second edition, including: Dwight French, Kathy Boles, and Juno Pandian of the Oregon Water Resources Department; Bernie Kepshire and Ray Hartelrode of the Oregon Department of Fish and Wildlife; Malavika Bishop of the Oregon Department of Environmental Quality; and Paul Cymbala of Oregon's Drinking Water Program.

I am also very grateful for the skilled editing and generous suggestions provided by the staff of Oregon State University Press, especially Jo Alexander, Mary Braun, and Tom Booth. Lastly, I thank my wife, Yvonne Bashor, for her patient support and encouragement (as well as the loan of her laptop) as I wrote the book, and to my son, Trenton, for many things, including his unsinkable enthusiasm for water in all its forms.

PART I

A Water Inventory

The clouds combed overhead and broke against the mountains
like waves breaking, and the water ran back toward the sea …
gullies bled into bigger gullies, bigger gullies into freshets dry
all summer, freshets into ditches choked full of Canada thistle
and buffalo weed, and these ran into Elk Creek and Lorain
Creek and Wildman Creek and Tyee Creek and Tenmile
Creek; sharp, steep, noisy creeks, looking like sawteeth on the
map. And these creeks crashed into the Nehalem and the Siletz
and the Alsea and the Smith and the Longtom and the Siuslaw
and the Umpqua and the Wakonda Auga, and these rivers
ran to the sea, brown and flat with the clots of swirling yellow
foam clinging to their surfaces, running to the sea like lathered
animals.

Ken Kesey
Sometimes a Great Notion

On my own place and the government land, where I pay for
pasturing, amounting to maybe fifty thousand acres, there is
not one stream, no lakes, no water on the surface at all. … We
measure humidity by the amount of sand in the air. When it
rains, we keep our hired man in—we want all the water on the
land.

R. A. Long
Water on the Desert in *The Oregon Desert*

CHAPTER 1

Clouds, Rivers, Rocks

OREGON REVERBERATES WITH RAIN AND RIVERS. The very name graced for a while the legendary River of the West (now called the Columbia). But how watery a place is it? Looking at rainfall or run-off totals, one has to conclude Oregon is a very watery place. Hidden within those statistics, however, are several Oregons, none of them a stranger to dusty fields or dry stream beds. This chapter explores where the water comes from and where, when, and why it goes—which turns out to be a matter of climate, streams, and lakes, and groundwater, or more simply: clouds, rivers, and rocks.

Climate

▣ Precipitation

In many ways, Oregon is built to catch precipitation. It is next to the earth's largest ocean, lies on the storm track a good part of the year and raises ramparts of mountains to intercept wet air masses. The precipitation that gives rise to Oregon's water supply usually has its beginnings in Pacific frontal systems. Storms are born when warm, subtropical air flowing north over the Pacific meets cold air flowing south from the Arctic. In winter, these "cyclonic" storms are swept toward Oregon by the jet stream. As they wheel over the state, the storms distribute their rain and snow as instructed by the mountains (Paulson et al. 1991).

When ocean air ramps up the mountains, it can cool as much as 5° for every 1,000 feet of elevation. So a trip over the Cascades can chill a Pacific air mass 25° or more. The water vapor condenses and falls out as rain or snow. Mountain-caused rain or snow is called orographic precipitation. Some of the heaviest annual precipitation in the continental United States occurs on the western slopes of the Coast Range and Cascades. Nearly two-thirds of the annual total falls from October through March (Pacific Northwest

River Basins Commission 1969). Generally, precipitation decreases from west to east and from north to south. Ultimately, Oregon's climate is most heavily influenced by cyclonic precipitation, with orographic precipitation responsible for the pattern of distribution across the state. In other words, without mountains, the state would still get a lot of rain (and it would still be heaviest next to the Pacific), but would lack the spice of today's contrasting alpine and desert environments (Dart and Johnson 1981).

The Cascade Mountains divide the state into two distinct zones of precipitation (Figure 1). West of the Cascades, average annual precipitation ranges from 40 to 140 inches. In eastern Oregon, annual precipitation drops to 10 to 20 inches (U.S. Army Corps of Engineers 1996a). The Cascade Range throws a rain-shadow across most of eastern Oregon. Air masses release much of their water as they pass up and over the Cascades. As they descend, they also warm back up, but at an increased rate thanks to their dryer condition (Jackson 2003). The result: sagebrush and Ponderosa pine instead of salal and Douglas fir. When the air masses encounter the Ochoco, Aldrich, Strawberry, Elkhorn, or Wallowa ranges of the Blue Mountains, they rise and cool again, increasing annual rainfall to 40 or 50 inches.

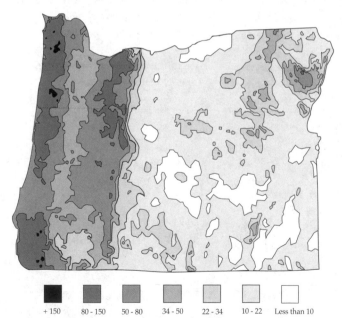

| + 150 | 80 - 150 | 50 - 80 | 34 - 50 | 22 - 34 | 10 - 22 | Less than 10 |

Figure 1: Mean annual precipitation. Based on records 1961 - 1990 (inches per year). (Source: Information from Oregon Climate Service)

Oregon has some of the wettest and driest locations in the United States. The Coast Range can receive over 10 feet of rain annually, while eastern Oregon scrapes by with eight or fewer inches. According to the State Climatologist, Oregon's wettest spot is somewhere in the higher elevations of the Coast Range, but because the measuring network is an imperfect weave, the exact location cannot be pinned down. However, we do know that northeastern Tillamook County, northeastern Lincoln County, and northeastern Curry County each receives approximately 200 inches per year. At the other end of the scale, the state's driest location is probably the Alvord desert, in the shadow of southeast Oregon's Steens Mountain, with an estimated annual average total of only 5 inches.

Even the driest places, however, have at least 50 days a year with some trace of precipitation, and wetter western Oregon may have 120 days of rain each year (Loy 2001). Western Oregon receives over three times as much precipitation as eastern Oregon. But whenever it falls, Oregon's precipitation is more often a consistent and determined drizzle than a driving downpour. Hourly intensities as great as an inch per hour are rare (Pacific Northwest River Basins Commission 1969). Overall, when Oregon's precipitation spikes and dips are averaged out, the state has an annual average of 28 inches—about the same as Texas, according to the World Book Encyclopedia (Figure 2).

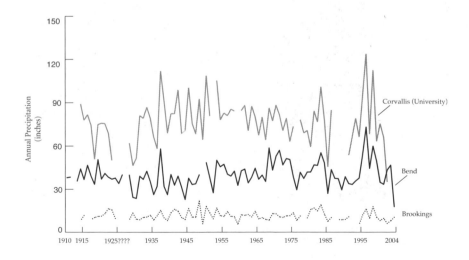

Figure 2: Oregon precipitation by region 1910 – 2004.
(Source: Data from Oregon Climate Service)

▣ *Snowpack*

Oregon's snowpack plays a crucial role in controlling the timing and amount of streamflows in many parts of the state. As snow falls in the Cascades and Blue Mountains, much of it accumulates through the winter. The snowpack is more than just deep snow. It is a crystalline reservoir. It has been melted, re-frozen, rained on, snowed over, and become a multi-layered water-storing wonder. Year after year of accumulating snows have formed glaciers on Mount Hood, Mount Jefferson, and the Three Sisters. While they cover a minuscule area of the state—less than five square miles—their local contributions to streamflow can be significant. Each square foot of glacier can generate about 13 cubic feet of average annual runoff. However, the runoff that really counts comes from the melting of the state's extensive snowpacks in late spring and early summer (Pacific Northwest River Basins Commission 1969).

Water in the form of snow and ice occupies more volume than its liquid form. The amount of water that results from melting snowpacks is measured by water content. Water content varies through the season, from year to year, and from place to place. Some snowpacks are drier than others. Skiers are familiar with water content, often bemoaning the "Cascade concrete"

Oregon Cascade's Mountains are one of the snowiest areas in the U.S. For much of Oregon, winter snows make for summer flows. (Oregon State Archives, Oregon Highway Division, OHD6718)

During the winter and spring months, government agencies closely track the depth and water content of mountain snows to help predict spring flooding and summer water supplies. (Photo by Ron Nichols; photo # NRCSOR00001; year 2000: "Snow survey near Mt. Hood, Oregon by NRCS employees Jon Werner and Sheila Strachen.")

that they would eagerly trade for Colorado powder. Similarly, slush-balls are one of the deadlier forms of snowballs by virtue of their water content. Heavy snow may make for restrained recreation, but it is great for water supply. Each winter, government agencies sample snowpack water content on established snow survey courses.

On the western slope of the Cascades, water content may reach densities of over 50 percent. That is, for every foot of snowpack, six inches of runoff is generated. In the Blue Mountains, however, the water content drops to around 10 percent. As in much of the western United States, snowmelt may account for as much as 80 percent of annual flow in many parts of eastern Oregon. In rainier western Oregon, at most 40 percent of the annual flow originates from melting snowpacks (Fox 1996). The Cascades are one of the snowiest places in the United States. Crater Lake has recorded 900 inches of total snowfall some years, and Timberline Lodge up to 700 inches. Snow depths in the Cascades may exceed 20 feet at the end of the season. Generally, snowpack water content grows from 25 percent in early winter to 40 percent and above by April (Pacific Northwest River Basins Commission June 1969; Loy 2001).

The Office of the State Engineer (now the Water Resources Department) started Oregon's snow survey program in 1928 by directing watermasters to measure snow in the Umatilla, Walla Walla, John Day, and Deschutes river basins. Measurements began the next year for the Malheur, Powder, Grande Ronde, Rogue, and Goose Lake basins (Water Resources Committee 1955). Today, most snow measurement is performed by the federal Natural Resources Conservation Service.

▣ *Climate Change*

Climate is always changing. About 45 million years ago, Oregon had a climate similar to that of Costa Rica today, with subtropical, vine-knotted forests of palms populated by crocodiles, three-toed horses, and cat-related predators. About 18,000 years ago, a 170-mile-long ice-cap crowned the Cascades, from Mt. Hood to Mt. McLaughlin, with ice that may have been a half-mile thick. (Bishop 2003) The climate change we hear about today, however, is not occurring in geologic time. Rather, it appears to be a sudden veering of the climate we knew to a climate we don't know—a climate that promises to have serious implications in our lifetimes, for water supply, as for many other aspects of life.

Scientists agree that the Pacific Northwest is warming. The annual average temperature in most of the region has increased by one to three degrees Fahrenheit over the last century. Since 1920, nearly every temperature recording station in the Pacific Northwest has shown a warming trend. Temperature rose steeply from about 1900 to 1940, dipped through 1975, and has been rising thereafter (Institute for Natural Resources 2004). By the 2040s, the average annual temperature is expected to increase by 4.1°F (Climate Impacts Group 2004). In addition, between 1950 and 1995, the snow-water equivalent of the snowpack in the Cascades decreased by about 50 percent and peaked earlier in the year, increasing March, and reducing June, streamflows (Institute for Natural Resources 2004).

These trends are expected to continue. By the mid-twenty-first century, in all likelihood, the Northwest will have drier summers and wetter winters that have more rain than snow. Scientists feel very confident with the estimates of temperature change, while the magnitude of the precipitation is less certain (Palmer and Hahn 2002).

The hydrologic impacts of such change are widespread and include greater risks of floods in winter and decreased late spring and summer streamflows (Climate Impacts Group 2004). In Oregon, climate change might increase storm severity, melt snowpacks faster, increase flooding, and decrease summer flows, making less water available for fish habitat, irrigation, drinking water, recreation, and pollution abatement (Oregon Department of Energy n.d.)

A 2004 report examining the economic impacts of climate change on Oregon identified eight key sectors of Oregon's economy that would likely encounter serious challenges: municipal water supplies, agriculture, forestry, snow-based recreation, coastal tourism and infrastructure, power generation, salmon recovery, and public health (Institute for a Sustainable Environment 2005).

Impacts on municipal water supplies are expected to be serious. To better understand the potential impacts of climate change, the City of Portland commissioned a study on its water supply. The resulting 2002 report indicated that by 2040, on average, the Bull Run system's winter flows will increase by about 15 percent and late spring flows decrease by about 30 percent. Because of climate change, Portland expects it will need to store roughly an additional 2.8 billion gallons per year (Palmer and Hahn 2002).

Similarly, Clean Water Services (a public utility providing municipal water and wastewater services in the Tualatin River watershed) also studied the impacts of climate change. In the Tualatin, annual precipitation will likely increase due to wetter winters, but summers will become drier. By 2040, summer flows will decrease by as much as 20 percent. The water yield of the water supply system is expected to fall by roughly 1.5 percent per year over the next 40 years. Refilling the system's reservoirs annually will become less easy, with multi-year drawdowns more common. Climate change will require the utility to expand its system five to eight years sooner than otherwise planned. Overall, climate change will add to the difficulty of achieving water quality standards and maintaining needed streamflows for fish in the Tualatin (Palmer et al. 2004).

Climate change would add another stressor to already-afflicted salmon runs. Increasing winter floods, reduced summer and fall streamflows, and rising summer temperatures in streams and estuaries will very likely hit salmon populations hard (Climate Impacts Group 2004). In the Columbia system, changes in runoff patterns will likely hamper the ability of the reservoir system to meet instream flow requirements for fish and to generate power (Climate Impacts Group n.d.).

Climate change may create improved growing conditions in the Pacific Northwest for many crops—if they get enough water. In some areas, however, this may become increasingly difficult as summer streamflows drop and reservoirs struggle to fill, even as the water demand of crops soars (Climate Impacts Group 2004). Drought frequency and severity will likely increase. The droughts we currently see as harsh, but episodic, acts of God may ultimately become known under a new word: "summer" (Institute for a Sustainable Environment 2005).

Rivers and Basins

▣ *Oregon's rivers*

The rivers of Oregon carry about 66 million acre-feet annually in four out of five years. An acre-foot is the amount of water needed to cover one acre one foot deep. Since Oregon is about 62 million acres in area, that runoff is roughly enough to put the entire state under one foot of water each year. About 55 million acre-feet comes from west of the Cascades and the other 11 million from eastern Oregon (State Water Resources Board 1969a). Ultimately, all of that water can be traced back to precipitation— whether recent (e.g., last week's rain) or more distant in time (e.g., last century's snowmelt absorbed and stored as groundwater for later release). However, not all precipitation ends up in rivers. In the Northwest, of the 28 inches of average annual precipitation, about 12 return directly to the atmosphere through evaporation and plant use (Pacific Northwest River Basins Commission 1969).

The water that is delivered to Oregon's surface water bodies occupies very little area—around 700 square miles, or less than 0.5 percent of the state (Todd 1970). However, Oregon has well over 100,000 miles of stream channel—roughly one linear mile for every square mile of land (Oregon Water Resources Department, n.d.[a]). Clearly, Oregon is stream-rich. There are over 12,000 named streams in the state, which is almost twice as many as Washington (with nearly 6,300) and significantly more than California (with about 9,800). Oregon also has 226 named waterfalls, the highest being Multnomah Falls at 620 feet. In addition, the state has more than 4,500 named springs, of which over 120 are geothermal hot springs (water temperatures greater than 15° F above the mean annual air temperature).

With 12,000 streams to keep track of, the Oregon Water Resources Department and the federal Environmental Protection Agency both have developed stream-coding systems to assure each stream has a unique

Table 1. Name origins of Oregon's river basins

Willamette: Derived from a Native American word, "Wal-lamt," for a place on the west bank of what is now the Willamette River near Oregon City.

Sandy: In 1805, Lewis and Clark observed the Sandy River throwing out "emence quantitys of sand" and compared it to the famously turbid Platte. They called the stream the Quicksand River, which by 1850 had been shortened to the present form.

Hood: Originally named Labeasche River by Lewis and Clark after one of their party, settlers called it Dog River after having been forced by starvation to partake of a canine supper. Later, a well-known pioneer resident who found the name objectionable, promoted Hood River, after the river's source, Mt. Hood.

Deschutes: On the way down river in 1805, Lewis and Clark referred to this stream by its native name "Towornehiooks." On their way home, they called it Clarks River. However, in the world of the fur trappers, it became known as "Riviere des Chutes," or River of the Falls — not because of any falls or rapids on the river itself, but because its mouth was just above the famous falls of the Columbia River.

John Day: Although Lewis and Clark named it Lepages River, a later incident at its mouth proved more name-worthy. John Day and a companion became separated from their trapping party during the winter of 1811-12. At the mouth of what is now the John Day River, the two were relieved by the natives of all their possessions, including their clothes. The naked men were ultimately rescued by other trappers.

Umatilla: The native name for this river was recorded by Lewis and Clark as "Youmalolam." Other explorers rendered it "You-matella," "Umatallow," "Utalla," "Ewmitilly," and "Umatilah." There is apparently no reliable interpretation of the name.

Grande Ronde: Named after the Grande Ronde valley, which impressed fur trappers with its fine surrounds and who described it in their native French.

Powder: Reputed to have its origins in Chinook jargon as "polallie illahe" — a sandy or powdery stream bank.

Malheur River and Lake: French for "bad hour," or misfortune, the name was applied by trappers who learned a cache of furs they had hidden near the river had been found and stolen. The name was in use as early as 1826.

Owyhee: Prior to 1800, this was the word used for Hawaii. The river was named after two Hawaiians who were killed by members of the Snake tribe in 1819.

Goose/Summer Lake: First named Pit Lake, under common usage of settlers it became Goose Lake , probably for the wild geese that frequented it. In 1843, from a chilly height Captain John Fremont observed the sun shining on a grass-rimmed lake far below, which he called Summer Lake, in contrast to his perch on Winter Ridge.

Klamath: The origins of the name are uncertain, some claiming a French root, others tribal. The trapper/explorer Peter Skene Ogden referred to the "Claminitt Country" in 1826. The name was used by local natives to describe the large lake they lived beside, and settlers used the term to describe the local natives.

Rogue: French trappers referred to this stream as "Riviere aux Coquins," or River of Rogues, having found their encounters with local natives difficult. Contrary to some stories, the name was not derived from the French "rouge" and does not mean "Red River."

Umpqua: "Umpqua" was the native name for the country around the river and came into usage as a word for both the River and local tribes.

Source: Information from McArthur 1992

number. Not only does this numeric system work better for computerized databases, it also resolves the problem of duplicated stream names. As anyone who has read the state's fishing regulations or, for that matter, done much driving in Oregon knows, there are lots of streams with the same names. For example, there are two John Day rivers (the lesser known near Astoria) and two Rogue rivers (the lesser known in Polk County). It appears that rocks, bears, and springs were especially popular inspirations for stream-namers, as each of these names claims over 90 creeks across the state (U. S. Board of Geographic Names 1996). Tables 1 and 2 display information relating to Oregon's stream names.

Oregon's 12,000 streams vein the state from top to bottom, collecting water from the land and returning it to the sea—or, in the case of Oregon's bake-oven/freeze-dried southeast, the sky. That journey may be very long, or quite short. Oregon's longest stretch of river is the Columbia, which borders Oregon for the final 309 miles of its 1,214-mile length. However, the longest river inside Oregon is the John Day, at 284 miles. Other two-

Table 2. Common Creek Names in Oregon	
Name	Number in state
Rock	93
Bear	91
Spring	90
Cedar	77
Beaver	62
Fall	62
Deer	60
Camp	51
Indian	48
Buck	45
Johnson	44
Elk	42

hundred-milers include the Deschutes (252 miles); the Snake (233 miles alongside Oregon); the Rogue (215 miles); and the Grande Ronde (209 miles). That is not a bad showing, considering the Water Encyclopedia shows the earth as having fewer than two hundred rivers more than 250 miles long. D River at Lincoln City is listed (along with Montana's Roe River) as the world's shortest river, with a length of about 120 feet (Guinness Book of World Records 1998) .

Length does not determine discharge, however. Oregon's largest-producing river, the Willamette, is only a modest 187 miles long, but yields over 24 million acre-feet per year (Moffatt, Wellman, and Gordon 1990), out-performing the John Day River by a factor of twenty. In fact, the Willamette is the thirteenth-largest river by volume in the continental United States (Kammerer 1990).

But the Columbia discharges far more water—a spectacular 198 million acre-feet annually. This flow is second largest in the United States (Bonneville Power Administration et al. 2001) and, according to the Water Encyclopedia, nineteenth in the world. However, Oregon represents only 22 percent of the Columbia's drainage area and contributes about 27 percent of its flow (Loy 2001). The bulk of the Columbia (44 percent measured at The Dalles) comes from Canada (Kitzhaber 1984). After the Columbia and Willamette, Oregon's next largest rivers are the Rogue and Umpqua, which both discharge about 7 million acre-feet per year, followed by the Deschutes at roughly 4 million.

To get a feeling for streamflow, it may be helpful to visualize different quantities. The rate of streamflow is normally expressed in cubic feet per second (cfs), the amount of water traveling at a foot per second through a full, one-foot-wide pipe—about 450 gallons per minute. So, a flow of one cfs would fill up a 5-gallon bucket in a little over half a second; a fair-sized refrigerator in 20 seconds; or a 2,000-square-foot ranch-style house (assuming 8-foot ceilings and strong windows) in about four and a half hours. If the pipe shot out water at one cfs for a year, it would cover 724 acres (a little over a square mile) with a foot of water.

▣ Measuring Streamflow

London, Rome, Waterloo, and Troy may conjure up images of great cities or great battles to most people, but to those who measure Oregon's streams, they are more like hometowns. That is because the roughly 350 measuring points (called gaging stations) maintained by the Oregon Water Resources

The short and long of Oregon's rivers: D River in Lincoln City is the state's (if no longer the world's) shortest river; the John Day is its longest. (top photo: author; lower photo: Oregon State Archives, Oregon Highway Division, OHD8378)

Department and the U.S. Geological Survey are named for nearby places— many times ghost-towns. Thus, we have the Coast Fork Willamette River measured at London; the Owyhee River measured near Rome; the South Santiam River measured at Waterloo; and the Grande Ronde River measured at Troy. No matter where they are, gaging stations usually are easily recognized as the outhouse-sized buildings that shelter streamside measuring instruments.

A gaging station sits on top of a shallow well next to a stream. The well is fed by pipes from the stream. Water in the well, rising and falling with the stream level, moves a float up or down. The float is connected to a recorder, which may be a pen that scribes over constantly moving graph paper (much like a seismograph), a device that punches holes in a paper tape, or a computer. This recorded height of the water column, called a "river stage" or "gage height," is used to determine flow. Generally, the higher the river gets, the more water is flowing by. State watermasters or other technicians refine this relationship by taking regular flow measurements and graphing them against the river level. Flow rates are calculated by crossing the stream and measuring water velocity and depth at between twenty and thirty points. Multiplying average depth and average velocity gives a flow.

It may be surprising how slow a fast river is. The highest river velocity ever recorded with a current meter in the United States by the U.S. Geological Survey was 22.4 feet per second—or about 15 miles per hour (Leopold 1994). In Oregon, the lower Willamette River's velocity has been estimated to range between a high of about 1 mile per hour to a low of 0.2 miles per hour (i.e., five hours to travel one mile). In contrast, the lower Grande Ronde River may reach velocities approaching 12 miles per hour. Even during floods, it may take about a full day for water to get from Grants

Gaging stations (left) are common sights along Oregon streams. Each contains equipment that measures the rise and fall of nearby water levels, such as the battery-powered tape-punching device on the right. (Left photo:Northwest River Forecast Center, Long Tom R. gage http://www.nwrfc.noaa. gov/photos/mnro3_g.jpg; right photo: Oregon Water Resources Department)

Pass to the mouth of the Rogue River, from Pendleton to the mouth of the Umatilla, or from Salem to Portland on the Willamette (Pacific Northwest River Basins Commission 1969).

With a lot of measurements matching streamflow with river levels, a relationship or "curve" can be seen, allowing experts to infer streamflows from recorded levels. Water specialists convert the river stage readings from gages around the state into a variety of river flow statistics such as mean daily or monthly flows, flow-duration figures (the period of time a stream can be expected to be at or above a certain flow), or flow-exceedance probabilities (the statistical probability that a given flow will occur in a specific time period).

It may be worth noting that the gage heights are arbitrary and have meaning only in reference to one gage. For example, on April 15, 1996, the Willamette had the following stage-readings, upstream to downstream: Eugene, 10.3 feet; Salem, 9.4 feet; Portland, 11.3 feet (*Oregonian* 1996). These readings do not mean the river was flowing uphill from Eugene to a point one foot higher in Portland; all that the Portland reading means is that the river at Portland rose 0.1 foot from the previous day's reading of 11.2. Gage heights are not linked to mean sea level or river depth. They are simply a relative marker for future river level references at that location.

By measuring stream width, depth, and water velocity, hydrologists calculate streamflow. (U.S. Department of Agriculture, Agricultural Research Service. Image Number K10787-1. Photo by Peggy Greb. "Rangeland scientist Chad Boyd (right) and Jess Wenick, Burns Paiute wildlife program manager, measure streamflow within the irrigated reach of Lake Creek in Logan Valley, Oregon.")

Streams have been measured in Oregon since 1890 when gaging stations were established on the Owyhee River at Owyhee and on the Malheur River at Vale. Over the next fifteen years, only a few stations were added to the network, including the Umatilla River at Gibbon (1896), the Deschutes River at Morrow, and the Hood River at Tucker's Bridge, from 1897 to 1899 (Water Resources Committee 1955). For any number of reasons, most early records have substantial gaps. The longest continuously running gage within Oregon appears to be the Willamette River at Albany, from 1894 to the present (Moffatt, Wellman, and Gordon 1990).

▣ Basins

Water from rain or melting snow runs downhill from every point in the landscape and collects in channels that join with each other to form creeks and rivers. Just as a roof collects water and passes it on to valleys and gutters, or a parking lot slopes water to drains, a watershed—also called a basin or drainage—is the land that collects water in a given area and passes it on to streams. Whatever its name, the surface area that drains to streams is a natural organizing unit for water supply and management. Basins meet and inter-finger with each other on ridge lines. Smaller basins (called subbasins) nest within larger basins—the Minam River subbasin is within the Wallowa River subbasin within the Grande Ronde River basin which, like most inland streams of the Northwest, is in the Columbia basin.

The State of Oregon recognizes eighteen major river basins (Figure 3). Many, such as the Willamette, Rogue, Umpqua, and John Day river basins, represent whole stream networks. Others are actually collections of separate drainages. For example, the North Coast basin consists of the Nehalem, Trask, Nestucca, and other basins combined. The Goose and Summer Lakes basin in southeast Oregon is a grouping of many closed basins that collect and evaporate water. Oregon's largest basin is the Willamette (Table 3). At over 12,000 square miles, it accounts for approximately 12 percent of the state's area. The next largest stream systems are the Deschutes basin at about 10,000 square miles and the John Day basin at roughly 8,000 square miles. (Although the Malheur Lake basin is larger than the John Day, it is a grouped basin, rather than a single interconnected stream system.)

Oregon's official basins are creations both of nature and bureaucrats. Why the Hood River basin is singled out for recognition, but not the Imnaha, or why the Nestucca River is assigned to the North Coast basin but the Siletz to the Mid-Coast seems lost in administrative history. Apparently, lines simply have to be drawn somewhere.

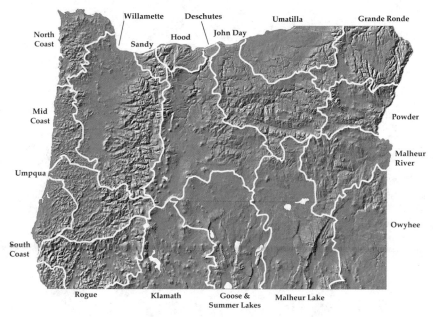

Figure 3. Oregon's river basins. (Relief image courtesy of Ray Sterner)

The map of Oregon would look very different if it followed the rule of rivers rather than the ruler of cartographers. As it stands now, water moves in and out of Oregon like wind through a woven-wire fence. The Grande Ronde, Walla Walla, Klamath, and North Fork Smith rivers all head in Oregon, but flow out of state. On the other hand, the Applegate and Illinois are born in California and move to Oregon, much in keeping with southern Oregon's human migration patterns. The Owyhee, Snake, and Columbia slide to home in Oregon, only after having rounded the bases in many other western states. As Portland-bound commuters sip their half-decaf skinny lattés and cross the Glenn Jackson bridge, a little bit of Utah and the Canadian Rockies glides by silently beneath them. Inside Oregon, few political boundaries respect the natural divisions of basins. Some exceptions include Washington, Douglas, and Baker counties, which coincide roughly (and respectively) to the Tualatin, Umpqua, and Powder-Burnt river basins.

It should probably also be noted that even within Oregon, basin boundaries are not inviolate. Given half a chance, people will move water wherever they want. The City of Portland diverts water from the Sandy basin and sells it to customers in the Tualatin subbasin. Washington County's growing population is partly served by pumping water out of the

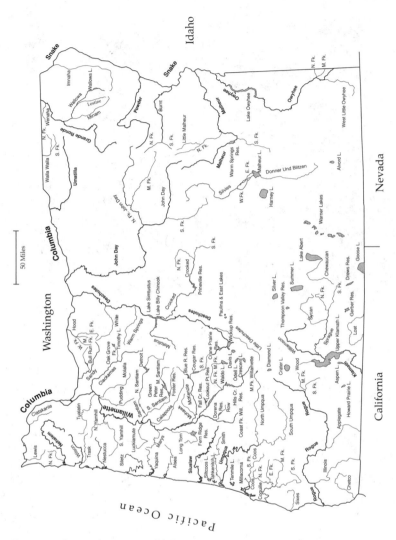

Figure 4. Oregon's rivers and lakes.

Trask River subbasin and over the Coast Range summit into the Tualatin River. Further south, the City of McMinnville takes the same approach with the Nestucca River. In a particularly cunning bit of plumbing, the Bureau of Reclamation's Rogue River project diverts water from high in the Rogue basin, transports it underneath the Cascade summit, combines it in a series of reservoirs with water collected high in the Klamath basin, and then routes it back under the Cascades where it is returned to the Rogue system by means of a storage reservoir.

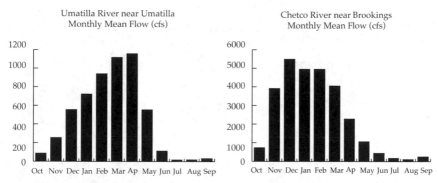

Figure 5: Water-year flow patterns. Eastern Oregon streams, as illustrated by the Umatilla River on the left, generally peak with snowmelt in late winter and spring. Western Oregon streams run fullest with winter rains in mid-winter, as shown by the Chetco River on the right. Another major difference is the dimension of flow: note that the vertical axis values for the Chetco are about five times greater than the Umatilla. (Source: Data from Moffatt, Wellman and Gordon 1990)

Every river that issues from a basin bears the signature of the local climate and landscape. (Figure 4 displays Oregon's major rivers and lakes.) The amount and timing of a watershed's runoff is a function of basin size, elevation, rainfall, geology, soils, temperature, vegetation, and many other factors. In fact, to estimate basin streamflow, the Oregon Water Resources Department uses a statistical model with thirty-one watershed characteristics (Oregon Water Resources Department 1993a).

However, most Oregon basins respond primarily to precipitation. Generally, streamflows are greatest when rain is falling (especially in western Oregon) or the snow is melting (particularly in eastern Oregon) (Loy 2001). But tracking flow using a January to December timetable would be like setting your VCR or TiVo to record at 9:15 when the program starts at 9:00. Thus, just as many companies have fiscal years that begin in off-months, hydrologists use a water year that runs from October 1 through September 30 (Figure 5).

The way in which basins release their waters may be characterized by their location within the state. Although this is far from the only way to view Oregon's streamflow patterns, the state's basins may be grouped into six regions: coastal basins; the Willamette and Sandy basins; Rogue and Umpqua basins; eastern Cascade subbasins; closed basins; and Blue Mountain basins. Each is briefly described below using some statistics that warrant a little caution. Generally, the streamflow figures reflect upstream storage and use, rather than untouched natural flow. Still, the

Table 3. Basin Summary

Basin	Average Annual Discharge (estimated acre-feet)[1]	Area (sq. mi.)[1]	% of State[2]	Length of Longest River (if different from basin name)[3]	Miles of Stream in Basin[4]
North Coast	10,227,000[5]	2,720	2.8	118 (Nehalem)	4,000
Willamette (inc. Sandy)	27,408,000	12,031	12.3	187	16,000
Hood	1,380,000	1,048	1.1	51 (Fifteenmile Cr.)	1,740
Deschutes	4,213,000	10,484	10.8	252	8,000
John Day	1,410,000	8,098	8.3	284	9,500
Umatilla	567,000	4,548	4.7	90	8,000
Grande Ronde	2,815,000	4,946	5.1	170 (Ore. Length)	6,000
Powder	700,000	3,266	3.4	144	5,000
Malheur R.	140,000	5,072	5.2	190	8,000
Owyhee	400,000	6,194	6.4	161	7,000
Malheur Lake	700,000	9,988	10.3	108 (Silvies)	7,000
Goose/Summer Lake	500,000[6]	7,953	8.2	53 (Chewaucan)	5,000
Klamath	1,205,000	5,702	5.9	92 (Williamson)	5,000
Rogue	5,661,000	5,022	5.2	215	6,500
Umpqua	6,700,000	4,690	4.8	112	6,400
South Coast	9,000,000	2,962	3.1	57 (Chetco)	5,000
Mid Coast	8,100,000	2,343	2.4	109 (Siuslaw)	4,500
Columbia R.	—	—	—	309 (Ore. Length)	—
Snake R.	—	—	—	233 (Ore. Length)	—

[1] State Water Resources Board, various basin reports, 1958-1971, except where indicated. [2] Oregon Water Resources Dept., 1984. [3] From inspection of Oregon Water Resources Dept. basin maps; Series x.6; various dates, 1964-1991. [4] Oregon Water Resources Dept., n.d. [a]. [5] Generated from per acre run-off estimates in State Water Resources Board, 1961b. [6] Estimated using run-off coefficient of 0.09 cfs/square mile identified in Pacific Northwest River Basins Commission 1969.

The Nehalem River on Oregon's north coast is the biggest river flowing out the Coast Range. (Oregon State Archives, Oregon Marine Board, OMB0029)

Table 4. Coastal Basins: Flow Regimes for Major Tributaries

Stream (measuring location)	Average Annual Discharge (acre-feet)	Highest Flow Occurrence Month	%	Lowest Flow Occurrence Month	%	Flow Range Factor[1]
Nehalem (near Foss)	1,956,000	Dec./Jan.	19.7	Sep.	0.7	28
Chetco R. (near Brookings)	1,713,000	Dec.	19.8	Aug.	0.4	50
Siuslaw R. (near Mapleton)	1,551,000	Dec.	20.9	Aug.	0.6	35
Siletz R. (at Siletz)	1,123,000	Dec.	18.7	Aug.	0.7	27
Alsea R. (near Tidewater)	1,097,000	Jan.	20.2	Aug./Sep.	0.7	29
Wilson R. (near Tillamook)	859,300	Dec.	19.1	Aug.	0.8	24

Source: Data from Moffatt, Wellman and Gordon 1990
[1]Ratio of annual discharge percentage in highest month to that of lowest month.

figures provided should be useful for understanding broad patterns and comparing one system with another.

COASTAL BASINS
The streams that drain into the sea from Oregon's rugged Coast Range are short and steep, but carry a lot of water—nearly all of it from rainfall. On the average, the coastal basins produce nearly 4 cfs of runoff for each square mile of area—a veritable water farm (Pacific Northwest River Basins Commission 1969). The largest coastal streams are the Nehalem, Chetco, and Siuslaw rivers, all with 1.5 million or more acre-feet.

The North, South, and Mid Coast basins account for only about 9 percent of the state, but over 30 percent of its flow. The highest flow month is usually December, which can produce about one-fifth of the total annual discharge. However, because the streams are so rainfall-dependent, many shrink to a fraction of their winter size during late summer and fall. The Chetco is the most variable in this regard, with its winter flow fifty times larger than its summer flow. The runoff patterns of some of the major coastal tributaries are displayed in Table 4 and illustrated in Figure 6.

WILLAMETTE/SANDY BASINS

The Willamette is Oregon's largest and most productive basin. It covers 12 percent of the state, occupying the area between the Coast and Cascade mountains in northwest Oregon. It drains both mountain ranges and combines the runoff imprint of both in a 27 million acre-foot discharge, about 35 percent of the state's total. The average discharge is an impressive 3-plus cfs per square mile (Pacific Northwest River Basins Commission 1969). The Willamette River's largest tributaries are the Santiam and McKenzie rivers, with 5.6 million and 4.3 million acre-feet of annual discharge, respectively (Moffatt, Wellman, and Gordon 1990).

The Willamette, Oregon's largest river, flows northward from its source in the mid-Cascades to its junction with the Columbia near Portland, a journey of nearly 200 miles. (U.S. Army Corps of Engineers; Image File: Sce0373.jpg; Date: 01JUN1988)

Table 5. Willamette/Sandy Basins: Flow Regimes for Major Tributaries

Stream (measuring location)	Average Annual Discharge (acre-feet)	Highest Flow Occurrence Month	%	Lowest Flow Occurrence Month	%	Flow Range Factor[2]
Willamette R. (at Portland)	24,130,000	Dec.	18.7	Aug.	2.1	9
Santiam R.[3] (at Jefferson)	5,647,000	Dec.	18.2	Aug.	2.0	17
McKenzie R. (near Coburg)	4,286,000	Feb.	13.1	Sep.	2.8	5
M. Fk. Will. R. (at Jasper)	2,981,000	Dec.	16.6	Jun.	5.5	3
Clackamas R. (near Clackamas)	2,656,859	Jan.	16.2	Aug.	2.3	7
Yamhill[4]	1,446,300+	Jan.	21.7	Aug.	0.2	109
Coast Fk. Will. (near Goshen)	1,188,000	Dec.	18.8	Jul.	1.5	13
Tualatin R. (at West Linn)	1,109,000	Jan.	24.1	Aug.	0.1	241
Pudding R. (at Aurora)	883,206	Jan.	18.7	Aug.	0.5	37
Molalla R. (near Canby)	842,600	Jan.	17.8	Aug.	0.8	22
Luckiamute R. (near Suver)	655,700	Jan.	21.4	Aug.	0.4	54
Long Tom R. (at Monroe)	557,900	Jan.	23.2	Sep.	0.2	116
Marys R. (near Philomath)	334,700	Jan.	22.2	Aug.	0.3	74
Sandy R. (near Bull Run)[5]	1,686,000	Dec.	14.0	Aug.	1.8	8

Source: Data from Moffatt, Wellman and Gordon 1990

[1] Where flow record adjusted for upstream dams, monthly data taken from adjusted record.

[2] Ratio of annual discharge percentage in highest month to that of lowest month.

[3] Tributaries indented.

[4] Combines North Yamhill near Pike with South Yamhill near Whiteson; monthly data reflects South Yamhill.

[5] Tributary to Columbia R., but often administered with Willamette Basin.

Highest runoff occurs during December and January, in response to rainfall. Although the lowest summer flows (typically in August) drop significantly, the big-bodied Cascade streams give up less of their total annual discharge during their high-flow months—in the range of 16 to 18 percent. The Middle Fork of the Willamette's winter flows, for example, are only three times higher than its summer flows. In contrast, the smaller, flashier streams that drain the eastern Coast Range proportionately give up far more in winter and drop far lower in the summer. In January, the Tualatin can discharge almost a quarter of its yearly flow and drop (by a factor of over 200) to 0.1 percent by August. The Long Tom and Yamhill rivers have similar patterns. While some of this range may be accentuated by water storage and diversion projects, the Willamette system is clearly Cascade-dominant. The U.S. Army Corps of Engineers operates thirteen storage projects in the Willamette basin that affect natural runoff patterns (see Chapter 7).

The runoff patterns of some major Willamette Basin streams are displayed in Table 5 and illustrated in Figure 6.

UMPQUA/ROGUE BASINS

The Umpqua and Rogue river basins together produce about 16 percent of Oregon's annual discharge and account for around 11 percent of its area. These basins are unique in Oregon in that they rise in the Cascades but

The John Day, Oregon's longest river, travels through nearly 300 miles of mountains and desert to deliver its fairly small annual 1 million acre-feet to the Columbia. The Willamette, two-thirds its length, carries nearly 25 times its flow. (Oregon State Archives, Oregon Highway Division, OHD6681)

punch through the Coast Range to the sea. Their runoff pattern combines the snowmelt signature of the Cascades and the rainfall-based contribution of the Coast Range. The Umpqua basin's largest tributaries are the North (2.7 million acre-feet annually) and South Umpqua rivers (2.1 million acre-feet annually) and the Rogue's Illinois and Applegate rivers (at roughly 3.0 and 0.5 million acre-feet annually, respectively).

January is the month of highest runoff for these basins, and August the time of lowest flow. The Umpqua is more variable than the Rogue, with a winter flow on the order of one hundred times greater than summer flow, compared to a more modest factor of forty to sixty for Rogue tributaries and only twelve for the Rogue itself. The Army Corps of Engineers operates two major flood-control reservoirs on the Rogue, which alter the flow regime to some extent (See Chapter 7).

The runoff patterns of some major Umpqua/Rogue streams are displayed in Table 6 and illustrated in Figure 6.

Table 6. Umpqua/Rogue Basins: Flow Regimes for Major Tributaries					
Stream *(measuring location)*	Average Annual Discharge *(acre-feet)*	Highest Flow Occurrence *Month*	%	Lowest Flow Occurrence *Month*	%	Flow Range Factor[1]
Umpqua R. (near Elkton)	5,442,000	Jan.	21.6	Aug.	0.2	108
N. Umpqua R.[2] (at Winchester)	2,736,000	Jan.	15.3	Sep.	2.2	7
Steamboat Cr. (near Glide)	538,300	Dec.	10.3	Aug.	0.7	15
S. Umpqua R. (near Brockway)	2,084,000	Jan.	24.3	Aug.	0.1	243
Cow Cr. (near Riddle)	641,900	Jan.	22.0	Aug.	0.3	73
Rogue R. (near Agness)	4,511,000	Jan.	20.8	Aug./Sep.	1.7	12
Illinois R. (near Agness)	2,966,000	Jan.	21.6	Aug.	0.5	43
Applegate R. (near Wilderville)	548,400	Jan.	19.4	Aug.	0.3	65

Source: Data from Moffatt, Wellman and Gordon 1990
1 Ratio of annual discharge percentage in highest month to that of lowest month.
[2] Tributaries indented.

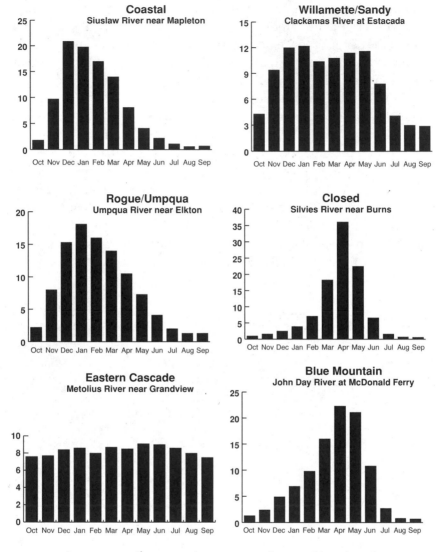

Figure 6: Flow patterns for Oregon basin regions illustrated by selected streams. Percentage of annual flow by month. (Source: Data from Moffatt, Wellman and Gordon 1990)

THE EASTERN CASCADES BASINS: HOOD/DESCHUTES/KLAMATH

The streams draining the eastern flank of the Cascades tap a bountiful and sustaining source. The rain-enriched snowpacks melt into porous volcanic rocks and charge aquifers to the point where streams like the Metolius can spring practically full-blown from the earth. On the average, groundwater accounts for about 97 percent of the Metolius's flow. The Deschutes River, supported by such groundwater contribution from this and other tributaries, is one of the least variable in the United States (Water Resources Committee 1955). On the whole, its peaks are only 1.7 times its summer

Stream (*measuring location*)	Average Annual Discharge (*acre-feet*)	Highest Flow Occurrence		Lowest Flow Occurrence		Flow Range Factor[1]
		Month	%	*Month*	%	
Deschutes River (at Moody)[2]	4,252,000	Jan.	10.8	Sep.	6.2	1.7
Crooked R.[3] (near Culver)	1,124,000	Apr.	13.4	Sep.	6.6	2.0
Metolius R. (near Grandview)	1,085,000	May	9.1	Sep.	7.5	1.2
Warm Spr. R. (near Kahneeta)	328,200	Feb.	13.0	Sep.	4.7	2.7
White R. (below Tygh Valley)	308,600	May	14.8	Sep.	2.4	6.2
Little Deschutes (near La Pine)	150,700	May	15.7	Oct.	3.5	4.5
Fall River (near La Pine)	108,700	May	9.0	Feb.	7.3	1.2
Klamath River (below Boyle Powerplant)	1,397,000	Mar.	13.8	Jul.	2.8	4.9
Williamson R. (near Chiloquin)	768,700	Apr.	15.7	Aug.	4.4	3.6
Sprague R. (near Chiloquin)	426,700	Apr.	18.1	Aug.	3.3	5.5
Wood R. (at Fort Klamath)	155,647	Mar.	19.3	Aug.	6.9	2.8
Hood River (near Hood River)	769,400	Jan.	13.2	Sep.	3.2	4.1

Table 7. Eastern Cascade Basins: Flow Regimes for Major Tributaries

Source: Data from Moffatt, Wellman and Gordon 1990
[1] Ratio of annual discharge percentage in highest month to that of lowest month.
[2] Monthly data after completion of Round Butte Dam.
[3] Tributaries indented.

Dry streams can result from seasonal variations or decades-long climatic cycles. In the drought of 1977 the John Day River essentially stopped flowing. (Oregon Water Resources Department)

lows. The Klamath and Hood basins are only slightly more variable, with winter/summer flow factors ranging from about three to five. High flows normally coincide with snowmelt in April and May. Low flow months are August and September (See Table 7 and Figure 6).

The Eastern Cascade streams contribute about 9 percent of Oregon's total yearly streamflow and represent approximately 18 percent of the state's area. The major tributaries that contribute to the Deschutes River's 4.5 million acre-foot annual discharge include the Metolius and Crooked rivers (both around 1.1 million acre-feet). The Williamson River provides about half of the 1.4 million acre-feet that pour over the border from the Klamath into California.

BLUE MOUNTAIN/OWYHEE

Unlike the Coast Range or Cascades, Oregon's Blue Mountains are not a narrow, continuous north/south line-up of mountain tops. Rather, they stretch from the center of the state north and east for several hundred miles. This area of geologic uplift has a complex topography and contains many individual ranges such as the Ochoco, Aldrich, Strawberry, Greenhorn, Elkhorn, and Wallowa mountains. The greatest topographic relief is to the east, where the highest Wallowa peaks exceed 9,000 feet and where the land falls away eastward to one of North America's deepest river gorges—Snake River's Hells Canyon, some 7,900 feet deep (McKee 1972).

The streams draining Oregon's Blue Mountains must often run a long way before picking up appreciable water. The John Day and Grande Ronde rivers are both over 200 miles long, and the Malheur nearly as long, yet by western Oregon standards they deliver only modest amounts of water. The Grande Ronde basin tops the bunch at about 3 million acre-feet per

Table 8. Blue Mountain//Owyhee Basins:
Flow Regimes for Major Tributaries

Stream (measuring location)	Average Annual Discharge (acre-feet)	Highest Flow Occurrence Month	%	Lowest Flow Occurrence Month	%	Flow Range Factor[1]
Grande Ronde R. (at Troy)	2,254,000	May	20.5	Sep.	2.1	10
Wallowa R.[2,3]	779,000	Jun.	29.7	Sep./Oct.	1.8	17
John Day R. (at McDonald Ferry)	1,524,000	Apr.	22.3	Sep.	0.7	32
North Fork (at Monument)	936,000	May	23.6	Sep.	0.8	30
Middle Fork (at Ritter)	185,500	Apr.	24.2	Aug./Sep.	1.0	24
Owyhee R. (at Owyhee)	759,300	Apr.	27.9	Aug./Sep.	0.3	93
Malheur R. (at Vale)	391,651	Apr.	28.3	Sep.	0.8	35
Powder R. (near Robinette)	386,600	May	23.0	Sep.	1.6	14
Imnaha R. (at Imnaha)	376,000	May	25.6	Sep.	2.3	11
Umatilla R. (near Umatilla)	335,400	Apr.	20.6	Jul./Aug.	0.4	52
Meacham Cr. (at Gibbon)	148,500	Apr.	22.3	Sep.	0.5	45
Pine Cr. (near Oxbow)	273,900	May	20.6	Aug.	1.2	17
Walla Walla R.[4]	163,270	May	14.6	Sep.	5.0	3
Burnt R. (at Huntington)	95,360	Apr.	23.7	Nov.	3.2	7

Source: Except where noted, data from Moffatt, Wellman and Gordon 1990
[1] Ratio of annual discharge percentage in highest month to that of lowest month.
[2] Because there is no long term gage on lower Wallowa, discharge is estimate taken from State Water Resources Board, Grande Ronde River Basin Report, September 1960. Monthly data from largest gaged tributary: Minam R. at Minam.
[3] Tributaries indented
[4] Discharge combines S. Fk. Walla Walla nr. Milton-Freewater and N. Fk. Walla Walla nr. Milton; monthly data from S. Fk. Walla Walla nr. Milton-Freewater.

year. The John Day (Oregon's longest stream) manages to eke out around 1.5 million acre-feet and the Malheur a trickling 0.4 million acre-feet. In other words, the Yamhill River gathers about the same amount of water from 550 square miles in the Coast Range as the John Day River does from the 7,600 square miles of its drainage. Overall, flows in the Blue Mountain region range from about 0.6 to over 0.9 cubic feet per second per square mile (Pacific Northwest River Basins Commission 1969).

However, comparisons to western Oregon streams, which are almost freakishly productive in a national context, are probably not very telling. The rivers draining the Blue Mountains, and the Owyhee, whose drainage area reaches into several other states, are major streams in their own right and are critical economically and ecologically to the region. These streams are most exuberant during the snowmelt months of April and May. They are most subdued in August and September. Their winter/summer flow range, however, is not as great as in some western Oregon streams, with factors ranging from the single digits into the nineties, but averaging around thirty (See Table 8 and Figure 6).

THE CLOSED BASINS

In nearly a fifth of Oregon, the little rain that falls or snow that melts finds not deliverance in the western sea, but instead an awesome purgatory— or is it a paradise?—of uplifted scarps and down-faulted sinks. Therein it remains, coursing down from the sky-hung rocks and collecting in big swamps and salt lakes, until taken back by the wind and sun.

The two closed basins recognized by the state are the Goose and Summer basin and the Malheur Lakes basin. Most runoff occurs from May snowmelt. Low flows generally occur in September. Interestingly, there is so little water conveyed in the stream systems that the variability is fairly low, with the highest flow months being only thirty-five to fifty times greater than the lowest flow months (see Table 9 and Figure 6). The

Lake Abert, like others in Oregon's closed basins, is an evaporating pan for all that flows into it, including the Chewaucan River. (Salem Public Library, Ben Maxwell Collection, #7425)

Table 9. The Closed Basins: Flow Regimes for Major Tributaries

Stream *(measuring location)*	Average Annual Discharge *(acre-feet)*	Highest Flow Occurrence *Month*	*%*	Lowest Flow Occurrence *Month*	*%*	Flow Range Factor[1]
Silvies R. (near Burns)	131,100	May	22.5	Sep.	0.6	38
Chewaucan R. (near Paisley)	107,200	May	28.6	Sep.	1.7	17
Deep Cr. (above Adel)	97,080	May	26.8	Aug.	0.8	36
Donner und Blitzen R. (near Frenchglen)	92,740	May	25.2	Sep.	2.8	9
Ana R. (near Summer Lake)	65,780	Mar.	8.9	Apr-Jun.	7.9	1.1
Drews Cr. (near Lakeview)	51,150	Apr.	25.4	Oct./Nov.	0.5	51

Source: Data from Moffatt, Wellman and Gordon 1990
[1]Ratio of annual discharge percentage in highest month to that of lowest month.

largest streams in the closed basins are the Silvies on the north and the Chewaucan on the west, both contributing in excess of 100,000 acre-feet annually. Most streams have small drainage areas—less than 1,000 square miles. The average annual discharge for the area is the lowest in the state: 0.09 cubic foot per second per square mile (Pacific Northwest River Basins Commission 1969).

This unique basin-and-range geologic province creates equally unique hydrologic settings. When viewed from space, Oregon's most noticeable water features are not its skinny rivers or pin-head mountain lakes, but the huge, flat, reflecting ponds of its closed basin lakes: Goose, Summer, Abert, Harney, and Malheur. (Admittedly, Upper Klamath Lake and Crater Lake also show up pretty well.) The water pouring into Harney and Malheur lakes creates 180,000 acres of wetlands—the largest freshwater marsh in the western United States. While it may seem odd, this driest part of Oregon can also be the location of serious, prolonged flooding due to heavy rain and snow—such as that periodically experienced by Malheur Lake (most recently in 1984) (U.S. Army Corps of Engineers 1996a). Lastly, when streams drop precipitously onto the flat fault-block floors, their hydraulics can change dramatically. For example, the Silvies River is one of the few major streams in Oregon to split downstream into east and west forks each with separate mouths, instead of having the forks join upstream to create a mainstem.

The deepest and the purest: both jewels of the Cascades. Crater Lake is the deepest lake in the United States, and Waldo Lake is one of the purest in the world. (Photo on right: author; photo below: U.S. Geological Survey. Photograph by D. Wieprecht on August 20, 1995.)

Lakes

Oregon may not claim, as Minnesota does, to be a land of 10,000 lakes. Still, among the state's 6,000 lakes, reservoirs, and ponds may be found the deepest lake in the United States (Crater Lake, at 1,932 feet) and one of the purest in the world (Waldo Lake). Oregon has nearly 1,800 named lakes and over 3,800 named reservoirs. However, in the named-lake department, Oregon is no match for neighboring Washington and California, which have about 3,000 and 2,600, respectively. Oregon's lakes range from near sea-level to the very top of the Cascades—the highest being an unnamed lake in the crater of the South Sister, about 10,200 feet above sea-level (U.S. Board of Geographic Names 1996; Johnson et al. 1985).

Oregon's largest water bodies are its desert lakes (see Table 10). With over 60,000 acres (about 96 square miles), Upper Klamath Lake has the greatest area of any Oregon lake and is one of the largest freshwater bodies in the nation (Johnson et al. 1985). Malheur, Abert, and Goose lakes are the next largest in size, with between 34,000 and 50,000 acres each, although their area varies considerably with wet and dry climatic cycles. While they may be the biggest lakes in the state, they are also among the shallowest. Upper Klamath has a maximum depth of 50 feet. The maximum depths of Malheur, Abert, and Goose Lakes range from 5 to 12 feet.

Crater Lake holds the most water of any in Oregon. It is the eighth-largest in area (13,000 acres) and this, combined with its startling depth, creates a spectacular volume of over 14 million acre-feet. Owyhee

---

Table 10. Surface Area of Oregon Lakes and Reservoirs (10,000 Acres or Greater)

Name	Acres
Upper Klamath	61,543
Malheur	49,700
Abert	36,538
Goose	34,087*
Harney	26,400
Summer	25,000
Owyhee	13,900
Crater	13,139
Wickiup	10,334

* Acreage in Oregon; total area is 97,391 acres
Source: Data from Johnson et al. 1985

Table 11. Volume and Depths of Oregon Lakes and Reservoirs, Ranked by Volume (100,000 Acre-Feet or Greater)

Name (reservoirs italicized)	Volume (acre-feet)	Maximum Depth (ft)
Crater	14,164,057	1,932
Owyhee	1,122,000	117
Upper Klamath*	849,290	50
Goose	798,606[1]	12
Waldo	787,395	420
Crescent*	566,556	265
Odell*	473,854	282
Lost Creek	465,000	322
Detroit	455,000	440
Lookout Point	453,000	234
Green Peter	430,000	315
Billy Chinook	399,952	415
Hills Creek	356,000	299
Warmsprings	285,260	140
Abert	268,186	11
Paulina	249,850	250
Wallowa*	243,517	299
Cougar	219,300	425
Wickiup	206,880	70
Prineville	160,020	130
Fall Creek	125,000	161
Gerber	110,000	65
Fern Ridge	101,200	33

* Small dams have been built at the outlets of these natural lakes.
[1] Entire lake, not just Oregon volume.
Source: Data from Johnson et al. 1985

Reservoir, residing in the sage-scented obscurity of southeast Oregon, is second largest with a little over one million acre-feet. The next set of large-volume lakes are Upper Klamath and Goose, each with around 800,000 acre-feet. High Cascade Lakes (Waldo, Crescent, and Odell) represent the next largest grouping, with volumes from around 500,000 to nearly 800,000 acre-feet. Next are U.S. Army Corps of Engineers flood-control reservoirs that catch runoff from the western Cascades: Lost Creek (Rogue basin), Detroit, Lookout Point, and Green Peter (Willamette basin). These projects each store on the order of 450,000 acre-feet. Thus, of Oregon's top ten lakes (in terms of volume), four are the result of bulldozers and cement mixers. Oregon has over sixty reservoirs of 5,000 or more acre-feet, twenty-three of which exceed 100,000 acre-feet (Johnson et al 1985). These artificial

Table 12. Major Oregon Lake Types

Cause	Description	Examples
Fault-block movement	Down-dropped or tipped blocks of land collect water	Klamath, Abert, Summer, Goose, Warner
Volcanic action	Craters or calderas capture water; lava dams streams	Crater, East, Paulina, Harney; Odell, Davis, Clear (Linn Co.)
Glacial action	Glaciers scour-out depressions in bedrock; glaciers deposit outwash that dams streams	Diamond, Waldo, Miller, Marion, Bull Run, Aneroid, Ice; Wallowa, Anthony, Elk (Marion Co.)
Landslides	Landslides dam streams	Pamelia, Strawberry, Triangle, Loon
Coastal shoreline processes	Sandbars dam stream-mouths; sand dunes block streams or catch water	Siltcoos, Tahkenitch, Tenmile; Mercer, Buck, Lily, Coffenbury, Cleawox
River processes	Rivers change course creating sloughs or oxbow lakes	Beaver, Colorado, Hayden, Humbug, Blue (Multnomah Co.)

Source: Information from Johnson et al. 1985

impoundments also create some of the deepest water in Oregon. Detroit, Cougar and Billy Chinook reservoirs are all deeper than 400 feet. Except for Crater Lake, Waldo Lake is the only natural lake that can match them (see Table 11). There are nineteen Lost, eleven Blue, and ten Clear lakes in the state (U.S. Board of Geographic Names 1996).

Oregon's lakes are hydrologic responses to a diverse landscape of volcanoes, glaciers, fault-blocks, sand dunes, and landslides. Lakes in Oregon represent seven of the eleven major categories used by scientists to classify lake-types (see Table 12). The combination of volcanic and glacial forces in the Cascades has produced over half of Oregon's lakes. Other lake concentrations occur in the Wallowa Mountains and along the coast. Lakes disappear when they fill with sediment or their outlets are lowered enough by erosion to drain them. Western Oregon basins tend to produce more sediment and supply their streams with more water than eastern Oregon basins. Consequently, lakes do not last as long west of the Cascades and are therefore more common in eastern Oregon. The cycling time of water through lakes (the theoretical average time required for a lake's entire volume to be replaced) is longest in the deep Cascade lakes (ten to twenty years is common, with figures as great as thirty-two years for Waldo, forty-six years for Paulina and one hundred fifty years for Crater Lake) and shortest in artificially regulated projects (such as three days for Toketee Lake) (Johnson et al 1985).

Droughts and Floods

Like all natural processes, precipitation and the streams it supplies are characterized by variation. Not just the geographic variation that is easily seen when driving from Eugene to Burns, or the seasonal variation that makes July a preferred vacation month over March. There is annual variation that imposes cycles of drought or drenching. Over the last hundred years (the timespan of Oregon's weather and climate observations), the number of wet and dry years is about equal. However, the record shows distinct wet and dry clusters. These wet and dry "cycles" generally last twenty to twenty-five years (Taylor 1999), while in the Portland area there appears to be an eighteen-year precipitation cycle (Figure 7) (Taylor 1997). Although the cycles are caused by a number of factors related to global ocean and atmospheric circulation, none are quite so influential as the "El Niño" and "La Niña" phenomena.

Roughly every three to seven years, the ocean off South America rapidly warms, while the western Pacific (i.e., around Japan, China, and the Phillipines) cools. This phenomenon is termed an El Niño (Spanish for the Christ child, for this often occurs around Christmas). The opposite

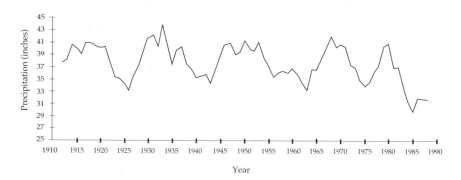

Figure 7: Portland precipitation cycles. Five-year running average. As shown by Portland's precipitation pattern, Oregon is prone to pronounced wet and dry cycles. (Source: Information from Taylor 1997)

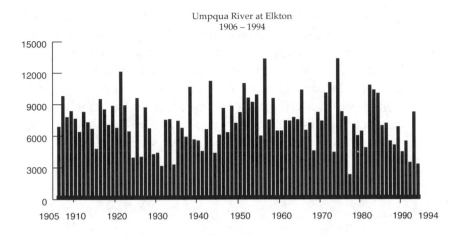

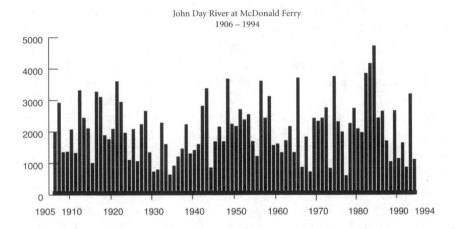

Figure 8: Historic average annual discharge comparison (cubic feet per second). Eastern Oregon streams, as shown by the John Day River, are subject to far more variable flow than western Oregon streams such as the Umpqua River. (Source: Data from Oregon Water Resources Department)

happens about as often and is called a La Niña. In the Pacific Northwest, El Niño events usually mean the following winter will be dry with mild temperatures; La Niña events signal wetter, colder winters (Taylor 1998). Interestingly, active Atlantic hurricane seasons are almost always followed by wetter than average winter conditions in western Oregon, and inactive years by dry winters (Taylor 1999).

These and other large-scale causes may be responsible for cyclical trends in Oregon's climate. These cycles spike and dip through millennia and are recounted through the stories of old-timers, tree rings, or fossilized pollen in ancient lake beds. However, to accommodate modern attention spans, they are usually reduced to statistics covering a thirty-year period—hardly a murmur in the earth's heartbeat. Two forms of natural variation that tend to stick in people's minds are when there isn't any precipitation or when there appears to be way too much—or, in different terms, when rivers are dry or high. And, as Figure 8 shows, rivers run a century's course with many lows and many highs.

▣ Droughts

Droughts are relative things, ranging from dry spells within years to a long string of dry years. Both dryness and the chances of drought increase from west to east across Oregon. West of the Coast Range, there is usually one completely dry month every year—but a two-month dry spell is generally a once-per-decade event. In the Willamette Valley and the Cascades, there is a dry month (usually July or August) nearly every year, a two-month dry period every third year, and a three-month drought period about once every ten years. In eastern Oregon, there may be two- or three-month dry periods every year. Four- to six-month dry spells can occur as often as once every four years. Similarly, dry years (with less than two-thirds of normal precipitation) occur about once every thirty years along the coast, every twenty years in the Willamette Valley, every thirty years in the Cascades, and every fifteen years in much of eastern Oregon (Pacific Northwest River Basins Commission 1969).

Drought conditions can occur in varying Oregon locales at different times. They can start one place and spread to another, or fade away in one basin while persisting in others. Seasonal droughts occur every year, especially east of the Cascades. Oregon has experienced four significant droughts since the 1920s. There was a prolonged drought that lasted from 1928 through 1941 in most parts of the state. However, during that period the north coast experienced three years of adequate moisture. The drought lasted a few years more in the Deschutes, Klamath, and closed basins. Eastern Oregon felt the effects of another drought from 1959 through 1964, and western Oregon from 1976 through 1981. The drought was statewide in 1977. Drought conditions returned to all of Oregon in 1987, 1988 (Paulson et al. 1991), and the early 1990s. The 2000-2001 water year was among the driest in Oregon in history, especially in Western Oregon, with Portland, Eugene, Astoria, and Corvallis setting new record lows in precipitation (Oregon Climate Service. n.d.). Although droughts have always been part of the Oregon landscape, as noted in the previous climate change section, their frequency and severity is expected to grow under most climate change scenarios.

⊞ *Floods*

Unlike droughts, most floods do not develop—they hit. Floods tend to be more immediate and dramatic than events at the other end of the hydrologic/meteorological spectrum. They occur relatively frequently and on most streams in Oregon. (See Table 13.)While the major flooding season runs from October through April west of the Cascades, more than half of the floods occur in December and January. Major floods are usually the result of rainfall combined with warm temperatures that melt snow in higher elevations. Many western Oregon streams experience one or more minor floods almost every year. Because they are strongly influenced by precipitation, most western Oregon floods are short-lived, with rivers reaching their peaks in a matter of hours (Water Resources Committee 1955).

Eastern Oregon floods are caused mostly by snowmelt. They usually occur in the late spring, but can come as early as November if temperatures are mild. The amount of flooding is more a function of how fast the snow melts than of how much snow there is. Often, high water lasts longer than in western streams, frequently up to sixty days. Late summer flash floods are also a noteworthy feature of eastern Oregon's hydrology (Water Resources Committee 1955). Eastern Oregon is also home to flooding that may take years to develop. From time to time, some of its closed basins receive enough water to swell basin lakes well beyond their normal size, as happened with Malheur Lake in the early 1980s. Following heavy rainfall and runoff in 1984, Malheur and Harney lakes were joined together for several years (U.S. Army Corps of Engineers 1996a). The inundation of homes, farms, highways, and railroads caused severe local economic disruptions. For the ultimate worst-case scenario of this type of flooding, one may observe the ancient wave-cut terraces of Pleistocene lakes in these basins at a level some 350 feet higher than today's shorelines (McKee 1972).

Oregon's worst flood in terms of lives lost was on Willow Creek in 1903. A cloudburst from a thunder storm caused a flash flood in Heppner. It killed

Table 13. Major Oregon Floods

Date	Location	Description
May - June 1894	Columbia River Basin	Rain on snowpack; highest flood stage ever recorded at Vancouver, WA (33.6 feet)
June 1903	Willow Creek	Flash flood in Heppner; 247 people killed
January 1923	Clackamas, Santiam, Sandy, Deschutes, Hood, McKenzie	Record flood levels
February 1927	Klamath, Willamette, Umpqua, Rogue, Illinois	Major flooding
Nov. – Dec. 1942	Willamette Basin	10 deaths; $34 million damage
December 1945	Coquille, Santiam, Rogue, McKenzie	9 deaths and homes destroyed in Eugene area
May - June 1948	Columbia River	Rain on snow; destruction of the city of Vanport
March 1952	Malheur, Grand Ronde, John Day	Highest flood stages on these rivers in 40 years
December 1955	Rogue, Umpqua, Coquille	11 deaths; major property damage
July 1956	Central Oregon	Flash floods
February 1957	Southeastern Oregon	$3.2 million in flood damage
December 1961	Willamette Basin	$3.8 million in flood damage
Dec. 1964 - Jan. 1965	Pacific Northwest	Rain on snow; record flood on many rivers
December 1967	Central Oregon Coast	Storm surge
January 1972	Western Oregon	Record flows on coastal rivers
January 1974	Western Oregon	$65 million in damages
Nov.- Dec.1977	Western Oregon	Rain on snow event; $16.5 million in damages
1979 - mid 1980s	Harney County	Cyclical playa flooding on Harney & Malheur lakes
July 1995	Fifteenmile Creek	Flash flood in Wasco County
February 1996	Nearly statewide	Damages totaling over $280million
November 1996	Southwest Oregon	Flooding, landslides, debris flows; 8 deaths in Douglas County
May - June 1998	Crook County, Prineville	Ochoco River
November 1999	Coastal rivers, Lincoln & Tillamook counties	Heavy rainfall and high tides
July 2002	Wallowa County	Flash flood above Wallowa Lake damaged Boy Scout Camp.
August 2003	City of Rufus	Flash flood (Gerking Canyon)
Dec. 2005 – Jan. 2006	Western and Central Oregon, Malheur County	Multiple heavy precipitation events on snow and/or saturated ground

Source: Oregon Office of Emergency Management 2006.

Eastern Oregon is subject to flash floods: the state's most disastrous flood was in 1903 when over 200 people were killed in Heppner (top photo). Western Oregon river systems are prone to frequent, broad-scale flooding like that of the Willamette (shown in the lower photo in the 1940s near Independence). (Top: Oregon Historical Society, OrHi24345; lower: Oregon State Archives, Oregon Water Resources, OWR0084)

225 of the town's 900 people. Eastern Oregon cloudbursts have caused other spectacular floods, including those in the John Day basin's Meyers Canyon (1956) and the Umatilla basin's Lane Canyon (1965). These are among the largest flash floods ever witnessed in the United States, with peak runoff on the order of 4,000 to 5,000 cfs for each square mile of drainage area. The 1948 flooding of the Columbia River wiped out Vanport, a community located near what is now Portland's Interstate Bridge (see Chapter 11). The largest floods in Oregon during the twentieth century occurred in late 1964 and early 1996. Both were caused by heavy, warm rainfall melting large snowpacks (Paulson et al. 1991). Major floods on the Willamette River also occurred in 1861, 1881, 1890, 1901, 1903, 1907, 1909, 1923, 1943, and 1945 (Water Resources Committee 1955).

Groundwater

Once rain or melted snow is accepted into the perfect darkness of subterranean Oregon, it begins an often long and poorly documented journey. All that is known for sure is that it will re-emerge, somewhere, sometime, in a stream by daylight or starlight, or up a well, out a faucet, and onto someone's dirty dishes. The journey can take a week or ten thousand years. Groundwater is just surface water waiting to happen—under the strict auspices of gravity and geology. How fast it goes and where it ends up is determined by a region's rocks and their history. Oregon's four major regional aquifer systems are summarized in this section. But first, a word about groundwater in general may be helpful.

⊡ *An Introduction to Groundwater*

Very simply, when water soaks into the ground it becomes groundwater. Gravity pulls it down through pores and fractures in soil and rock. Porosity is a measure of how many holes there are in a rock. Permeability is a measure of how well those holes interconnect and therefore of how easily water moves through them. (Some rocks, like pumice, can be extremely porous, but not very permeable.) Thus, geology and groundwater are inseparable. (In fact, the professionals who study groundwater are called hydrogeologists.) And Oregon's geology is a fascinating mess. It has been fractured, squeezed, exploded, extruded, seared, and smoothed-over. It is a geology of volcanoes, erosion, and sedimentation. Except for a few lava or limestone caves, it just doesn't supply passageways for "underground streams." Groundwater movement is more diffuse than that, having more in common with membranes than pipes—or with electricity, always finding the path of least resistance.

After traveling through pores, joints, fractures, or between beds, water accumulates at some depth determined by the rate of soaking (called recharge) and the type of rock (see Figure 9). It can come from rain, melting snow, and rivers, or from human activities such as irrigation. The level where water first completely saturates the surrounding material is called the water table. In most parts of Oregon, the water table is higher in winter than in summer. Like its surface counterpart, groundwater moves downhill—only, because there are no hills beneath us, "down gradient" is the term most often used. Some water can get caught and held (say by a clay layer) above where the rest of groundwater goes to. This is called a perched

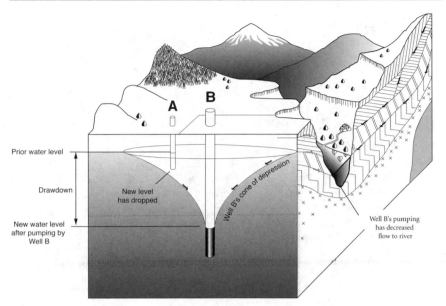

Figure 9: Dynamics of groundwater flow. Layered rock formations in much of Oregon route groundwater to streams and wells. Water from rain or melting snow moves downward between layers or through fractures until it emerges in creeks or is pumped by wells. In the diagram above, a new well (B) is drilled near a shallower, smaller well (A). When B is pumped, a "cone of depression" develops around it, like a milkshake's dimple around a straw. As more water is pumped, this cone can expand under other wells causing water tables to drop or it can intercept water otherwise bound for creeks and springs.

water table. Where a rock formation holds enough groundwater in a way that allows wells to yield usable amounts, it is called an aquifer. Oregon's most productive aquifers are the sands and gravels of the Willamette Valley and the basalts of northern Oregon.

The act of groundwater surfacing is called discharge. In Oregon, recharge typically happens in the mountains and discharge in lowlands. Discharge can be dramatic, such as when groundwater picks up enough momentum during its travel to spurt out under pressure, giving rise to artesian wells or springs. For example, the Metolius River, which jumps nearly full-sized from the base of Black Butte, is a discharge point for Cascade recharge. Or discharge can be more hidden, such as the quiet feeding of streams through the summer. Depending on the time of the year or the location of the channel, streams can feed groundwater or vice versa.

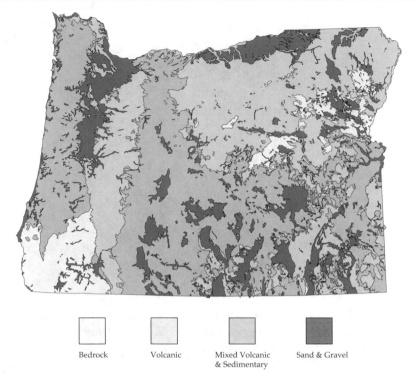

Bedrock   Volcanic   Mixed Volcanic   Sand & Gravel
                                    & Sedimentary

Figure 10: Oregon's major aquifers. Though not scientific, and surely an oversimplification, one can still get some sense of Oregon's aquifer character by visualizing the dark areas of the map above as being more, and the light areas less, "absorbent." (Source: Information from Gonthier 1985 and McFarland 1983)

It may also be worth noting that, compared to surface water, groundwater is poorly understood. No one knows the total quantity of groundwater in Oregon, let alone how current or future use will affect water tables or streams. Unlike streams, which drain to convenient channels and are visible and easily used, groundwater can be all over the place, hidden, and relatively hard to get at. The Oregon Water Resources Department maintains a statewide observation network of nearly 350 wells—roughly one for every 300 square miles of Oregon, if they were evenly spaced. Besides not being very dense, the network was originally established to track complaints of water users, not to provide an independent, scientific measure of Oregon's aquifers. Consequently, some aquifers are well represented in the network, others are skipped altogether. A comprehensive statewide groundwater resource assessment has fallen victim to budget cuts, as has the state's hope of adjusting the network to monitor each major aquifer's recharge and discharge (Lissner 1996). However, the department is working with the U.S. Geological Survey, local governments, and other state agencies on more site-specific regional assessments that are yielding lots of new information on Oregon's groundwater resource.

⊡ *Oregon's Major Aquifers*

This section deals with aquifers in big brush strokes. It is entirely possible to encounter local areas of abundant groundwater in water-stingy aquifers, or to have miserable well yields in areas of plenty. The data are simply not good enough for predicting water conditions at a specific location. The information offered in this section, unless otherwise noted, is based on or developed from the work of the U.S. Geological Survey in the mid-1980s (McFarland 1983; Gonthier 1985). In a broad sense, Oregon has four major aquifer groupings, as described below and shown in Figure 10.

BEDROCK AQUIFERS

Oregon's oldest rock formations occur in the southwest and northeast corners of the state—namely, the Klamath and Blue mountains, respectively. Most of the rocks have been folded, compressed, or otherwise subjected to great pressure. This has ruined them for water-transmitting purposes. Unless they have been weathered into something more spongy, the rocks have very low permeability. There is a limited amount of water movement through fractures, but these close up with depth. Wells generally yield less than 10 gallons per minute and consequently serve mostly domestic and livestock uses.

VOLCANIC AQUIFERS

Much of Oregon is built from material spewn, strewn, or spouted from different kinds of volcanic eruptions. Oregon's volcanic past has had profound effects on its groundwater present, creating some of the state's most productive aquifers. The Cascade Mountains range is probably the most obvious volcanic feature of the state, due in no small part to the majesty of snow-capped volcanoes like Mt. Hood or the Three Sisters. However, a feature that covers even more area is both less visible and more important for groundwater—the basalt flows covering large parts of eastern Oregon and a portion of the northern Willamette Valley.

CASCADE AND OTHER VOLCANICS. Although unnoticed by most of us, Oregon has two Cascade Mountain ranges—at least geologically speaking.

One stands new, shiny, and high; the other is older and more worn down. The older range is called the Western Cascades; the newer (the one ornamented with snowcaps) is called the High Cascades. The Western Cascades is a layer of formations, with older explosively produced material capped by younger lava flows. Altogether, these rocks can be as much as 10,000 feet thick. However, they do not amount to much in the way of groundwater production; most wells have domestic-sized yields of 20 gallons per minute or less.

The High Cascades aquifers on the average have the same small yields. However, with volcanic domes, cinder cones, glacial deposits, ash and lava flows, the High Cascades are more diverse volcanically and can have large variations in water-bearing character. These aquifers have enormous volumes of debris from volcanic explosions. Recharge from rain and snow is very high. Ultimately the groundwater discharge feeds many major river systems, but is especially important in the Deschutes and Klamath. The water table is deeper than in the Western Cascades, on the order of hundreds rather than tens of feet. Springs are common and some of those are quite large—like Medford's water source which can supply 26 million gallons per day. Also, most of western Oregon's hot springs are found where the High Cascades and Western Cascades zones meet.

Some other noteworthy volcanic aquifers are located in Wheeler and Grant counties. The Clarno and John Day formations are more famous for their fossils than their water, with yields of generally less than 10 gallons per minute. However, this limited resource is locally important for domestic use.

While none of these volcanic aquifers will win any prizes for production, it should also be noted that the country they underlie is high, wide, and lonesome. There just are not very many wells from which to draw a highly detailed portrait of these mountain and desert aquifers.

BASALT FLOWS. Compared to other lavas, basalt is positively runny. If Oregon were a pastry, most of its mountains would be made from a stiff dough, while its basalt landscapes would be from a batter. These landscapes are mostly in eastern Oregon where, about 15 million years ago, the basalt flows erupted simultaneously along many extended fissures. Although there are basalt flows throughout eastern Oregon, the most evident (and hydrologically important) are those of the Columbia River Basalt Group.

A typical flow of the Columbia River basalts covered the ground at a pretty good clip—some estimate 30 miles per hour—and pooled to a depth

Basalt flows, like these exposed in the John Day basin, can store a lot of water between layers and in fractures—but they can also be emptied quickly if pumped heavily. (Photo by author)

of 100 feet thick. Thick flows could take decades to cool. This slow cooling led to the basalts' characteristic columnar jointing. The period between flows could be tens or thousands of years (McKee 1972). When one flow topped another, a contact zone was developed. Both the jointing and the contact zones are important for storing and transmitting groundwater.

Some of the basalt erupting in northeast Oregon and eastern Washington flowed down the channel of the ancient Columbia River and into northwest Oregon. This basalt is visible as the walls of the Columbia Gorge and crops out in valley hills such as Cooper and Bull mountains, the Chehalem Mountains, the Eola-Amity hills, and the South Salem hills. It also shows up around Molalla and in the Clackamas and Bull Run river drainages. Its maximum thickness (all flows added together) is about 1,500 feet. While basalt itself is not very permeable, water can travel readily through fractures, joints, and interflow zones. Where wells penetrate several of these flows (a controversial practice given state regulations about commingling, explained in Chapter 10), well yields may exceed 1,000 gallons per minute. More typically, yields are on the order of 100 gallons per minute or less, with water tables at 300 feet or deeper.

Most of the basalt stayed in eastern Oregon where it flooded, and now buries, the pre-existing topography primarily north and west of the Blue Mountains. Oregon's share of the 50,000-square-mile basalt block known as the Columbia River Plateau is called the Deschutes-Umatilla Plateau. Here, the basalt flows form a single unit that can be several thousand feet thick. The upper 1,500 feet, however, is the zone tapped for groundwater. Some of Oregon's deepest water wells occur in the Umatilla Basin and approximate this depth (Lissner 1996). In the uplands, the basalts are cut deeply by major rivers such as the Deschutes, John Day, Grande Ronde, and

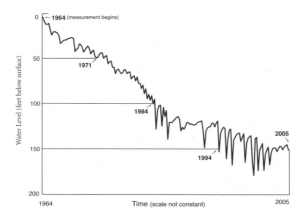

*Figure 11: Water level declines in state observation well 502, Linn Co., Oregon.
Wells decline in the well-watered Willamette Valley, too. This example shows a drop
of 150 feet over 40 years. Water levels go up and down with wet and dry seasons
— the problem often being that they go down further than they come up the next
year, creating a downward trend. (Source: Data from Oregon Water Resources
Department)*

Imnaha. Beneath these uplands, the water table is usually several hundred
feet below the surface. Irrigation is the primary use of the Columbia River
basalt aquifers, and wells for this purpose may produce between 500 and
2,000 gallons per minute. The basalts are a particularly important water
source in parts of Wasco, Morrow, and Umatilla counties.

In some ways, basalts present the best and worst of aquifer characteristics.
On one hand, wells that tap them can have high yields of good quality
water and be economic drivers of local communities. On the other, when
it comes to water, the basalts are more battery than generator. They have
a low recharge rate, low permeability, and low storage capacity. When too
many big wells start pumping too close together, water tables can decline
dramatically (Figure 11). Most of the state's critical groundwater and
groundwater limited areas (see Chapter 4) involve declines in Columbia
River basalt aquifers. In some areas, such as around Hermiston, water
tables have been dropping 10 to 15 feet per year (Oregon Water Resources
Department 1993b). The Water Resources Department has measured total
declines of over 400 feet in some locales—which throws an interesting
light on a nationally acknowledged trouble spot: the Mid-West's Ogallala
aquifer which dropped, on the average, 10 feet between 1940 and 1980
(U.S. Department of Energy 1996).

MIXED AQUIFERS (VOLCANIC AND SEDIMENTARY)

Just as volcanoes or vents deposit material, so too can water and wind. Sometimes all work together to create aquifer systems. The Coast Range is a good example of a joint project between volcanoes and water. So is the area just east of the Cascades which broadens to include most of southeast Oregon.

Much of the Coast Range consists of beds of sand, silt, or clay washed off the continent into the ancient ocean. In some areas, these beds are more than 15,000 feet thick. Lava was also erupted in this submarine environment and congealed in a bubble-wrap pattern called "pillow" basalts. Today, recharge of these tight-grained rocks is low. Well yields are small, ranging from 5 to 20 gallons per minute, though an occasional gusher might produce up to 200. Water can be found at depths between 50 and 200 feet. However, these aquifers can provide some of the poorest quality (that is, mineralized) water in Oregon, with up to 1,000 milligrams per liter of dissolved solids (which is about five times greater than most good quality water elsewhere in the state). In some areas of the Willamette Valley where these formations occur, the concentration may exceed 10,000 milligrams per liter at depths of only a few hundred feet.

A band of volcanic and sedimentary rocks parallels the Cascades to the east and extends across most of southeast Oregon. These formations consist of layers of lava flows, sand, gravel, pumice, ash, clays, and silts. The lava flows predominate in uplands and the sedimentary deposits in lowlands. Across southeast Oregon, water may be found at depths of several hundred feet. Older sedimentary deposits are often topped by younger material washing off today's uplands. Wells can yield more than 250 gallons per minute where these aquifers are tapped, such as in Harney, Fort Rock, and Christmas valleys, as well as the areas around Klamath Falls, Lakeview, and Vale-Ontario.

In north Central Oregon, volcanic and sedimentary rocks are the primary aquifer for the Bend-Redmond area. Lava beds cap several hundred feet of sediments and volcanic debris. The water table at Bend is typically 500 feet or more deep, but only 200 to 300 feet near Redmond, because the land surface slopes downward to the north.

A recent study of the Upper Deschutes River basin suggests groundwater moves through young, permeable volcanic deposits first eastward from the Cascades and then northward. North of Madras, these permeable deposits become increasingly thinner as they meet more impermeable formations, much as a lake shallows out as it encounters the shore. When groundwater runs into this relatively water-repellant layer, it is forced to the surface where it creates an enormous and very stable discharge to the Deschutes River system—over 2,000 cfs near the confluence of the Deschutes and Crooked rivers (Sherrod, Gannett, and Lite 2002).

Annual recharge appears to be swift, and water moves quickly down and through the aquifer. Groundwater levels respond rapidly to changes in river levels and to irrigation canal seepage—even at depths greater than 600 feet. Groundwater seems to flow primarily from the Cascades to discharge areas near Lake Billy Chinook through basalt fractures and coarse-grained volcanic and sedimentary materials (Gannet and Lite 1996). Wells in this area most often tap the volcanics and sediments, but some are developed through lava beds. Yields may be as much as 500 gallons per minute.

SAND AND GRAVEL AQUIFERS

The sands and gravels of Oregon are probably the single most important component of groundwater supply in the state. Sands and gravels typically fill lowlands, stream valleys, and coastal benches. To a greater or lesser degree, they are usually accompanied by other sediments such as silts or clays. Their biggest concentration is in the Willamette Valley, where at Portland they reach a thickness of over 1,000 feet.

This incredible amount of sediment is an inheritance from an equally incredible series of events known as the Missoula Floods. Between thirteen and nineteen thousand years ago at the close of the last Ice Age, ice dammed streams in western Montana, forming huge lakes. The dams gave way, creating inconceivable floods, which flowed on the order of 17 million gallons per second. The floods thundered across eastern Washington, raged down the channel of the Columbia, and swept up the Willamette Valley, creating a murky lake 100 miles long, 60 miles wide, and 300 feet deep from which sediments settled out (Bishop 2003).

Valley well yields seem to increase from south to north, from less than 50 to around 300 gallons per minute, though some wells may produce up to 2,000 gallons per minute. Water tables vary from around 25 to 50 feet deep. South of Oregon City, the uppermost component of the Valley's fill is lake silt and sand as much as 130 feet thick. However, this layer is not present in major flood plains or above 500 feet. The flood plains of the McKenzie, North and South Santiam, and Willamette rivers are underlain by up to 50 feet of highly permeable sand and gravel which freely exchange water with these streams. In general, the sediments on the west side of the valley are finer grained than on the east.

In western Oregon outside the Willamette Valley, the sand and gravel is rarely more than 150 feet thick. While more silt and clay is present, the hydraulic connection with streams is good, resulting in respectable well yields of between 20 and 100 gallons per minute. Other locally important aquifers in this grouping are the sand dune formations along the coast. Most extensive between Astoria and Seaside and from Heceta Head to Coos Bay, dunal aquifers can be 200 feet thick and support wells with yields up to 250 gallons per minute—but 50 is more the average.

Eastern Oregon also has significant sand and gravel aquifers. To the north, sands, gravels, and glacial sediments hold the most water and are typically hundreds of feet thick, but in the Grande Ronde Valley can exceed 2,000 feet in thickness. In some places, the sand has been consolidated into sandstone beds. Important sand and gravel aquifers are located in the Tygh, Grande Ronde, and Wallowa valleys, and near Milton-Freewater and Hermiston, where they are heavily tapped for irrigation supplies. In southeast Oregon, sands and gravels have been deposited on the floors of the interior basins. These materials, which also include clay, silt, and volcanic ash, have been carried by streams, worked by lakes, or dropped by the wind. The best water-bearing materials are usually the coarse deposits found where streams empty onto the basin floors, such as the Silvies River in the Harney Valley, where wells can produce from 300 to 1,000 gallons per minute.

PART II

Dividing the Waters:
The Decision-making Process

Whiskey's for drinking; water's for fighting.

Attributed to Mark Twain

Chapter 2

Prior Appropriation:
The Code of the New West

Today, Oregonians are fighting over water. Farmers and ranchers feel under siege by environmentalists who want more restrictions on agricultural water use. Environmentalists are feeling increasingly desperate as populations of wild salmon and other water-dependent species—never fully worked into Oregon water law—plummet. Booming cities, pinched for water supplies to serve more people and industries, are jockeying to secure long-term water supplies. Sounds a lot like 1909.

Prior Appropriation in Oregon

By 1909, the Oregon country had already been plowed or paved for three generations. Settlement had been guided by water. Fishing, farming, ranching, cities, and industry all drew upon and grew upon Oregon's water. Across Oregon, people were elbowing each other out of the way to keep using "their" water and to get more for the new century. The twentieth century promised big dividends in growth, but growth required water. Oregonians needed to stop fighting and come up with a new way to share water—some way to both settle the disputes and make room for new development. As one irrigator put it, "It is imperative that, before the waters of any stream within the State of Oregon can be put to the most economic use, ... there must be an administrative system. ... There is now ... no way provided to enforce [the law], except the farmer's weapons—the pitchfork and the shotgun" (*Oregonian* 1909).

Getting things straight water-wise was not just a matter of good government acting on sound principle. It was also a matter of capitalists jumping at a great offer. Around the turn of the century, the U.S. Reclamation Service was prepared to plumb the West with irrigation projects, greening both wasteland and wallets. It was not prepared, however, to lay any pipe unless the states could guarantee a sound, up-to-code water right structure. The Service became a powerful advocate for the creation of a workable system of water right laws (Dunbar 1983).

Wyoming was the first state to respond, passing a new set of water right laws in 1890. This legislation served as the legal template for all western states (except Colorado) many of which adopted state water codes before Oregon (Gould 1990). Wyoming-style provisions were submitted to the Oregon Legislature in 1905 and 1907, but were not passed. So a group of business and government leaders from the Oregon Conservation Commission, the Portland Board of Trade, the State Grange, and the Oregon Irrigation Society lobbied the 1909 Legislature to pass a new law declaring water a public resource and requiring a permit for anyone to use it. Most important for the times, the proposed legislation included a court-based process for settling existing water right disputes. The introduction of the courts as decision makers (rather than agency engineers or administrators) was a new twist in the field of water rights determination, and became known as the Oregon system (Dunbar 1983; State Engineer 1910).

The legislature bought the idea and enacted a sweeping new law on February 24, 1909, to the enthusiastic reaction of old-style and apparently learned bureaucrats that were allowed the luxury of opinion:

> Under the old law, no foundation existed for titles to water.
> Utter confusion prevailed as to the legal status of a water right.
> Litigation among water users grew to such an extent as to
> prove a serious burden upon irrigated agriculture. Dams and
> flumes were destroyed and lives threatened in community
> quarrels to secure a proper division of streams. Under such
> conditions capital declined to invest, and home-seekers went
> to other states where the purchase of a water right did not
> mean the purchase of a lawsuit. To remedy these conditions,
> a complete code of law was enacted which became effective
> February 24, 1909.
>
> The ultimate object of this law is to secure a proper division
> of streams among those entitled to its use. It is not supported
> by any constitutional provisions relating to water, but rests
> entirely upon the police power of the State to preserve the
> public peace and safety.

> Its enactment is of as great importance to Oregon as was
> the making of the 'Doomsday Book' in 1085, by William
> the Conqueror, which was the first attempt in England to
> systematize land titles. (State Engineer 1910)

The Oregon Water Code established four general principles which still hold today:
- Water belongs to the public
- Any right to use it is assigned by the State through a permit system
- Water use under that permit system follows the "prior appropriation doctrine"—older water uses are entitled to water before newer uses
- Permits may be issued only for beneficial use without waste

The primary aims of the new approach were to control chaos (through both the state's permit system and its maintenance of a central bank of water right records), and to prevent the assertion of excessive and often speculative claims to water. These principles meant a big change from previous practice. Prior to 1909, two somewhat competing approaches to water use co-existed in Oregon. One invoked the riparian system based on the common law of England. Under a riparian system, land owners have rights to use water from streams running through their property. In England, that right was to use water as long as flow was left undiminished for the next downstream landholder. In America, the riparian system evolved to allow use of water as long as enough is left for downstream property owners to make reasonable use of the streamflow (Gould 1990).

Riparian systems are designed for regions with humid growing seasons, such as England or the eastern United States, where off-stream landowners can get along fine without access to flowing water. However, it is not one that works well where crops need more water than can be supplied by rain, or where extensive land areas are far-removed from water sources—such as the American West. Without a system to share water, many nineteenth-century citizens feared the West would remain more dust than destiny, manifesting not bustling cities and wholesome farms, but panoramas of prickly pear interrupted only occasionally by shantytowns shimmering in the heat.

To get the most out of the West (including Oregon) in the shortest time possible, a system was needed that optimized use of a scarce resource for development. Thus, a new approach grew out of necessity: the prior appropriation system. This system not only allowed, but encouraged,

Riparian   **Prior Appropriation**

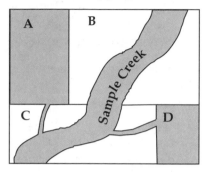

Figure 12: Comparison of riparian and prior appropriation systems. Under the riparian system, property owners B and C are entitled to water from the creek, as long as their use does not diminish downstream landowners' ability to make use of the creek also; but property owners A and D are out of luck. Under the prior appropriation system, B and C have no automatic right to use the creek, while A and D can get water rights to use the creek—even to the point of leaving no water in the creek for downstream users.

diverting water from streams and delivering it by ditch to wherever it was needed (Figure 12). At first diversion was mostly for mining; later, for farming; still later, for cities. Relying on the riparian system would have confined development and settlement to narrow Western river corridors and left vast expanses of unwatered wasteland—at least to the development-minded. Also, leaving water in-stream riparian-style (i.e., largely undiminished for someone else to use) hardly seemed a prescription for rapid development, either.

The prior appropriation doctrine got around this problem by allowing the transport of water far from river courses. It also provided an important incentive for doing so: certainty. Those who took the trouble to dig ditches or build diversion dams would be protected from others who might demand water later. No new user could take water away from a pre-existing user; new users would have to wait their turn and divert water only after older uses were satisfied. In short, prior appropriation is a doctrine of "first come, first served"—and it is still at work today: every summer state watermasters visit hundreds of water users and order them to "shut off" so that older water rights can be served.

The federal government recognized the value of this approach in at least two laws vital to the settling of the West. In the Mining Act of 1866,

Oregon's water rights system, like those of other western states, has its roots in mining. To these hydraulic miners near Medford in the 1880s, "beneficial use" meant hosing hillsides (and, as we know today, ecosystems). Today, with increasing demands, water is becoming as valuable as the gold it once dislodged. (Oregon Historical Society, OrHi13222)

Congress honored rights for mining and agriculture claimed under local laws by private parties already using water on public lands. In the Desert Land Act of 1877, it went a step further by not just allowing such claims, but by declaring any rights to use water must be grounded in "bona fide prior appropriation." The first appropriative water rights statutes were passed by California in 1872 and allowed for the establishment of rights by posting claims and proof of actual use. This type of legislation was enacted Westwide in the 1880s and '90s (Johnson 1992).

In Oregon, through the late 1800s, while the courts upheld riparian rights when they were disputed, the legislature was favoring prior appropriation, in part through laws requiring water claims to be publicly noticed (Clark 1983). This notice was served by literally "posting" at the proposed diversion site a sign describing a person's intent to use the water. This gave other users a chance to counter the claim if it would interfere with their rights. The state's objective then, was pretty much as it is today: to protect older ("senior") users. However, some riparian rights still exist, as they were established and recognized prior to the passage of the 1909 Water Code (U.S. Bureau of Land Management 2001).

Unlike some other western states whose water code is contained in their constitutions, Oregon's is established in state law: Oregon Revised Statutes (ORS) Chapters 536 through 543 and ORS 545. (Note: here, as elsewhere in this book, the statutes refer to those as found in the 2003 Oregon Revised Statutes. These statutes are the laws of Oregon, which are amended and maintained by, and are available on the Web or in print from, the Oregon Legislature.) These statutes, along with the administrative rules implementing them (Oregon Administrative Rules, Chapter 690), reflect the principles of Oregon's Water Code and lay out the thousands of factors and regulations which drive today's water management decisions. The most fundamental of these decisions is how the state gives permission for use of public water: the water rights process.

Water Right Basics

"Water from all sources of water supply belongs to the public" (ORS 537.110). Permission to use the public's water is granted through a state permit called a water right. Requests for water rights are approved or denied by the Oregon Water Resources Department. As of 2006, the department had records of over eighty-two thousand rights. (Table 14.)

The act of deciding whether or not to issue a water right is at the very core of Oregon's water management system. For better or worse, it is still the primary way the state attempts to meet increasing water demand. This decision-making process is frequently the first place that citizens, eager for a little water for a prized project, are exposed to the confusing mosaic of water law. Increasingly, it is also where thorny policy issues are raised and hotly debated. Understanding what a water right is, when it is needed, and what it allows is a necessary first step in grasping Oregon's water management system.

Given the importance of water to practically every Oregonian, it is surprising how poorly understood water rights are—even by those lucky

Table 14. Water rights in Oregon (primary only).	
Type	Total
Surface water	51,683
Groundwater	15,754
Storage	14,819
Total	82,256

Source: Oregon Water Resources Department. 2006d

enough to have them. Despite chronic water difficulties through the decades, the perception of Oregon as a water-rich state has prevailed. For the most part, water supply has simply not been worth worrying about. Getting a water right has been a formality. Water use requests were not heavily scrutinized, and neither the public nor the state monitored water rights very closely. All that has changed.

The rest of this chapter explores key provisions of Oregon's water right laws, which were developed both to control and to promote taking water out of streams (or the ground) and putting it somewhere else, usually for profit. Recently, the state has been attempting to bring some of the law's mechanisms into play to keep water in streams or stabilize groundwater levels. These attempts are described in Chapter 4. However, because most of the laws were written for traditional consumptive uses—the kind that cities, farmers, or property owners most often request—so too is the remainder of this chapter (unless otherwise noted). Because in water law there seems to be an exception for everything, what follows should be approached as a general overview rather than the last word in water use legalities.

▣ Water Rights

A water right is the legal authorization given by the state to a party to use public water in a specific way for a specific purpose. It is not a title to the water itself. Only the public owns the water. That means that the creek running through the back forty, a llama-crammed 5-acre hobby farm, or a county park is not anyone's to use as they please. Nearly always (see the exceptions sections, below), they have to ask for and receive state permission first. The same goes for any substantial use of groundwater.

The bottom line on a water right is that it has to be for a beneficial use without waste (ORS 536.310(1)). While, on its face, beneficial use may seem an obvious starting point, it has some problems as a central organizing principle of an entire system of water law.

First, the very concept of beneficial use is predicated on there being "good" uses and "bad" uses. But our understanding of good and bad can change: during the goldrush, dissolving eastern Oregon hillsides with hydraulic mining was considered a good use, while keeping water instream was bad. Second, a proposed use is almost always beneficial, if to no one other than the proposer. Benefit is in the eye of the beholder and therefore doesn't call into sharp relief the true character of the use. Third, reliance on the concept has created lists of hallowed Beneficial Uses, shifting the system away from analyzing the impacts of a proposed water withdrawal, and toward a bureaucratic obsession with nomenclature. Naming what a water right would be used for ("agriculture") is far less important than nailing down what it would do ("take 0.25 cfs from the 5 cfs flowing in Drift Creek"). Getting hung up on the names of uses creates pointless debates about which use is better than another: a strawberry farm is not inherently any more, or any less, beneficial than a fish hatchery. Both are valid uses, worthy of consideration. However, the effects of either on other users or the public can vary a great deal—and that's where the debate should be.

While failing to provide a definition, the statutes do offer a list illustrating beneficial uses (ORS 536.300), many of which are described in greater detail in Oregon Water Resources Department administrative rules. The "without waste" mandate has long been the fine print of every right—though usually overlooked, or at best poorly understood. Waste is not well defined in statute, by rule, or in court decisions. Consequently, enforcing against waste has been very rare (see Chapter 6).

A water right establishes:
- the beneficial use;
- the specific source that water will be taken from and the point of taking;
- the priority date, which determines where the holder stands in relation to other water right holders;
- the maximum amount of water that can be used;
- exactly where the water will be put to use; and
- any other use-related requirements imposed by the department.

These elements bound the right. That is, a water right holder is prohibited from doing anything other than using water exactly as laid out in black and white (at least without permission; see Transfers in Chapter 5). For example, deciding to water a new field not identified in the right—even if an old field of the same size is retired and/or no additional water is used—would technically be illegal. Similarly, it would be illegal for a water right holder to start using water for a recreational vehicle campground when her right specified the use was for dairy purposes.

▣ Basic Rules

There are a number of other basic rules for water rights. The first is implied by the assignment of the priority date: even if you get a water right, you will have to stop using water when people with older water rights need it—not because you did anything wrong, but because that's "rule number one" of the prior appropriation doctrine (see Figure 13).

The second basic rule is "use it or lose it." A water right must be exercised at least once every five years to be valid. If it is not, it is lost, or "forfeited." The rule of forfeiture reflects the development-oriented underpinnings of Oregon water law: water is too important to be tied up by slackers or speculators. Either use it or get out of the way of those who will. The sole exception to this is for cities, whose certificated rights are generally not subject to forfeiture. When the Water Resources Department discovers forfeiture, it is required to begin proceedings to cancel the right—that is, to get it off the books and free the water up for someone else. Usually these proceedings end up in a contested case hearing (see glossary) where water right holders object to—and often their feuding neighbors cheer— cancellation of the right.

Third, under most circumstances, the right attaches to the land where water is used (presumably exactly as detailed in the right). This legal attachment is called appurtenancy. When the land is sold, the right goes with it. A little-known Oregon law requires that when sellers accept an offer, they inform buyers in writing whether there is a valid permit and deliver any valid permit at closing (ORS 537.330). Water rights are only infrequently recorded in deeds. Buyers should be absolutely sure that sellers have been legally using water before making financial commitments. It is not uncommon for landowners to believe they have the proper permits

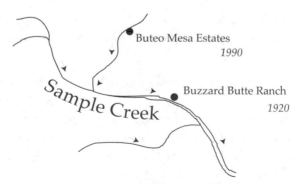

Figure 13: How priority dates work. On Sample Creek, Buteo Mesa Estates must stop using water anytime Buzzard Butte Ranch doesn't get the amount allowed under its water right. The reason: Buzzard Butte's priority date is seventy years ahead of Buteo Mesa's. Even though Buteo Mesa is on a tributary, the water is on its way downstream to Buzzard Butte and beyond. So the former shuts off, whether the use is for a fish hatchery, city, or corn field and despite the consequences.

when in fact they do not. Buyers should ask to see copies of the valid permit or check with the Water Resources Department.

Fourth, water rights are forever. In only a few recent instances have water rights been issued for time periods other than perpetuity. Unlike other state permits, such as for grazing, discharging pollutants, or for that matter, fishing or driving, once a water right is issued, it never expires. While this offers the utmost in certainty to water right holders and their community economies, it means there is next to no opportunity to adjust the water use to reflect evolving societal needs. This is a remarkable benefit bestowed by the state—and for a pretty good price. Aside from handling fees at the beginning of the process, a water right holder pays the state nothing for the use of public water.

Another basic of water rights is understanding when one is needed—an understanding that warrants its own section.

When Is a Water Right Needed?

Except for a short (but growing) list of exceptions, a person wishing to divert any water from a creek or pump significant amounts from a well must first get state permission:

> … no person shall use, store, or divert any waters until after
> the department issues a permit to appropriate the waters. ORS
> 537.130(2)

> … the use of ground water for any purpose, without a permit
> … or registration … is an unlawful appropriation of ground
> water. ORS 537.535(2)

For most uses, the state grants permission through a water right (though there are other types of authorizations, as described in Chapter 5). The safest and simplest rule before using water is to assume that some kind of permit is required for any use and to check with the Oregon Water Resources Department. The actual test of whether a right is required involves three factors.

First, is the source included in "waters of the state"?

> "Waters of this state" means any surface or ground waters
> located within or without this state and over which this state
> has sole or concurrent jurisdiction. ORS 536.007(12)

What kinds of sources would *not* meet the definition? Ocean water (while the state has never required a permit to use sea water, a strict reading of the definition above may cast doubt on this landlubber perspective), rain water, and springs that rise on but do not leave a property offer some examples. For instance, you would not need a water right to capture water in a rain barrel. At that point the water has not yet entered into the domain of waters of the state.

Second, is there an act of physical control (statutorily waived in the case of instream water rights; see Chapter 4)? In other words, is there damming, diverting, pumping, piping, or any other action that would steer water from where it naturally would be to where somebody wants it? There are some important water uses that do not involve physical control and that therefore do not require water rights. For example, when livestock drink directly from a stream, no water right is needed because there is no physical control of water.

Third, is the water to be put to a beneficial use? Because most people would not take control of water if they did not somehow benefit, this factor may seem a question hardly worth asking. However, benefit (or lack of it) can be in the eye of the beholder, and differing eyesight can have unexpected consequences. Take, for example, a city that would like to pump 100,000 gallons per day from a 300-foot-deep well to serve its growing population: groundwater is clearly public water, pumping is definitely taking control, and providing houses and industries with water is undoubtedly beneficial. Conclusion: the city would need to apply for a water right. Now take an open-pit rock quarry right across the road from the well. As the operators excavate the pit, they encounter groundwater and pump it out of the pit at a rate of 100,000 gallons per day. They are also taking control of public water, but the intent is not to make beneficial use of the water. Just the opposite. The quarry operators want to get rid of it. Conclusion: no water right needed—no matter what consequence the pumping may have on the groundwater table or nearby wells. (However, in 2003 the Legislature gave the Oregon Department of Geology and Mineral Industries [DOGAMI] the authority to limit mine operations to protect groundwater resources [ORS 517.835].)

This illustrates an important characteristic of Oregon water law: it can often have fairly nutty outcomes. Its design is based largely on a nineteenth-century code of conduct: get cracking and use all that water otherwise wasting to the sea by taking charge, putting it where it is needed, and transforming wasteland into Eden. A system true to its time and successful in its day, but proving an odd fit for today's needs.

And as growing demand leads more and more people to first-hand encounters with the system, either to gain approval of new uses or legalize "historic" (historic sounds better than "illegal") uses, there is often one common reaction: "You've got to be kidding."

▣ *Exemptions and Other Good Intentions*

Most systems by their very operation generate their own exceptions. Often the older the system, the more loopholes: witness the federal tax code or health care regulations. When the unpredictable spectrum of individual need meets the letter of the law, the results are governed by the physics of human behavior. Whether based on the momentum of special interests, the inertia of past practice, the mass of public opinion, or the tug of heartstrings for hard cases, legal provisions are waived, modified, or stayed. Water law is no different.

Most exemptions fall into three general categories: (1) uses too small to worry about; (2) uses that protect the environment; and (3) other uses that represent a grab-bag of special needs. A definition may be in order here. Exemptions are uses that would require a water right if they were not otherwise singled out in statute for special treatment. For the purposes of this book, they do not include uses that fall short of the three-point threshold described above, or uses that require some other form of formal authorization (Chapter 5). Table 15 displays a summary of exemptions.

USES TOO SMALL TO WORRY ABOUT

This category is composed of the specific groundwater exemptions found in ORS 537.545, most of which were first adopted in the 1955 Groundwater Act. It is believed these uses were exempted because they were too small to worry about at the time (though whether that is still true is open to debate). In some ways they are souvenirs from a simpler time and reflect a type of homesteading ethic. This exemption category favors homes, gardens, small industry, and livestock husbandry:

> ... no registration, certificate of registration, application for a
> permit, permit, certificate of completion or groundwater right
> certificate ... is required for:
> (a) Stockwatering purposes;
> (b) Watering any lawn or non-commercial garden not
> exceeding one-half acre in area;

(c) Watering the lawns, grounds and fields not exceeding 10 acres in area of schools located within a critical ground water area ...

(d) Single or group domestic purposes in an amount not exceeding 15,000 gallons a day;

(e) Down-hole heat exchange purposes; or

(f) Any single industrial or commercial purpose in an amount not exceeding 5,000 gallons a day.

While they do not require a water right in the paper sense, once established, these groundwater uses are recognized as full rights under the law. They are assigned a priority date and other wells may be shut off to keep them going. They are also subject to the standards of forfeiture and beneficial use without waste.

By far the most widespread exemption is that for domestic groundwater use. Buy a house and you are entitled to a lot of groundwater—15,000 gallons per day per household. No water right required, no questions asked. In fact, this amount would easily serve the needs of at least fifteen families and probably more like thirty. While any family trying to exercise its right to 15,000 gallons would find it hard not to exceed some unquantified threshold of waste, this exemption is now proving to be a problem in some areas.

Table 15. Exemptions from Requirement to Obtain a Water Right	
Type of Exemption	Statutory Basis
Small Groundwater Uses	ORS 537.545
Domestic purposes ≤ 15,000 gallons a day;	
Watering any lawn or non-commercial garden ≤ one-half acre	
Any single industrial or commercial purpose ≤ 5,000 gallons a day	
Stockwatering	
Watering school grounds ≤ 10 acres within a critical ground water area	
Down-hole heat exchange	
Environmental Water Uses	ORS 537.142
Salmon/Trout Enhancement Program (STEP) projects	
Fish passage structures	
Enclosed, self-limiting diversions for livestock	
Emergency firefighting	
Other Water Uses	ORS 537.141
Forest management requiring water for mixing pesticides or controlling slash burning	
Diverting water to tanks from authorized reservoirs	
Using rain water collected from an impervious surface	

No water right is needed to use groundwater for livestock use. It is one of the more common "exempt" uses enumerated in Oregon's water laws. (Oregon State Archives, Oregon Dep't. of Agriculture, OAG0079)

In a number of (usually rural residential) areas around the state, development is contributing to quite a drawdown of the water table. But it is an impact that is mostly off-limits to state control. First, most wells can be drilled anywhere any time in Oregon without a permit, as long as they meet certain construction standards and notice requirements (see Chapter 10). So there is little state leverage in terms of drilling controls. Second, with no water right application to deny or permit to condition, if development is allowed under the local government's comprehensive land use plan, the groundwater use can just happen. The Oregon Water Resources Commission can prohibit exempt uses altogether in an area (and in a practical sense stop most, if not all, development), but it cannot adjust the quantity thresholds set in statute. In other words, state control becomes an all-or-nothing proposition.

Prohibiting exempt domestic uses altogether would often be going too far. It could preclude all new residential development and even single homes, including farm dwellings (held nearly sacred under Oregon's land use planning system). It would have very serious impacts on property owners, while over-protecting the groundwater resource. On the other hand, doing nothing can put at risk the future water supply of a residential area and its neighbors. Lowering the amount of the exemption and closer coordination between the state and local governments have both been proposed as solutions, but whether planning commissions, the legislature, or the public will demand solutions remains to be seen.

Other types of groundwater exemptions also have their peculiarities (e.g., no limit whatsoever on how much groundwater can be pumped for livestock without a water right; the school exemption that was pushed by one legislator to solve one specific situation in one county but is now the law of the land statewide). In summary, these groundwater exemptions have made life a lot simpler for quite a few people, but in some cases are beginning to turn into water management soft spots.

USES THAT PROTECT THE ENVIRONMENT
Another set of water uses deemed special enough to be relieved from the perceived agony of obtaining a water right is that for projects benefiting the environment—for example, salmon and trout enhancement projects. Such projects usually involve citizen volunteers caring for salmon eggs in hatch-boxes fed by streams. No water right is required for any such project certified by the Oregon Department of Fish and Wildlife. Similarly, certain structures (such as side channels or fish ladders) that help fish get over or around instream obstacles do not need water rights either (ORS 537.142).

Until recently, any entity that pumped water from a lake or stream for fire-fighting purposes (a beneficial use) technically had to get a water right. No one ever did, because the system just was not built to accommodate such requests. But, to do away with any confusion or liability on the state's part, the legislature exempted "emergency fire-fighting" from water right requirements. A lot of land management activities have also been freed from water right requirements, such as the construction of terraces, dikes, or retention dams and certain tilling practices, all of which involve slowing or stopping water for a certain period. Another management activity forgiven the need for a water right are projects that deliver livestock water away from streambanks, thus helping protect these ecologically sensitive areas. By exempting projects with enclosed delivery systems and automatic shut-off devices, the state hopes to promote sound riparian area management (ORS 537.141).

OTHER USES
There are a handful of other exemptions inspired, no doubt, by other encounters with the water right system. For example, in recent years, the legislature exempted forest managers' use of water for mixing pesticides or controlling slash burning; diverting water to tanks from authorized reservoirs; and using rain water collected from an impervious surface (ORS 537.141).

It may be worth noting that the trend seems to be for more exemptions. The natural reaction to every difficult encounter is to try to make it go away. But in a practical sense, every exemption lessens protections for senior users. While most exemptions in theory cannot make a "call" on water, and must yield to bona fide water right holders, the state's enforcement structure simply is not up to the task of policing lots of exempt uses in a big state like Oregon. Many exempt users, inadvertently or otherwise, will pump or divert water whenever they need it, despite the consequences to other, technically more deserving, water users. Ultimately, if exemptions do not become the rule and undercut the system altogether, then a growing list of exemptions means less water for water right holders. This seems to be a concept lost on many water use sectors who often champion easing regulations for their own kind. While this birds-of-a-feather behavior is understandable or even normal in most situations, when it comes to water rights, what would make more sense is alliance by age, not type. Senior users have far more in common with each other whatever their type than they do with newer users of their own kind. This is how the system is designed—to protect the club of established users from newcomers.

Oregon and Neighboring States

It may be worth looking at neighboring states to get a sense of how Oregon rates in allocating one of the West's most precious commodities. (Many of the processes and provisions spotlighted below are defined in the glossary and covered in later chapters.)

A short answer to Oregon's grade would be "average" (Table 16). Oregon was neither the first nor last to adopt a comprehensive water code, though it broke new ground in assigning adjudication to courts, was the first to pass long-lasting legislation creating a groundwater appropriation system, and through legislative water withdrawals in the 1920s offered some of the earliest forms of instream flow protection (Dunbar 1983; Shupe 1989). Oregon's deference to the public interest is on a par with (that is, as vague as) most of its neighbors'. Like adjacent states, Oregon employs a three-step water right process and recognizes five years of non-use as the forfeiture threshold. All the far western states have well-developed systems for transferring water rights, though California and Nevada consider the public interest or instream impacts in their evaluation, whereas Oregon, Washington, and Idaho consider only injury to existing water right holders. And, unlike Idaho, which has water allocation provisions in its constitution,

Oregon and its other neighbors rely on a statute-based system of water allocation and management. Finally, like Oregon, all of its neighbors (except California) afford appropriative standing for instream flow protection (i.e., link priority-dated flow amounts with specific beneficial uses through water rights and/or administrative action).

There are a number of differences among the states, however. Keying off California, with its nation-state economy and bellwether reputation in the West, may offer a good starting point for contrasts. For example, unlike most other far west states, California has a plural water rights system, recognizing both riparian and appropriative surface water rights, and has established no groundwater appropriation code, per se. Rather, it divides its groundwaters into three somewhat antique legal classifications. Significantly, most groundwater is presumed to fall into the "percolating" class, and is subject to use without any state permit (Johnson 1992). The tens of thousands of Californians who resettle across the West every year may be surprised that in their new homes permits are required for all but the smallest groundwater uses.

As noted above, California does not recognize water appropriations for instream use. Rather, instream flows are protected by rejecting or conditioning other proposed uses in order to best preserve the public interest, considering instream needs. To instream flow advocates intent on gearing up to change the system, this may seem like trying to drive and shift with one hand tied behind your back, but California has been at the forefront of exploring a potent protection mechanism that may have profound ramifications for all western states—the public trust doctrine.

The public trust doctrine is based on English common law, which held that the public's interest was so bound up with certain activities (e.g., commerce, navigation, and fishing) that it was the sovereign's duty to keep tidal and submerged lands in trust for the people and never sell the lands out from under the public who depended on them. This principle was carried over into the American judicial system and for many years operated mainly to keep states from transferring title in coastal lands to private parties. However, in 1983 the California Supreme Court applied this principle for the first time in the water rights arena (Shupe 1989; Glick 1995a).

The City of Los Angeles holds water rights to the tributaries of Mono Lake in northern California. In diverting water to its thirsty residents, the City caused lake levels to drop, creating both habitat and scenic problems. The National Audubon Society challenged the legality of these diversions

Table 16. Water Allocation Characteristics of Oregon and Neighboring States

Allocation Characteristic	California	Nevada	Idaho	Washington	Oregon
Comprehensive Water Code Adoption Date[1]	1914 (SW)	1905 (SW) 1939 (GW)	1903 (SW) 1963 (GW)	1917 (SW) 1945 (GW)	1909 (SW) 1955 (GW)
Allocation Agency	Water Resources Control Board	State Engineer	Dept. of Water Resources	Dep't. of Ecology	Water Resources Dept.
Source of Authority	Statutes: Cal. Water Code -Sect. 1200 - 1675	Statutes: NRS 533 & 544	Constitution: Article 15, Sect. 3; Statutes: Title 42, Id. Code	Statutes: RCW 90.03, & 90.44	Statutes: ORS 536—543
Public Interest Provisions for Proposed Water Uses	Must "best conserve" public interest. Public trust requires supervision of permitted uses to protect public values.	May not "threaten to prove detrimental to the public interest."	Must be in "local public interest."	May not "threaten to prove detrimental to the public interest."	May not "impair or be detrimental to the public interest."
Water Right Sequence	Application Permit License	Application Permit Certificate	Application Permit License	Application Permit Certificate	Application Permit Certificate
Primary Instream Flow Protection Strategies	Public interest review & conditions on out-of-stream rights; Public trust actions.	State Supreme Court has upheld Engineer's authority to issue water rights for instream uses.	Water Resources Board may apply for and hold instream water rights on behalf of citizens.	Statutory mandates to preserve fish & wildlife flows; administratively established instream flow appropriations.	3 state agencies may apply for instream rights to be held in trust by WRD.

Groundwater Allocation	Most groundwater presumed "percolating," and no permit is required for use.	Permit required except for domestic uses.	Permit required except for domestic-type uses of 13,000 gpd or less.	Permit required except for domestic-type uses of 5,000 gpd or less.	Permit required except for domestic-type uses of 5,000 gpd or less.
Transfer Criteria	May not injure other rights or unreasonably affect instream uses.	May not impair existing rights or threaten to prove detrimental to public interest.	May not injure other rights.	May not injure . other rights	May not injure other rights.
Non-use period for forfeiture	5 years	5 years	5 years	5 years	5 years

1 SW=surface water; GW=groundwater;
Source: Information from: Johnson 1992; Shupe 1989; Western States Water Council 1986

on the basis of the public trust. The court ruled (33 Cal 3d 419) that the diversions threatened public values and that, under its public trust responsibilities, the State of California had no right to approve diversions that broke that trust. Thus, the court found the doctrine not only applied to water allocation generally, but applied retroactively to existing rights (if the public trust was not considered at the time of issuance). It also concluded the public trust doctrine does not merely secure state control of tidal and submerged lands, but extends that control to flowing navigable water.

The Mono Lake decision essentially unlocked the doors to the heretofore members-only club of western water rights, and invited the general public to sample the bar and stroll the grounds. However, there has not been a rush on the system; the public trust in water has only been tested by the courts in a couple of other states. Aspects of the public trust in flowing water have been addressed by the courts in both Washington and Idaho. In 1993, the Washington Supreme Court ruled that the public trust did not extend to non-navigable waterways. In 1995, the Idaho Supreme Court held that the public trust applies to Idaho water rights, but is not at play during adjudication. The public trust doctrine is an element of the body of Oregon law, but no court has yet defined its relation to Oregon water rights (Glick 1995a; Glick 1995b).

In summary, Oregon's water management rules are about the same as other western states. Becoming familiar with others' systems might be likened to Americans motoring in Canada: although the culture may be different, aside from the arithmetic itch of kilometers, imperial gallons, and exchange rates—driving is driving.

CHAPTER 3

To Issue or Not to Issue: Water Rights and the Public Interest

GETTING A WATER RIGHT IN OREGON is not as easy as it used to be. But that goes hand-in-hand with allocating a limited resource. The more that has been claimed, the more care is taken with what remains. With about a hundred years of water allocation and over eighty thousand water rights "under the bridge," more care means many more hoops for applicants to jump through. Many of these hoops involve making sure the public's interest in its water is protected. This chapter explores some of those protections, why they were created, and how they are applied step-by-step in the permitting process.

The Prime Directive

Reduced to its simplest form, getting a water right is a three step process (Figure 14). First, you fill out an application provided by the Water Resources Department, the state agency responsible for carrying out Oregon's laws on water supplies and water use (see Figure 15). Second, if the request is approved, a permit is issued. The permit allows actual diversion and use of the water and requires that construction of most diversion works be completed within five years. Third, once water is used according to the terms of the permit, a confirming certificate of water right is issued.

The toughest part of the process is the decision whether or not to issue a permit. It is a difficult balancing act, pitting the individual benefit gained from using the public's water against any injury to the public's interest posed by that use. The decision is to be based on two sections of statute:

> ... the Water Resources Department shall approve all
> applications made in proper form which contemplate the
> application of water to a beneficial use, unless the proposed
> use conflicts with existing rights (ORS 537.160(1);

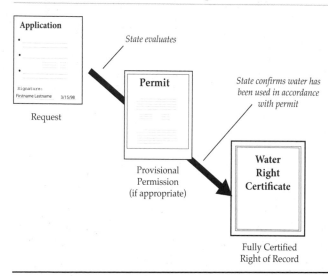

State evaluates

State confirms water has been used in accordance with permit

Request

Provisional Permission (if appropriate)

Fully Certified Right of Record

Figure 14 (left): Three steps to a water right certificate

Water Resources
Commission

• 7 citizens appointed by the Governor and confirmed by the Senate to serve staggered 4-year terms.

• By statute, the authoritative body charged with carrying out state water law and policy—but like most such bodies, has delegated more of its authority to the Department.

• Sets new policies, adopts administrative rules, hears water right appeals, and otherwise acts as a board of directors.

Water Resources
Department

(150 - 200 staff, depending on biennial budget)

• Issues authorizations to use water (permits, water rights, hydroelectric licenses, transfers, registrations).

• Enforces water laws (especially through state watermasters).

• Measures streamflows and groundwater levels.

• Verifies "grandfathered" water uses (adjudication process).

• Administers construction code for wells and dams.

• Responsible for developing an "integrated, coordinated" program the use and control of all water.

Water Management Agency Timeline*	
1905	Office of State Engineer established.
1909	Water Code enacted. State Board of Control (with State Engineer as president) created and charged with code administration.
1913	Board of Control becomes State Water Board.
1923	Water Board functions consolidated within State Engineer's Office.
1955	Water Appropriations Act passed, creating State Water Resources Board to work with State Engineer in establishing an integrated water program.
1975	State Engineer and Water Resources Board merged to form Water Resources Department, with oversight by a Water Policy Review Board.
1985	Water Policy Review Board replaced with Water Resources Commission to do away with irregularities in authority between Board and the Department's director.

Figure 15: Water resources agency purpose and history. (Information from Oregon Water Resources Department 1993a)

> ... if the director determines that the proposed use ...
> would impair or be detrimental to the public interest, the
> director shall issue a final order rejecting the application or
> modifying the proposed final order to conform to the public
> interest. (ORS 537.170(6); paralleled in ORS 537.625(1))

In other words, the prime directive is to issue unless there is injury to other rights or the public interest. This might be contrasted with a more conservative approach—arguably one more appropriate for allocating a limited public resource—which would be to deny applications unless it can be shown no harm results to other rights or the public interest. The existing starting point of the process is important because it forces the state to prove it should not issue a water right, rather than requiring an applicant to demonstrate that the state should.

In theory, the approach to the public interest is different for groundwater than for surface water (Clark 1983). It is in the affirmative and must determine "whether the proposed use will ensure the preservation of the public welfare, safety and health." (ORS 537.621(2)) This is arguably a tougher standard, though in practice it has not been applied.

Thus, the state (specifically, the Water Resources Department) must issue a water right unless it shows reason that it should not. This is the crux of the water rights evaluation process, which may be viewed as a series of filters applied to a request to determine if there are reasons not to approve it.

The Public Interest

Oregon's water is owned by the public. In theory, no use of water that injures the public's interest is allowed. But what is the public interest? Oregon's water laws frequently refer to it, or public benefit, or public welfare. However, none of these terms is ever fully explained. The statutes list, but do not define, public interests. The list most often consulted consists of some vague, but competing, "considerations."

> (a) Conserving the highest use of the water for all purposes,
> including irrigation, domestic use, municipal water supply,
> power development, public recreation, protection of
> commercial and game fishing and wildlife, fire protection,
> mining, industrial purposes, navigation, scenic attraction or
> any other beneficial use to which the water may be applied for
> which it may have a special value to the public.

(b) The maximum economic development of the waters involved.

(c) The control of the waters of this state for all beneficial purposes, including drainage, sanitation and flood control.

(d) The amount of waters available for appropriation for beneficial use.

(e) The prevention of wasteful, uneconomic, impracticable or unreasonable use of the waters involved.

(f) All vested and inchoate rights to the waters of this state or to the use of the waters of this state and the means necessary to protect such rights.

(g) The state water resources policy ... (ORS 537.170(8); 537.625(3))

More sayings than standards, they offer little guidance in making a specific water right decision. Too often, the public is unconsciously reduced by the department to the living, (sometimes fire-) breathing person on the other side of the water rights counter. The interests of the other 3.6 million Oregonians who don't have (and for the most part aren't aware of) water rights become bloodless legal abstractions. It may be unrealistic, even unfair, to expect a term so broad and important as "public interest" to find concise expression in law, let alone water law. Its counterparts may be concepts like "morality," "pornography," "a well-regulated militia" or "free speech"—lofty concepts well worth debating by each generation. However, in the real world, decisions need to be made for each permit request. And the public interest presumably has real meaning for each of those decisions. The Oregon Water Resources Department includes a rather awkward "definition" of public interest in its statewide water policy rules:

As a standard for reviewing new uses of water, ... (the public interest is) a beneficial use which is consistent with state law and includes providing the greatest good for the people of the state based on current values, protecting water rights and conserving water resources for present and future generations. (OAR 690-400-0010)

While more a recipe for apple pie than a definition, the concept it describes seems to value a broad public good and the preservation of opportunities for future Oregonians. But it falls short of an operational instruction for the day-to-day decisions which shape how the public's water is managed.

Perhaps the public interest is something that can only be understood, but not defined (though readers will find an attempt has been made in the glossary), if that understanding is constantly refreshed through broad and informed public participation in the process. Through time, the state has articulated its understanding of the public interest through its policies, regulations, rules, and individual decisions. A number of these are brought into play explicitly in the water rights process.

▣ Public Interest and the Process

The water rights evaluation process is laid out in statute (ORS Chapter 537) and rule (OAR Chapter 690, Divisions 2, 3, 300, 310 and 77). When the Water Resources Department receives an application for a water right permit, it is required to check on five primary public interest factors. (The specific language varies slightly depending on whether the proposed source is surface or groundwater.)

- Is the use prohibited by statute?
- Is the use allowed under the applicable basin program?
- Is there water available for the use?
- Is the use prohibited under any other rules of the agency?
- Would the new use hurt any existing water right holders?

Exploring these factors one by one will not only help in understanding the water rights evaluation process, but shed some light on how the state has grappled with public interest issues over the years.

IS THE USE PROHIBITED BY STATUTE?

One of the earliest protectors of the public interest was the legislature. ORS Chapter 538 lists many streams that the legislature has "withdrawn" from appropriation. Usually this either means that no diversion is allowed from the stream or that it is allowed only by a specific entity. For example, some streams are off-limits to diversion in order to protect their waterfalls, including thirty-one creeks in the Columbia River gorge. Other streams are withdrawn to preserve municipal supplies. The most famous (if fame and water law are not mutually exclusive) of these is the statute (ORS 538.420) giving exclusive rights to the City of Portland for the use of the Bull Run and Little Sandy rivers. Some water bodies, such as Diamond Lake or the Clatskanie River have been nearly totally withdrawn except for use by the State Fish and Wildlife Commission to protect fish life. If an application for a water right is filed on any of these sources and the applicant is not specifically identified in the statute, the application will be rejected. So far, these legislative protections have not been applied to groundwater.

IS THE USE ALLOWED UNDER THE APPLICABLE BASIN PROGRAM?
Basin programs (sometimes called "basin plans") are state administrative rules which set down future allowable water uses in, and other water regulations specific to, Oregon's eighteen major river basins. Basin programs do not affect previously issued water rights or permits.

Basin programs were the state's first effort to formulate what the 1955 Legislature mandated—a coordinated, integrated state water resources policy (ORS 536.220). The programs combined basin water-supply inventories with long lists of findings supporting the final outcome: regulations delineating which waters were "open" for specific beneficial uses. This act of specifying the allowable beneficial uses (that is, uses for which the Water Resources Department would accept future water right applications) is called "classification." ORS 536.340 states:

> Subject at all times to existing rights and priorities to use
> waters of this state, the (Water Resources) commission ...
> may ... classify and reclassify the lakes, streams, underground
> reservoirs or other sources of water supply in this state as to
> the highest and best use and quantities of use thereof for the
> future in aid of an integrated and balanced program for the
> benefit of the state as a whole.'

Thus, a program is a menu of uses composed by the state to balance existing and future basin water demands toward a goal of benefiting the entirety of Oregon. The bottom line is that if an application requests a type of water use not listed in the basin program, the statutes imply it is not in the public interest. The department in almost all cases would deny the application on that basis. However, there is a limited variance process (ORS 536.295) that allows some types of use to be approved, even if they do not appear in the basin list.

The state experimented freely in the 1990s with different forms of basin programs to achieve better integration and balance. Whereas early programs allowed almost all uses, later programs have struggled to close the door on further appropriation without slamming it in the face of local communities. The results have been mixed: while some innovative management concepts have been espoused, few have met with successful implementation; but all have added to the complexity of basin programs (see Chapter 11).

Basin programs are hard to understand, which, owing to their fundamental importance to permitting, tends to burden rather than clarify the decision-making process. Their basic structure is not straightforward: they first allow every use basin-wide, then back out of that general allowance stream by stream, dropping one or two allowable uses from a list of a dozen or more, creating a numbing repetition. Not only are they hard to follow but, depending on the basin and the year they were adopted, can offer very different water management objectives, definitions, and regulatory approaches. In the face of these difficulties and changing budget priorities, and given the impacts of the water availability policy (explained below), the Department abandoned basin programming in the late 1990s.

IS WATER AVAILABLE FOR THE USE?

In the past, whether water was actually available for new users was not much of a worry. Many thought (and for that matter, still think) that the prior appropriation system relieves any concern about availability. They argue that people will take turns using water in an orderly fashion based on the priority dates of their rights. When there is no longer enough for junior users, they simply shut off. Therefore, letting more people on the system really is not a liability, because the ability of a new user to injure an old user is a theoretical impossibility under the prior appropriation doctrine. Thus no harm is done during periods of shortage, and use can be maximized in times of plenty.

But Oregon has rejected this "no harm" argument. In reality, the actual workings of the system just about guarantee harm if water availability is not taken into account. This was clearly acknowledged by the State Engineer as early as 1907:

> If there is unused water in the streams of this State available for
> new appropriations, it is exceedingly important to the state's
> development to have this fact officially known and recorded,
> and the conditions clearly set forth under which it can be
> secured. If all the water in a stream has been utilized, it is just
> that the appropriator and would-be appropriator should be
> apprised of this fact, and it is not right that the homes of those
> who have used the water should be jeopardized by a new use.
> Either the new claimant must fail or the old must suffer and
> both be involved in controversy. (Lewis 1907)

Junior users do not always and automatically know it is their time to shut off. Even if they somehow did know, to voluntarily stop watering crops or serving paying subscribers is a tall order. And on the face of it, to be able to issue an infinite number of rights on a finite resource would not only seem illogical, but ultimately shady. A fiscal equivalent might be bouncing checks or printing twenties in your basement—practices that are easily understood but that most upright citizens nonetheless steer clear of. In any event, the statutes now direct the state to consider water availability in its permitting decisions (ORS 537.150(4), 537.153(2), 537.620(4)(b), 537.621(2), 537.170(8)(d), and 537. 625(3)(d)).

So when is enough enough? For a time, it was assumed if water was available for a new user more often than not, that was good enough. A "50/50" approach had a reasonable ring to it. But what it actually meant was the state was accepting fees in return for granting perpetual rights that would not be satisfied half the time. While 50/50 may represent pretty good odds for some endeavors, when it comes to water those odds can spell mischief for private property protection, local economies, and sound resource management.

In 1992, the Oregon Water Resources Commission adopted a new water allocation policy by rule which defined "over-appropriation" and established water availability standards for permit decisions for out-of-stream, groundwater, storage, and instream uses.

> "Over-appropriated" means a condition of water allocation in which:
>
> (a) The quantity of surface water available during a specified period is not sufficient to meet the expected demands from all water rights at least 80% of the time during that period; or
>
> (b) The appropriation of groundwater resources by all water rights exceeds the average annual recharge to a groundwater source over the period of record or results in the further depletion of already over-appropriated surface waters. (OAR 690-400-0010)

The allocation policy ties into this definition by stating surface water "shall be allocated to new out-of-stream uses only … when the allocations will not contribute to over-appropriation" (OAR 690-410-0070). Thus, agricultural, municipal, industrial, or other types of use that take water from streams are held to a month-by-month 80 percent standard: water must be available 80 percent of the time for each month of proposed use.

This can get confusing. People seem to understand the idea of 50/50. However, once state regulators begin talking about "80 percent," especially if they call it the "80 percent exceedance probability" standard, many people get lost. This is the curse of statistics, for that is what the 80 percent standard is. It is a statistic picked by the Water Resources Commission to help make decisions about permits. Eighty percent was selected because it provides new water users with pretty good assurance they will get a crack at using water most of the time, but it does not over-do it by requiring water to be available 100 percent of the time. It respects the natural variability of stream systems and the resilience of many users.

But what does the 80 percent flow mean? It is that flow which, based on historic patterns of streamflow, is met or exceeded 80 percent of the time. Table 17 illustrates how the statistic is calculated for one month on a hypothetical stream. When the years of monthly measurement are listed in order of magnitude, the flow that was met or exceeded eight out of ten years (20 cfs, shown in bold) is easily identified as the 80 percent flow. Statistically there is only a 2-in-10 chance that the flow would drop below this point.

This 80 percent figure is the starting point for calculating water availability. After any consumptive uses are accounted for, the amount of all instream water rights issued on the stream is totaled and subtracted. The result is how much water is left to appropriate. If the result is zero or less, that means the stream is fully, if not over-, appropriated for that month (see Figure 16).

Table 17. Illustration of 80 Percent Exceedance Probability			
Year	Flow (cfs)	Flow (low to high)	Ranking	Chance of meeting or exceeding the flow (%)
1915	60	10	0	100
1916	45	15	1	90
1917	35	20	2	80
1918	40	25	3	70
1919	30	30	4	60
1920	20	35	5	50
1921	25	40	6	40
1922	15	45	7	30
1923	10	50	8	20
1924	50	55	9	10
1925	55	60	10	0

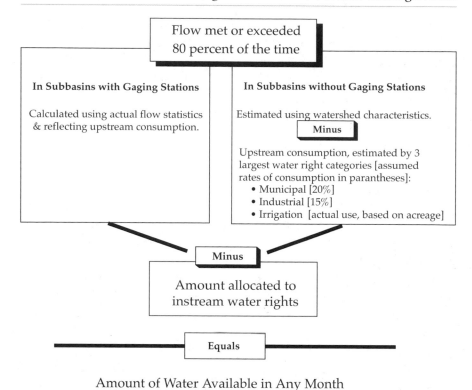

Figure 16: Method for determining surface water availability

If the stream is over- or fully appropriated, that is reason enough to deny or condition an application. Most streams are over-appropriated only for certain summer or fall months. Thus it is possible, though perhaps not always practical, to approve a permit for some months and not others. However, the state's water allocation policy does allow an exception to this standard: "… when a stream is over-appropriated, some additional uses may be allowed where public interest in those uses is high and uses are conditioned to protect instream values" (OAR 690-410-0070 (2)(a)).

The 80 percent limit is well known in some circles because it can torpedo people's requests to use the most obvious and visible of sources—running water (Figure 17). But preventing over-appropriation in other sources is also a state objective, as explained below. For these sources, different availability methods are employed.

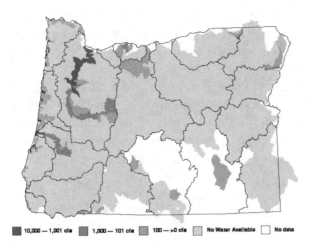

Figure 17: Surface water availability in late summer. By late summer (August shown at left), most of Oregon's surface water supplies have already been allocated for use (based on an 80% standard)— there's no more for new water rights. (Source: Oregon Water Resources Department)

■ 10,000 — 1,001 cfs	■ 1,000 — 101 cfs	■ 100 — >0 cfs	No Water Available	No data

GROUNDWATER USES. Instead of using a statistical standard, groundwater availability is determined using a performance test—namely, whether the proposed use, when added on top of all other groundwater uses, would exceed the annual recharge to the groundwater table, and thus cause the water table to drop. Another measure of over-appropriation has to do with effects on streamflow. Sooner or later, nearly all groundwater finds its way into streams. In Oregon, with its dry summer months, it is groundwater that provides the base flows of many streams. If groundwater uses end up hijacking water from already over-appropriated streams, then the groundwater source is also considered over-appropriated.

STORAGE USE. Filling storage reservoirs is excepted from the 80 percent standard. This amounts to tacit recognition that if streams are over-appropriated, it is generally during the summer and fall, while reservoirs fill during winter and spring. The Water Resources Department uses a 50 percent standard as a measure of water availability, perhaps because the effects of over-appropriation (e.g., a reservoir does not completely fill) are more manageable when compared to a water diversion context (e.g., a growing crop being cut off from water). It also offers some encouragement for new storage projects by opening the availability window slightly. Furthermore, effects of storage on the stream system and downstream users can be addressed through storage seasons set in basin programs or specific permit conditions tailored to the use.

INSTREAM USES. In recognition of Oregon's having overlooked instream flow needs for much of its history, the state has chosen to use a different approach for determining the amount of water available for instream water rights—rights held by the Water Resources Department for keeping water flowing in-channel. Instead of limiting new instream water right requests to what is left in a stream, the Water Resources Department allows agencies to request up to the natural flow of the stream prior to any out-of-stream use (OAR 690-77-0015). This is seen as a mechanism to bring instream interests more fully into the water right system. Where the amount approved in an instream water right exceeds what may currently flow in the stream, it is official acknowledgment of flow needs that may eventually be satisfied as old uses are abandoned and the instream rights move up the line in terms of priority dates. (See Chapter 4 for a discussion of instream water rights.)

WATER AVAILABILITY AND STATE SCENIC WATERWAYS. Another important water availability consideration involves protecting streamflows in state scenic waterways. A 1988 Oregon Supreme Court Decision (*Diack v. City of Portland*) affirmed that, before issuing any water right, the state must make explicit findings about scenic waterway impacts. In order to issue a permit, the state must find that the proposed use would not diminish scenic waterway flows below those needed to support fish, wildlife, and

Because state scenic waterways, such as the Clackamas River just east of Portland, must remain free-flowing, upstream water uses may be restricted. (Oregon State Archives, Oregon Highway Division, OHD6354)

recreation, the highest and best water uses identified in Oregon's Scenic Waterways Act (ORS 390.805–390.925). This has the effect of creating an implied instream reserve for all state scenic waterways.

In response to this court decision, the department identified flows needed to protect fish, wildlife, and recreation in scenic waterways. The flows were based on fisheries studies, recreational publications, and expert testimony—qualitatively through observations, rather than quantitatively through site-specific measurement. These "Diack" flows were approved by the Water Resources Commission to be used in the water right application process as rebuttable presumptions of flow needs (that is, parties are free to challenge their validity). Accordingly, Diack flows are basically subtracted like instream water rights from total streamflow to arrive at a final water availability figure.

This created additional permitting difficulties for applicants in some areas of the state. About half of Oregon is upstream of one state scenic waterway or another and therefore subject to Diack findings. Also, overlaying a scenic waterway designation—with its mandatory focus on water needed for fish, wildlife, and recreation—over a stream that may have had a very different focus for the last century, such as irrigation or municipal supply, can help generate an over-appropriated condition in a hurry. Thus, in many cases, scenic waterway designation has had the effect of closing upstream basins to further surface appropriations and, where the sources connect, groundwater appropriations as well. In 1993 and 1995, the legislature loosened restrictions in the scenic waterway statutes to allow the permitting of human consumption, livestock, and groundwater uses under certain circumstances (ORS 390.835 (5)–(12)).

IS THE USE ALLOWED BY OTHER RULES?

According to ORS Chapter 537, any requested ground or surface water use must be checked for compliance with the rules of the commission. With over seventy divisions of rules, that single requirement opens up a lot of regulatory territory. However, only a few of those divisions emerge at the forefront of protecting the public interest during the allocation of water.

STATEWIDE POLICIES. Since 1990, the Water Resources Commission has adopted "statewide water policies" by rule on eight topics—water allocation, instream flow protection, water storage, conservation and efficient water use, hydroelectric power development, groundwater management, interstate cooperation, and protection of water resources on public riparian lands (OAR Chapter 690, Division 410). These statewide water policies compiled directives existing at the time of adoption in statute or other rules or implied by customary practice. They also established new policy interpretations for the agency. The statewide policies were an attempt to lay down a common set of principles to guide agency decisions and activities, including basin planning, permitting, and conservation.

The statewide policy most pertinent to permitting decisions is that on water allocation. Besides establishing the definition and the standards relating to over-appropriation, as previously discussed, the policy includes other principles related to the water right application process. For example: if surface or groundwater is contaminated, new rights may be issued only if the commission determines the use does not pose a significant hazard to human health or the environment; the state does not have to wait for an instream water right application to protect instream needs, but can actively protect them by limiting new out-of-stream permits; conservation, storage, and water right transfers and leases—not new permits—are held out as the means to meet future water demands.

LAND USE. Another important checkpoint in evaluating a proposed water use is whether or not it would serve a land use allowed under the local comprehensive plan. Under Oregon's land use planning system, every city and county has a comprehensive land use plan that has been acknowledged by the state. Thus, Oregon's land use plan is actually a mosaic of 277 separate local plans (Rohse 1987). Under rules of both the Department of Land Conservation and Development and the Water Resources Department, the state's water rights process must comply with the nineteen statewide planning goals and be compatible with local plans. ORS 537.153(3)(b) directs the department to address land use compatibility in making its permit decision. The bottom line is that the department will not authorize a water use that violates a local plan, unless there is an extraordinary, over-riding public interest in doing so. The compatibility procedures differ between out-of-stream and instream uses.

For out-of-stream permit requests, applicants are required under OAR Chapter 690, Divisions 5 and 310 to submit a land use information form for their local government to fill out. The Water Resources Department takes this local "sign-off" seriously: it will not accept an application unless it is accompanied by a completed form (unless local governments somehow fail to complete it). If the associated land use is not allowed outright in the land use plan, or has already been denied at the local level, the department will not issue a permit.

The compatibility procedures for instream water right requests (OAR Chapter 690, Divisions 5 and 77) are a little different primarily because an instream right involves no land. That is, there is no associated land use that would end up violating a local land use ordinance. The object of an instream right is to keep water running in local creeks, much as it always has (only three state agencies are authorized to request instream rights; see Chapter 4). Although the risk of potential incompatibility is small, it is still possible. For example, an instream right might take up the last water from a stream reach identified in a local plan as a site for a future city water in-take. Thus, an instream right applicant is required to give local governments a "heads up" about the application. As with out-of-stream applications, if the department does not hear from the local government within certain time periods, it may presume the instream right would be compatible with the local plan.

If a local government asserts a proposed water use would violate its plan, certain land use dispute resolution procedures are triggered. These can get very involved and have a variety of potential outcomes. In addition, if there is a land use concern, it can also be raised in the water rights process as a comment or protest for resolution through a contested case hearing, or ultimately, the courts.

ENDANGERED SPECIES PROTECTIONS. There are additional public interest standards for evaluating permit applications in light of threatened or endangered fish species. OAR Chapter 690, Division 33 establishes definitions and adds procedures to protect fish species that have been listed as threatened or endangered by either the State of Oregon or the federal

Requests for water rights come under special scrutiny where endangered or threatened fish species, including many stocks of salmon, might be harmed by the proposed water use. (Portland Water Bureau)

government (Chapter 9). These specialized rules apply to surface water, hydraulically connected groundwater (see Division 9 discussion, below), storage, and groundwater recharge applications in basins tributary to the mainstem Columbia and Snake rivers.

For applications in these basins and reaches, the Water Resources Department must determine whether the proposed use would be detrimental to the protection or recovery of threatened or endangered species. If so, the use is presumed detrimental to the public interest and the application is denied. In determining impacts on listed species, the department consults a number of agencies and environmental documents. One of the most important of these is the Northwest Power and Conservation Council's Columbia River Fish and Wildlife Program, a regularly-updated plan that protects and enhances fish and wildlife on the Columbia River and its tributaries affected by hydroelectric projects (see Chapter 11).

In order to be judged non-detrimental, proposed uses in waters above the Columbia River's Bonneville Dam must:
- use water only from October 1 through April 14 of any year;
- comply with any mandated fish screening, water quality, or water use measurement requirements; and
- adhere to any conditions requiring the restoration of riparian areas disturbed by constructing a point of diversion.

However, these standards may be waived for domestic uses; projects that provide net benefits to fish; emergency uses necessary to protect public health and safety; certain existing small ponds; multi-purpose storage projects; or "other projects with measurable public benefits."

Proposed uses in waters below Bonneville Dam must comply with the same standards listed above, except for the seasonal restriction.

While not placing a moratorium on new uses in these basins, as the Fish and Wildlife Program urges, the standards do lop off a fair amount of water use potential in a large block of Oregon, especially eastern Oregon basins where the seasonal restriction applies. In practice, most of the standards become automatic conditions. Where the conditions bite too deeply into applicants' plans, they may propose mitigation measures or argue their project provides measurable public benefits.

HYDRAULICALLY CONNECTED GROUNDWATER. "If it acts like surface water, treat it like surface water." That is the basic directive contained in OAR Chapter 690, Divisions 8 and 9. Although all groundwater at some point finds its way to the surface, of primary concern to the department is the immediacy of the connection between surface and groundwater. Consider a 10-foot-deep well dug into loose gravel 10 feet from the bank of a stream. Clearly, the stream's water and the well's water are actually the same source. Pumping the well would have the same effect on stream users as pumping from the stream itself. It makes sense to manage the well by the same rules that apply to the stream. The question becomes one of determining where to draw the line in defining immediate connections.

These rules lay out a detailed process for drawing the line. The department reviews each groundwater application in a two-step analysis. First, it determines if the proposed groundwater source is hydraulically connected to surface water. If it is, then the department evaluates whether the application has "the potential to cause substantial interference" with the surface water body. Substantial interference occurs when the effects of a well's pumping reaches a surface water body, depriving surface appropriators of their legal rights to water. These rules establish certain distance-from-stream thresholds and aquifer characteristics for determining when there is potential to cause substantial interference. If the proposed use has this potential, the application is processed in accordance with surface water rules and may also be denied or conditioned to control interference.

ORDERS OF WITHDRAWAL. Neither statute nor rule, these withdrawals are a collection of orders closing or limiting streams to further appropriation. Orders at one time provided a slightly simpler way to achieve restrictions now primarily established through basin classifications. Many of the orders are quite old, having been issued by the State Engineer in the 1920s and '30s, often to reserve water for specific irrigation projects or for exclusive use by certain cities. For example, in January 1931 the State Engineer issued an order reserving all future use of the South Fork of the Clackamas River and Memaloose Creek for the cities of Oregon City and West Linn. Newer orders have been issued by the Water Resources Commission, often at the request of local citizens wanting to protect remaining flows. A number of these orders have been collected and incorporated into rule by reference in OAR Chapter 690, Division 80. Orders of withdrawal are consulted during the water rights evaluation process, much as basin program classifications or legislative withdrawals.

WILL THE USE INJURE EXISTING WATER RIGHT HOLDERS?

The last of the explicit public interest checks required by statute in the water rights evaluation process is determining whether the proposed water use would injure other water right holders. Protecting existing legal water users from encroachment by new users may appear primarily a private, rather than a public, interest. After all, most water rights in Oregon are the private property of individuals. However, there is a public interest in upholding the fundamental precept of the water rights system: first in time is first in right. Failing to do so would be to return to the chaotic conditions of the late nineteenth century, putting at risk both private and public investment.

But given the importance of protecting existing water rights, it is interesting that the state's assessment of injury is quite limited. It consists of preventing irrigation rights from being superimposed on each other. Because these water rights are appurtenant to the land, there should not be more than one right covering a given parcel. If there is, it means one of two things: 1) there already is an active right available to the applicant for immediate use or transfer; or 2) there is an old right that has fallen into disuse that needs to be canceled before the new right can be established. But there is no analysis of other potential impacts of the proposed use on existing rights. This is mostly because it is assumed that the prior appropriation doctrine will automatically protect older uses: new uses have to cease when older uses need the water. Perhaps one of the biggest reasons the analysis is limited is because the statutes offer absolutely no guidance on what injury the department should be checking for. Even if the statutes were explicit, the time allotted for the department to assess injury is very brief (see Steps in the Permitting Process, below) and there may be little opportunity for far-reaching analysis, anyway.

Steps in the Permitting Process

The business of reviewing an application to decide whether to issue a permit has been getting more complicated. As more and more people line up to use the same, finite water sources and as society increasingly values previously overlooked water uses, pressures on the system increase. One measure of the stress involved was the Oregon Water Resources Department's application backlog of the mid-1990s. At that time, there were about five thousand applications awaiting a decision. Some were decades old. Although the department had been trying for several years to reduce the backlog, little progress was made. In an attempt to legislate a solution, the 1995 Legislature made far-reaching changes to the water use permitting process. By 2006, the application backlog had dropped to about three hundred (French 2006).

The rest of this chapter highlights the evaluation process put in place by the 1995 legislation (Figure 18). Although it also applies to instream water right requests, it was probably designed—and is best visualized—with out-of-stream applications in mind.

⊞ The Application

Except for the exemptions described in Chapter 2, people must first ask the public's permission to use water. Applications must be submitted to the Water Resources Department before anyone begins construction of water works (ORS 537.130(1)). For the most part, the application form is fairly predictable, requiring the applicant's name and address; identification of the proposed water source; and the nature and amount of water use. Most applicants must also submit a land use information form with their application (OAR Chapter 690, Divisions 5 and 310). Usually a map of the proposed point of appropriation and location of use is also required. Beyond that, specific requirements vary based on particular types of use.

One other important submittal requirement is money. Unlike many other state agencies, in general the Water Resources Department cannot raise (or lower) its permit application fees. The fees are set in statute, subject to change only by the legislature, and vary by the type of use. Fees are intended to help cover the cost of having staff look over the application, make sure it is complete, compare the request to all the statutes and regulations that affect it, and document a decision. A second fee is required for filing and recording a permit, if one is issued.

The department estimates that it costs around $3,000 to process an average permit request all the way from application to certification. The fees paid by the applicant typically cover from 25 to 35 percent of that cost, with the taxpayers picking up the rest (French 2006).

In 2003 the Legislature authorized the department and applicants to enter into alternative service delivery agreements to expedite processing, allowing the department to hire additional temporary staff or contract with an outside service provider to process applications, and charge the applicant for the cost (Oregon Water Resources Department 2003). This arrangement provides applicants who are willing to foot the cost of

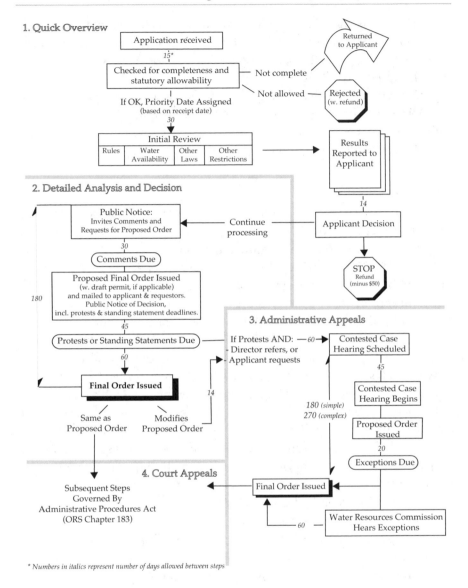

1. Quick Overview

Application received

15*

Checked for completeness and statutory allowability — Not complete

If OK, Priority Date Assigned
(based on receipt date)

30

Initial Review
| Rules | Water Availability | Other Laws | Other Restrictions |

Returned to Applicant

Not allowed — Rejected (w. refund)

Results Reported to Applicant

14

Applicant Decision

STOP
Refund
(minus $50)

2. Detailed Analysis and Decision

Public Notice:
Invites Comments and Requests for Proposed Order ← Continue processing

30

Comments Due

Proposed Final Order Issued
(w. draft permit, if applicable)
and mailed to applicant & requestors.
Public Notice of Decision,
incl. protests & standing statement deadlines.

180

45

Protests or Standing Statements Due → If Protests AND: —*60*→ Contested Case Hearing Scheduled
- Director refers, or
60   - Applicant requests

Final Order Issued   *14*

Same as Proposed Order   Modifies Proposed Order

3. Administrative Appeals

45

Contested Case Hearing Begins

180 (simple)
270 (complex)

Proposed Order Issued

20

Exceptions Due

4. Court Appeals

Subsequent Steps Governed By Administrative Procedures Act (ORS Chapter 183)

Final Order Issued

60 —

Water Resources Commission Hears Exceptions

* Numbers in italics represent number of days allowed between steps

Figure 18: Water right application evaluation process

processing to get a determination from the department more quickly than waiting in line with everyone else.

To use the public's water, applicants must pay $300 up-front plus a sum based on the amount of water they are requesting. (See Table 18a and b.) For example, if an irrigator or a city wants to divert 2 cubic feet per second (about 900 gallons per minute), they would pay the $300 base fee, $200 for the first cfs, another $100 for going beyond 1 cfs, and, finally, a permit recording fee of $250 (Oregon Water Resources Department 2006a). So, $850 (plus other nominal processing costs for even larger uses) secures the free use of a large quantity of Oregon's water. Forever. While Oregonians would likely agree that water is a priceless treasure, they might stop short of employing that chestnut as a pricing formula for a valuable state asset.

Raising fees is likely to be a subject of continuing attention at the legislature in the future. Whether they ever will be raised to fully cover the cost of processing is a perennial question, and the answer anyone's guess.

⊞ *The First Cut*

Upon receiving an application, the department checks to see if it is incomplete, inaccurate, or on statutorily withdrawn waters. If it is, the application is rejected. If not, the application is accepted and time-stamped to establish the priority date of any right issued in response to the application. The department then conducts an initial review of the application.

This is supposed to be a quick and dirty assessment of what chances the applicant has of getting a permit. The department looks at water source restrictions, water availability, conditions placed on similar requests in the area, and any other factor that might block, restrict, or shape the issuance of a permit. The department also consults with other agencies, such as the state departments of Fish and Wildlife or Environmental Quality to flag environmental concerns. It then reports its findings to the applicant, who then has fourteen days to bail out and get all money back, save for a small non-refundable fee.

If the applicant decides to stay in the process, the department lists the requested water use in a public notice. This notice is sent to state agencies and county planning offices, as well as to individuals or organizations that have paid to receive it. In contrast to other public processes, neighbors do not have to be (and seldom are) notified. The notice requests comments within thirty days on the proposed water use. It also gives people information on how to obtain copies of future decision documents on the application. This

Table 18a. Water Use Authorization Fees

Authorization	Base Fee	Recording Fee	Additional Fees	
			First Increment	Additional Increments
I. Application for A Water Use Permit (OAR Chapter 690, Division 310)				
A. To Appropriate Groundwater or Surface Water	$300	$250	First cfs* or fraction: $200	Each additional cfs or fraction: $100
If also proposing to appropriate stored water under the same application:			Each acre foot or fraction up to 10 acre-feet: $20	Each additional acre-foot or fraction: $1
B. To Appropriate Groundwater or Surface Water for STORAGE	$300	$250	Each acre foot or fraction up to 10 acre-feet: $20	Each additional acre-foot or fraction: $1
C. To Appropriate ONLY Stored Water	$150	$250	Each acre foot or fraction up to 10 acre-feet: $15	Each acre foot or fraction up to 10 acre-feet: $20
If appropriating ground water for above ground storage, direct ground water use, and use of above ground storage under a single application:				
To Appropriate Ground Water	$300		1st cfs or fraction: $200	Each additional cfs or fraction: $100
ADD: Above Ground Storage			Each acre-foot or fraction up to 10 af: $20	Each additional acre-foot or fraction $1
ADD: Appropriation of Stored Ground Water			Each acre foot or fraction up to 10 acre-feet: $20	Each additional acre-foot or fraction: $1
II. Alternate Permit Application for Qualifying Reservoirs				
(Storing less than 9.2 acre-feet or with dams less than 10 feet in height) ORS 537.409	$40		Each acre-foot or fraction: $10 (not to exceed $250 total)	

III. Water Use Authorizations (OAR Chapter 690, Division 340)

A. Limited License $150 (1st point of diversion)   Each additional point of diversion: $15

B. Registration of Water Use for Road Maintenance and Construction $300 (initial application)   Annual renewal: $50

IV. Transfer of Water Rights (Changes To Water Rights)*(OAR Chapter 690, Divisions 21 and 380)

A. Regular Transfers, Permit Amendments and Water Right Exchanges (A type of change is any of the following: place of use, point of diversion/appropriation, or character of use.) $350 (for one change)   Each additional type of change requested: $350   Each change in place of use or type of use or water exchange: $175 for each cfs or fraction beyond 1st cfs

B. Temporary Transfers $175   Non-Irrigation uses—Each cfs or fraction beyond 1st cfs: $75   Irrigation uses—$1 per acre of land irrigated

C. Permanent Irrigation District Transfers $350   Each cfs or fraction beyond 1st cfs: $175.

D. Temporary Irrigation District Transfers $175   Non-Irrigation uses—Each cfs or fraction beyond 1st cfs: $75   Irrigation uses—$1 (or $0.25 if submitted in a Department approved digital format) per acre of land irrigated

E. Substitution of Surface for Groundwater $250

V. Allocation Of Conserved Water (OAR Chapter 690, Division 18: fee may be reduced under ORS 536.050)
   $700

*cfs = cubic foot per second

Table 18b. Water Use Authorization Fees

VI. Instream Leases (OAR Chapter 690, Division 77)
  Involving 4 or more landowners or 4 or more water rights: $200
  All other leases: $100
  Lease renewal: $50

VII. Water Management And Conservation Plan (OAR Chapter 690, Division 86)
  Examination of a water management and conservation plan
    submitted by an agricultural water supplier: $250
    submitted by a municipal water supplier serving a population of 1,000 or fewer:
      $500
    submitted by a municipal water supplier serving a population of more than 1,000:
      $1,000

VIII. Other Fees
  Application for an extension of time to develop a water use (municipal and other):
    $250
  Filing a protest: $250
  To request standing in a contested case hearing: $50
  To participate in a contested case hearing: $200
  To receive a copy of a proposed and final order on a pending water right application:
    $10
  Surface water application for stock watering outside of riparian areas: $40
    examination fee and $10 permit recording fee

Source: Oregon Water Resources Department 2006a

---

is one of two public participation opportunities (and the only free one) during the process.

▣ *Making the Decision*

If an applicant keeps the request in play, the department begins preparing a detailed decision document called an order. One of the changes made by the 1995 Legislature was to normalize water right decision making with other state processes governed by Oregon's Administrative Procedures Act (ORS Chapter 183). The APA sets up a decision structure consisting of proposed orders, protest opportunities, contested case procedures, and final orders. Orders are state actions "expressed orally or in writing directed to a named person or named persons" (Kulongoski 1995).

Based on agency consultation, review of water regulations, public comments, and any other information it has on hand, the department makes a decision about the requested water use. It documents the decision through "findings of fact and conclusions of law" in a written order. This kind of documentation has a familiar, and often comforting, ring to process-veterans across the state, especially lawyers.

As discussed earlier, the basic decision boils down to determining if the proposed use would have adverse impacts on other water right holders or the public interest.

If all the public interest criteria detailed earlier in the chapter are satisfied, then the use is presumed to be in the public interest, but this presumption can be voided by a finding in the department's order. In this case, the state would have the burden of proof in demonstrating that even though a proposed use may have passed all the tests, there is still some very good public interest reason why it should not get a permit. In any event, the department is required to include this key public interest determination in its order, along with a number of other decision elements, including:

- a statement explaining the decision criteria, including basin programs and land use plans; and
- assessments of water availability; the amount of water necessary for the use; and injury to other water right holders. (ORS 537.153)

If the decision is to issue a permit, the order must include a draft copy of the permit, including any conditions of use. Otherwise, the order must include a recommendation to deny the application. Either way, the order must be issued within sixty days of the close of the "bail-out" period provided the applicant.

◧ *Announcing the Decision and Taking Flak*

At this stage, the decision document is called a proposed final order. It is mailed to the applicant and anyone who has paid to receive it. The applicant and the public have forty-five days to protest the department's decision. A protest costs anyone but the applicant $250. If no one protests, the order becomes final.

Typical grounds for a protest might be that the department made a mistake and that the public interest presumption criteria were not met— maybe the proposed use violates a rule overlooked by the agency. Or, perhaps, the presumption criteria were met, but there is an over-riding public interest reason not to issue a permit. For example, it might be that water is available for a use allowed in the basin program, no other water right holder would be injured, and the use may not violate any other department rule. However, an environmental group may assert that a rare amphibian would be harmed, and therefore that it would not be in the public interest to issue a permit allowing the use.

On the flip side, an applicant might protest the department's decision to deny the application, or might take issue with a proposed condition of use.

A protest must be in writing and be very specific to the application in question. If a protest is filed, the department has sixty days to evaluate it and decide whether to issue a final order or to schedule a contested case hearing. Unless otherwise arranged with the applicant, the Water Resources Department must reach this decision point within 180 days of the close of the applicant "bail-out" period described above.

A contested case hearing is a trial-like proceeding undertaken by an agency to determine the facts of a case or to apply agency policies to a particular situation—in this instance, whether the department made a decision consistent with the law, rules, and facts of the case. Opposing sides present arguments to the presiding officer, called a hearings referee. Just as in a trial, witnesses are sworn, testimony is taken, and there may be cross-examination. The hearings referee's decision on whether the agency should issue or deny a permit is issued through a written order. A contested case hearing can only be held if the department finds a non-applicant protest raises a "significant" dispute of the proposed order. However, if an applicant protests and requests a contested case hearing, the department must schedule one. The department cannot schedule a contested case on its own motion.

The department's final order will be identical to the proposed order if protests were not judged "significant." Alternatively, either based on new information or a protestant's argument, the final order could modify the proposed order. In other words, the department could be talked out of its original decision. If the final order does modify the proposed order, the applicant can request and then must receive a contested case hearing on the changes. Similarly, if outside parties were satisfied with the original proposed order—and therefore had nothing to protest—but wanted to lock in their ability to participate just in case a contested case hearing is held, they can file a statement of standing for a small fee to keep their options open. If a contested case hearing is scheduled, they must pay $200 to participate and present their argument.

Although modeled on the state's Administrative Procedures Act, the water rights process includes a number of far stricter provisions. These were added in 1995 to limit participation in water right decisions. Some people believed the pre-existing process suffered from too many interests intervening in a matter best kept between the state and an applicant—and the occasional serious protestant.

The first of the stricter provisions is that contested case hearings are closed to broad participation. The hearing is limited to the applicant, protestants, and people who have filed for standing—in other words, people willing to pay $200. Unlike in many other state decision-making processes, other people and organizations cannot petition for party status and be allowed to duke it out in the hearing. Second, there are limits to the types of court appeals allowed during the process. Third, all issues and arguments that can reasonably be raised must be raised in the protest or hearing. The court is precluded from considering any new issues raised in an appeal, unless it was impossible for them to be raised earlier in the process. Fourth, strict time-lines are imposed on conducting contested cases: hearings must be concluded and an order issued within 180 days (unless there are three or more parties to the hearing, in which case 270 days are allowed).

Finally, just in case there is some more liberal provision in the APA that has been overlooked, ORS 537.140(7) states:

> Notwithstanding any provision of (the APA), an application
> for a permit to appropriate water shall be processed in the
> manner set forth in ORS 537.120 to 537.360. Nothing in
> (the APA) shall be construed to allow additional persons to
> participate in the process. To the extent that any provision in
> (the APA) conflicts with a provision set forth in ORS 537.120
> to 537.360, the provisions of ORS 537.120 to 537.360 shall
> control.

Clearly, the legislature was dead serious about limiting who may take issue with the one-at-a-time permit decisions by which the state appropriates the remainder of the public's water. Certainly, allowing each permit decision to be dragged into a contested case proceeding does not make for efficient government. However, it may be an open question whether the limitations imposed represent the answer most consistent with Oregon's citizen participation and conservation traditions.

⊞ Consequences

In the realm of water resource decision making, it is a given that few will be satisfied. The course of appeals for permit decisions varies according to the routing of a permit request. If the department issues a final order without holding a contested case hearing, parties may appeal to the local circuit court. Decisions stemming from contested case hearings may be appealed to the Oregon Water Resources Commission, and beyond that, to the state Court of Appeals.

There is one other consequence worth noting added by the 1995 Legislature. In the event that the department misses the 180-day deadline for issuing a final order or scheduling a hearing, the applicant may obtain a writ from the Marion County Circuit Court to force the agency to make a decision. If the applicant is requesting an out-of-stream use, the writ can go one step further. It can compel the department to issue a permit, unless the department shows to do so would injure another water right holder.

One can imagine many reasons for slipped deadlines: staffing cuts; loss of a key worker familiar with the request; a rush of new permit applications triggered by drought or enforcement actions, all converging on one specific deadline; or simple human error. This provision would essentially debit the public's remaining store of water and reward it to a private party, all as a penalty for administrative tardiness. This is a double-whammy for the public interest: not only is this kind of writ unavailable for instream rights or storage reservoirs, but to nullify the writ the department has to show injury to a water right holder, rather than to the interests of the public at large. To say the least, this is an unusual set-up for the allocation of a public resource.

CHAPTER 4

The Ifs, Ands, and Buts of Water Rights

THIS CHAPTER DESCRIBES THE PECULIARITIES OF the major types of use and some of the conditions commonly attached to permits these days.

Types of Use

⊟ Irrigation

According to OAR Chapter 690, Division 300, irrigation is "the artificial application of water to crops or plants by controlled means to promote growth or nourish crops or plants." Examples include watering crops, commercial gardens, tree farms and orchards, and golf courses as well as washing away alkali salts that can build up in some eastern Oregon soils. Furthermore, the 1995 Legislature decided to recognize reality by allowing incidental use of irrigation water (like tractor washing) for other on-farm purposes, as long as the amount and timing of water use was within the bounds of the permit (ORS 540.520). The state structures irrigation rights through major controls on season, rate, and duty. These vary by location within the state.

The season is the period of the year in which the right can be exercised, normally corresponding to the growing season. The rate is the maximum instantaneous amount of water that may be diverted or pumped (normally expressed in cubic feet per second ([cfs]). The duty is the volume of water that can be applied over the course of the season (expressed in acre-feet). The maximum rate, however, cannot typically be sustained on a full-time basis without exceeding the duty. Rate is to duty what spending is to a credit card limit. Interestingly, few users measure their rates, let alone their duties. To monitor the complete scope of use would require a totalizing water meter (not necessarily a cheap proposition), and would imply a willingness to quit using water when the limit of the right is met. Both

Water rights for irrigation, which withdraws more water than any other use in Oregon, are limited—legally, if not in practice— by both "rate" (how much water can be pumped at any time) and "duty" (how much can be used altogether over the course of a season). (Oregon Department of Agriculture)

of these would demonstrate an admirable public spiritedness, given that measurement is only rarely required (Chapter 7).

The three controls are usually determined by rule of thumb during the application review process. Applicants had long been required to say how much water they need for their use. But 1995 changes to the permitting process required the department to perform its own assessment of the water needed for the use before approving a permit. Historically, the department has relied on rates and duties contained in local court decrees. In a good part of the state, many claims to pre-1909 surface water uses (before a permit was needed) have been verified through an adjudication. This is a process where people present evidence of pre-1909 water use to a local court and the court then issues a decree establishing who has a valid claim. The decree also sets out how much water can be diverted under those claims, usually by applying customary or reasonable (for the climate, crops, and economies of the time, anyway) seasons, rates, and duties.

Although many of these decrees date from early in the twentieth century, the department has felt they represent a consistent standard of reasonable water use limits for a locality, and therefore applies them to new uses. The department may establish new basin-wide standards. However, unless an applicant agrees otherwise, to issue a permit for rates or duties less than the standard in a particular case, the department must make specific findings stating why the lesser amount is appropriate (ORS 537.621). The department is the final authority on the amount of use to allow under a permit (although the applicant may challenge its decision in the previously described contested case process.)

Irrigation uses west of the Cascades usually are assigned a rate based on 1/80th of a cfs (0.0125 cfs) per acre and a duty of 2.5 acre-feet per acre (i.e., over the course of the irrigation season an irrigator can apply 2.5 feet of water to each acre). East of the Cascades, the rate doubles to 1/40th of a cfs (0.025 cfs) per acre, and the duty to 3 or 4 acre-feet per acre. The Deschutes Basin's Trout Creek drainage wins the prize for the largest duty, hands down, with a sopping 6 acre-feet per acre seasonal limit (if that term can be used to describe such a volume). It should be noted that some rights have been issued with rates but no duties, and vice versa.

Much like any other water use limit or stipulation, irrigation seasons are contract terms of a water right. But, like a vacuum, nature seems to abhor a contract. A typical irrigation season may run from April 1 to October 1, but the need for water in a given year may start before April and run past October. However, crop and/or economic need notwithstanding, water use outside of the irrigation season has never been legal. If they do not shut off precisely on the last day of the season, most users taper off sooner or later—much as traffic reduces speed from 65 to 55 mph to comply with highway speed limits. Compared to the dictates of the market, water right limitations often represent a weak force.

The Water Resources Commission may extend the irrigation season of an area if asked to do so by the Oregon Department of Agriculture, as long as water is available and water quality and instream flows are not harmed (ORS 537.385). Not all areas or rights have been assigned irrigation seasons. In such cases, for field enforcement purposes the default irrigation season is March 1 through October 31 (OAR 690-250-0070).

Other than these three primary controls, some common conditions placed on water uses (see conditions section, below), and the ever-present legal bottom lines of priority and appurtenancy, the department has not generally gotten into much permitting detail about how irrigation is to take place. That has been left primarily to the operator. Although it would be within its authority, as a general rule the department does not require specific technologies (e.g., sprinkler versus furrow irrigation) or otherwise set irrigation efficiency standards. However, as competition for water supplies heats up, irrigation permits are likely to get more detailed (which is to say, rigorous or picky, depending on the perspective) through the conditioning process—either as a result of a dispute negotiation process or enlightened self-interest on the part of applicants.

▣ *Municipal Use*

Municipal water use enjoys a number of enviable preferences under Oregon water law. First, unlike other water users, they do not have to complete construction of water works within the normal five-year period. Municipalities have twenty years to begin and complete construction for new surface water and groundwater permits for municipal use. All permit extensions issued during this time must be conditioned to require a water management and conservation plan prior to diverting more water than currently being used (ORS 537.230). If the permit is to store water, municipalities (as well as counties and districts) must both begin and complete work within ten years (ORS 537.248).

Second, a municipality can get a water right certificate for part of its permit and keep the remainder in permit status. When other permittees request certification, if they have not fully developed all the water allowed under their permit, they lose the undeveloped part. but as long as cities and towns "perfect" at least 25 percent of the water use, they qualify for a water right certificate in that amount and can keep the rest in a type of permit-reserve (ORS 537.260(4)). Third, if and when a municipal permit finally ripens by use into a water right certificate, it is difficult for that use

Municipal water rights are frequently more open-ended than others, allowing greater flexibility in construction deadlines (such as for intake facilities, shown here for Salem's North Santiam water plant). (U.S. Geological Survey. Photo from U.S. Geological Survey Fact Sheet 2004-3069, The North Santiam River, Oregon, Water-Quality Monitoring Network, by Heather M. Bragg and Mark A. Uhrich)

ever to be lost. Although if unused for five consecutive years it is presumed forfeited, the presumption is easily overcome by showing the use was for a municipal purpose ((ORS 540.610(1) - (2)).

Fourth, municipal uses are not subject to strict appurtenancy requirements. As city boundaries are adjusted or deals struck to supply other cities or districts, water can be moved far and wide, the original configuration of the city notwithstanding (ORS 540.510(3)).

Fifth, municipal uses can take preference over an already-established instream water right established through the permitting process (as opposed to "conversion" or acquisition; see instream use discussion, below) if the department determines that this would comply with the public interest considerations in ORS 537.170. Sixth, in the event of competition between a municipality and a private interest seeking a hydroelectric permit, the municipality always receives preference (ORS 543.270).

Taken together, all the preferences described above give cities and towns a powerful ability to reserve significant quantities of water for the future through the permit system. It is not uncommon for cities to have undeveloped rights. As available supplies diminish, the value of valid rights will presumably increase either in some future water right market or by providing a competitive edge in attracting industries by offering a secure water supply.

Interestingly, undeveloped municipal uses are held "off-book" by the department in terms of water availability calculations (Chapter 3). If a city were to start diverting water under a heretofore undeveloped permit, it could come as quite a surprise to newer users of the system who may have thought they had at least an 80 percent chance of being served. All applications for municipal use, however, must be tied to some reasonable expectation of growth: "the application shall give the present population to be served, and, as near as may be, the future requirements of the city" (ORS 537.140(1)(e)).

▣ *Hydroelectric Uses*

"Hydro" is a world unto itself. It is legally and technically complex; one of the few state water permitting functions with a federal counterpart; subject to a different fee structure; and authorized under an additional set of statutes (ORS Chapter 543) than most other uses (ORS Chapter 537). What follows should be viewed as a summary of the major elements of Oregon's hydroelectric permitting program, rather than an in-depth treatment of the entire field.

Oregon recognizes two categories of hydro developers: municipal applicants and non-municipal entities. The evaluation process for hydro applications is laid out in ORS 537.282-299; ORS Chapter 543; and OAR Chapter 690, Division 51.

Municipal corporations (e.g., cities, towns, water districts) are eligible to receive water use *permits* to appropriate water for hydroelectric power purposes (ORS 537.282-537.299). The process for granting any permit or certificate is like that for other types of water uses as described in ORS Chapter 537, but the mandatory processing timelines do not apply. Like other water rights, a municipal right for hydroelectric purposes lasts forever unless otherwise conditioned.

Citizens or private corporations (that is, non-municipal applicants) may receive time-limited *licenses* to appropriate water for hydroelectric power purposes (ORS 543.050(2)). Those interested in obtaining a license must first apply for a state preliminary permit, as must anyone applying

Hydroelectric water uses may require either permits or licenses from the state and/or the federal government, depending on who operates them and the nature of the streams where they are sited. As a federal facility, Bonneville Dam (completed in 1938, it was the first on the Columbia) is not subject to state permitting requirements. (Oregon State Archives, Oregon Highway Division, OHD5648)

to the Federal Energy Regulatory Commission for a preliminary federal hydro permit (ORS 543.210). Federal hydro permits are required when the project would be on navigable waters of the United States; occupy lands or reservations of the United States; use water from a federal government dam; or, in specific instances, affect interstate or foreign commerce (Hayes 1996).

A state preliminary permit lasts at most three years. It gives the holder "priority of right to make application for a license covering the project for which the preliminary permit was issued." A license lasts a maximum of fifty years (ORS 543.260) and project construction must begin within two years (ORS 543.410). A public hearing must be held on preliminary permits and all licenses for projects of more than 100 theoretical horsepower (thp). A public hearing may be held on smaller projects, if the commission concludes it is in the public interest (ORS 543.225). Unlike the federal government, the state presently has no process to re-license projects once their fifty years are up.

The fees are the same for either a municipal or non-municipal applicant. They represent a very different approach than fees for other water uses. First, they are significantly more expensive: $1,000 for projects less than 100 thp; for bigger projects, $5,000 plus $1,000 for each megawatt in excess of 5 megawatts, up to a maximum of $100,000. Second, in addition to the "project fees" just listed, if the development is to be sited where anadromous fish (such as salmon or steelhead) or threatened and endangered species are present, the applicant must pay a surcharge of 30 percent of the total project fee (ORS 543.280). Third, a mandatory condition of any license is that the licensee pay to the state annually "not more than $1 for each horsepower covered by the license." This is as close as Oregon comes to a water use fee.

Whether for municipal or private purposes, any hydroelectric project must comply with strict environmental standards. First, upon receiving an application, the department must conduct a cumulative impact analysis of all existing hydro projects and hydro applications pending in a given river basin. If the department determines there is a potential for cumulative impacts, it must then conduct a "consolidated review" in the form of a contested case hearing. In the resulting final order denying or approving the application, the agency must make findings on the individual and cumulative effects of the proposed project (ORS 543.255).

Oregon law requires the department to deny an application for a project "that may result in mortality or injury to anadromous salmon and steelhead resources or loss of natural habitat of any anadromous salmon and steelhead resources." In addition, on non-anadromous fish streams, the agency may not approve an application that results in a net loss of wild game fish or recreational opportunities. Furthermore, the department cannot approve projects that may result in a net loss of natural resources, generally. As a matter of fact, other natural resources in the project vicinity—water quality, wildlife, scenic values, cultural sites—must be at least maintained, if not enhanced. It is interesting that these natural resource protections are prominently labeled in the statutes as "minimum" standards (ORS 543.017(1)(a)-(d)).

Overall, any proposed project must comply with the Northwest Power and Conservation Council's Columbia River Basin Fish and Wildlife Program, which addresses the protection and enhancement of the fish and wildlife resources of the region. The Northwest Power and Conservation Council has designated certain areas (mostly stream segments) that are off-limits to hydro development. In Oregon, the protected area designation applies to over 9,000 stream miles (StreamNet n.d.). Under the Water Resources Department's Division 51 rules, the agency may not accept an application for hydro power in these areas, nor in any of the following: national parks and monuments, wilderness areas, federal wild and scenic rivers, state scenic waterways, and state parks or wildlife refuges (OAR 690-51-030).

In addition to these environmental standards, there is another standard for developers to contend with. In determining whether it is in the public interest to allocate water for a proposed hydro project, the commission must make a finding on the need for the power, considering present and future power needs. In making this finding on big projects (those that would generate 25 megawatts or more of power), the commission consults with the state's Energy Facility Siting Council (ORS 543.017(1)(e)).

Thus, on its face, Oregon is not a very hydro-friendly state. First, there has to be a need for the power, not just a desire on the part of an applicant to build a project. Second, because of what has come to be called the "one dead fish" standard, it would appear to be all but impossible to construct a new project on an anadromous fish stream, and difficult to do so anywhere else. Third, even if the fish protection criteria can be met, a developer must maintain or enhance other nearby natural resources. Lastly, hydro licenses are subordinated to all other uses (OAR 690-51-0380). On the other hand, under certain circumstances the statutes do allow the commission to waive or modify any of the protective requirements (ORS 543.300(7)).

⊞ *Instream Uses*

For a good half of the twentieth century, keeping water in Oregon streams to protect fish or control pollution was not considered beneficial—at least not in terms of the water rights system. Deep in the programming of the state's water laws was the idea that water needed to be taken up and put to use. Leaving it alone was not use, but waste. Instream flows were therefore far from being a beneficial use and very much outside the system. In fairness to the times, a water rights-based affirmation of instream flows seemed a legal absurdity. It might be compared today to trying to maintain open space by requesting a building permit to not build a house.

Holding back on issuing rights in order to protect a few select streams, however, was not a concept altogether lost on the state. As explained in Chapter 3, many of the legislative and administrative water withdrawals were to protect fish propagation or scenic values. But withdrawals were never common enough to assure widespread protection for streamflows— a fact evidenced by the impacts water uses have had on Oregon's wild fish (Table 19).

The most dramatic effect of putting water to work for agriculture and cities has been the complete blockage of rivers by dams, which have wiped out fish species or isolated and depressed fish populations. For example, chinook salmon are now extinct in the Snake and Deschutes rivers above the Hells Canyon and Pelton dam complexes, respectively. Dams have also closed the book on chinook in the upper Klamath basin and in portions of the Sandy (including City of Portland dams on the Bull Run River),

A century of water use passed before instream water uses (such as for fish, wildlife, recreation, and pollution abatement) were given a place in the water rights system through minimum perennial streamflows (first allowed in 1955) and instream water rights (authorized in 1987). (U.S. Bureau of Reclamation. Image Number C448-100-238 . Date Taken: 1997-6-27. "Steelhead jumping ladders at Gold Hill Dam, OR. Rogue River Basin Project, OR." Photo by Dave Walsh)

Table 19. Water-Use Factors[1] Affecting Wild Fish in Oregon

Basin(s)	Chinook	Coho	Sockeye	Steelhead	Redband Trout	Cutthroat Trout	Bull Trout	Other[2]
Lower Columbia	DD	WW DD		WW DD		DD		
Willamette / Sandy	WW DD/✖	DD		WW DD		DD	DD /✖	FC
Coastal	DF	Dr		WW				
Umpqua	WW DD/✖	WW DD		DF				
Rogue	WW DD/✖	WW DD		WW DD				WW
Klamath	WW DD/✖			DD/✖	WW			WW DD,Dr
Deschutes	WW DD/✖	WW DD/✖	DD	WW DD/✖	WW		DD/✖	
John Day	WW	WW DD/✖		WW	WW	WW DD	WW	
Umatilla	WW DD/✖	WW DD/✖		WW	WW		WW	
Grande Ronde	WW	WW DD/✖	DD/✖	WW DD	WW DD		DD	
Snake & tributaries	WW DD/✖	WW DD/✖		WW DD/✖	WW		DD/✖	
Closed					WW DD,Dr	DD WW		WW DD

[1] Frequently listed in combination with other factors (e.g., land management or other fish species introductions). WW = water withdrawal; DD = interference from dam/diversion structures; DF = decreased flow (no cause listed); Dr = drainage or diking; FC = flood control activities; ✖ = species extinction has occurred in all or parts basin
[2] Other: includes various suckers (Goose Lake, Warner, Shortnose, Lost River, Jenny Creek), lampreys, and chubs.
Source: Data from Oregon Department of Fish and Wildlife 1995

Willamette, Umpqua, Rogue, Umatilla and Walla Walla rivers. Pelton and Round Butte dams have also blocked migratory sockeye salmon, steelhead, and lower river redband trout.

Some dams are scheduled for big changes. Many of the major Willamette flood-control dams will likely be required to provide for fish passage under final Endangered Species Act decisions expected from the NOAA Fisheries Service and the U.S. Fish and Wildlife Service. Under the Federal Energy Regulatory Commission relicensing process, a plan was approved in 2004 to allow salmon, steelhead, and bull trout to migrate past the Pelton-Round

Butte project for the first time since 1968 (Portland General Electric. n.d. [a]). In addition, a number of smaller dams are slated for removal to help fish runs, including Portland General Electric's Marmot Dam on the Sandy River and the Grants Pass Irrigation District's Savage Rapids Dam on the Rogue. As of early 2006, discussion is also underway concerning fish passage needs (including possible dam removal) at the Klamath River dams in Oregon (J.C. Boyle dam project) and California.

Where populations have not been extinguished, they are subjected to other stress. For example, the Columbia's white sturgeon population is now fragmented and cut off from productive marine environments and bull trout in many eastern Oregon streams can no longer move up and down the stream systems as their biology demands.

The actual use of water can also dewater streams and diversion dams can take a toll on the necessary mobility of fish populations. While far from the only cause of wild fish difficulties (others include poor land management practices and damaging introductions of non-native fish species), water withdrawals are injuring coho salmon in the Rogue and Umpqua basins,

Water use has hit many fish runs hard. When Portland built its water supply dams in the 1890s (above, horses are shown hauling pipes to the site), salmon runs were cut off on the Bull Run River. (Oregon Historical Society, B.C. Towne, OrHi35389)

steelhead in the Illinois and Umatilla basins, redband trout in many areas of eastern Oregon, bull trout in rivers draining to the Snake, and an assortment of specialized fish in the closed basins (Oregon Department of Fish and Wildlife 1995).

MINIMUM PERENNIAL STREAMFLOWS

In 1955, after a century of settlement and a half-century after the initiation of Oregon's water permitting system, the legislature recognized the need to bring instream flows into the system:

> The maintenance of minimum perennial stream flows sufficient to support aquatic life, to minimize pollution and to maintain recreation values shall be fostered and encouraged if existing rights and priorities under existing laws will permit. (ORS 536.310(7))

Minimum perennial streamflows are administrative rules in basin programs that establish (usually monthly) flow levels for fish, water quality, and recreation. Water rights newer than these flows must yield to these "minimum flows." Thus, these flows are vested with water-right-like qualities—they receive a priority date and can make a call on water from junior appropriators. But this also means that minimum flows cannot demand water from earlier (senior) water rights. So "minimum" does not mean a real physical water level that must be maintained at all times regardless of the cost to other water users. Rather, a minimum flow reserves from further appropriation an amount of water that will conservatively serve the fish, water quality, or recreation needs.

Originally, the Water Resources Board and the Water Policy Review Board (predecessors to the current seven-member Water Resources Commission, which oversees the Water Resources Department) adopted minimum flows based less on biological needs than on the physical characteristics of streams during low-flow periods—namely, the mean monthly flows of the lowest three consecutive months of record (Oregon Water Resources Department 1981). Thus, early minimum flows were adopted at very low levels with little monthly variation (Borden 1989).

During the 1960s, what is now the Oregon Department of Fish and Wildlife identified minimum and optimum flows based on stream surveys and a (at that time cutting-edge) habitat model, called the Oregon Method. This approach considered the passage, spawning, egg incubation, and rearing needs of salmon and trout, based mostly on water depth and velocity requirements at key points along a stream. Surveys were completed for fifteen of Oregon's eighteen river basins (Oregon Department of Fish and Wildlife 1984).

As data on streamflow needs became more available, the state moved toward an approach where other agencies applied to the Board for certain flows (ORS 536.325). These flows tended to be higher than those previously considered, and in many cases were hard for the board to swallow. The mission of the agency was still strongly aligned with the development, rather than the conservation, of water resources. The board inclined toward minima rather than the higher flows contained in many requests. Rather than fostering and encouraging minimum flows as the statutes required, the state was seen by many as resisting them.

This resistance came to the attention of an environmentally active legislature, which in 1983 made it "the policy of the State of Oregon that establishment of minimum perennial streamflows is a high priority of the Water Resources Commission and the Water Resources Department" (ORS 536.235). The legislature also directed the Departments of Fish and Wildlife and Environmental Quality to submit minimum flow requests on seventy-five of their highest-priority streams, and it required the Water Resources Department to take action on them within three years. Furthermore, in a novel approach to the executive branch of government, it instructed the governor to "guide and assist ... the Commission in performing duties ... to insure compliance with the time limitation" (ORS 536.325).

The first minimum perennial streamflows were adopted in 1958 in the Umpqua basin and the last in 1988 in the Umatilla basin, for a total of 547. Although the authority to adopt minimum flows is still on the books, this method of providing for instream flows is a thing of the past. Which is just as well, for it has major drawbacks. First, minimum streamflows may only be adopted to protect some of the many benefits provided by instream flows (primarily fishlife and clean water). Second, although recreation values are recognized as an important objective of minimum flows, the state's Parks and Recreation Department is not allowed to request them. Third, minimum flows are rules, not rights, and rules can be changed. For example, in bone-dry 1973 when push came to shove, the Water Policy Review Board temporarily suspended some minimum flows (Sherton 1981). In addition, in 1981 the commission suspended minimum flows to allow municipal uses on the South Umpqua River, Cow Creek, and Canyon Creek (Oregon Water Resources Department. n.d. [b]).

INSTREAM WATER RIGHTS

In response to some of the shortcomings of minimum flows—and following in the footsteps of other states such as Colorado, Wyoming, and Utah—the Oregon Legislature created appropriative instream water rights in 1987 (Trelease 1990). The instream water right statutes offer an improved means for protecting instream flows because they broaden the kinds of uses eligible for instream flow protection; declare these "public uses" to be beneficial uses once and for all; recognize the State Parks and Recreation Department as a key player in instream matters (ORS 537.332-537. 336); and grant instream rights the same legal status as other types of rights (ORS 537.350). But the statutes also reiterate that instream rights cannot impair the rights of senior water users (ORS 537.334). Instead of seeking to undo the appropriation system that had ignored instream values, the legislation bought into that system and sought full participation under it for instream water rights.

There are about fifteen hundred instream water rights in Oregon. (Oregon Water Resources Department. n.d. [c].) There are three ways instream rights can be established in Oregon: conversion, application, and acquisition.

CONVERSION. In 1987 the legislature directed the department to review and convert all minimum perennial streamflows existing at that time to instream water rights. (Seventeen additional minimum flows were adopted in the Umatilla basin in 1988 after the legislative direction and thus were not included in the conversion requirement.) Some of the conversion process was complicated by difficulties interpreting the storage component (i.e., the amount of the instream flow dependent upon reservoir releases)

The Umpqua River basin was the first in the state to receive instream flow protections under the appropriation system.(Photo by author)

of some Willamette basin minimum flows. All but twenty-four minimum flows have been converted to instream water rights.

APPLICATION. Only the state departments of Fish and Wildlife, Environmental Quality, and Parks and Recreation may apply for new appropriations for instream uses. This means no private parties, environmental groups, or federal agencies can do so. It also means that, unlike setting of minimum flows, the Water Resources Department cannot take action on its own motion. It needs to receive a request. Instream rights can be requested to protect a wide variety of purposes as shown in Table 20.

ORS Chapter 537 and OAR Chapter 690, Division 77, set out the evaluation process for instream right requests. These are processed in much the same way as out-of-stream applications, with the same range of outcomes—the department can approve an instream right as requested, reduce it, or reject it. The department is the final authority in determining the instream flows necessary to protect public uses (ORS 537.343). However, there are some things unique to instream right processing.

First, the water availability standard is different from that used for other types of water right requests (see Chapter 3). Second, if an instream water right request is approved, the department issues a water right certificate rather than a permit. The reason for the permit phase for an out-of-stream use is to provide time to divert the water, to prove it can be put to beneficial use. In theory, that time or proof is not needed for an instream use, because the water is intended to stay where it is, providing the benefits it always has. Third, when an instream water right certificate is issued, it is not in the name of the applicant, but of the Water Resources Department as trustee for the people of the State of Oregon. Fourth, the department does not collect a fee for instream water right processing.

Table 20. Instream "Public" Uses

• Recreation
• Pollution abatement
• Navigation
• Conservation, maintenance, and enhancement of:
- aquatic life
- fish life
- wildlife
- fish and wildlife habitat
- any other ecological value

(ORS 537.332)

Fifth, in a unique exception to the prior appropriation doctrine, certain new uses can take water away from some older instream water rights. Multipurpose storage projects and municipal uses may take precedence over agency-requested instream rights if the Water Resources Department considers it in the public interest (ORS 537.352). In addition, when an instream water right and a hydroelectric request are competing for the same water source, the hydro applications get first priority (ORS 537.360).

ACQUISITION. In a far-sighted provision, the Oregon Legislature created a mechanism that not only includes instream flows in the water rights system, but allows instream interests to reach back in time to claim water long-since allocated to other uses. Under ORS 537.348, "any person may purchase or lease an existing water right or portion thereof or accept a gift of an existing water right or portion thereof for conversion to an instream water right (which) ... shall retain the priority date of the water right purchased, leased or received as a gift."

In other words, old out-of-stream uses (which, by virtue of their age, have more power to demand water under the prior appropriation doctrine) can be acquired for instream purposes. This ability, however, is to be exercised in accordance with the transfer process which is designed to prevent injury to existing water right holders (Chapter 5).

Acquisition may hold the greatest promise of any mechanism in restoring instream flows through the water rights system. The conversion and application processes may represent potent forms of instream flow protection, but only if the stream was not already over-appropriated at the time the protection set in.

For those nervous about taking a permanent transfer plunge, the state's instream leasing program (also addressed under OAR Chapter 690, Division 77) offers a way to get their feet wet. The first lease agreement was signed in 1994, with three others following that same season. By 1995, there were thirteen agreements, keeping an estimated 17,000-plus acre-feet instream in four southern and eastern Oregon (especially the Malheur) basins. By 2005, 335 leases were in effect that protected 458 cfs instream for some part of the year—or about 105,000 acre-feet (counting only primary, not supplemental, water rights). Over three-quarters of these were leases for five years (Oregon Water Resources Department 2006[b]). Originally, instream lease regulations mandated water rights holders to forego use of any leased water for the entire season or year. In 2001, the legislature allowed split season leases, so that a water right holder can apply water for part of the season and then stop, leasing the remainder of the use for instream purposes, usually during the low-flow season. The water right holder must measure and report use to the Oregon Water Resources Department (ORS 537.348).

Acquisition opens up a number of intriguing avenues and interesting questions. First, it recognizes a market in water rights by clearly and explicitly using the term "purchase," thereby creating a promising context of private action. In other words, if you have the money and a heart for instream uses, you can make somebody an offer for their water right. Or if you have a water right and feel environmentally inspired, you can donate all or part of it to the cause. Second, the statutes recognize pragmatism. If water right holders are flirting with forfeiture (that is, nearing five consecutive years of non-use), they could lease the right for instream purposes without breaking the chain of use (ORS 537.348 (2)). When once again in a position to return the right to its original use, they could pick up legally where they left off. Third, the language of ORS 537.348(1) opens up the possibility of individuals or organizations holding the purchased/leased right in their own name. While the state currently disputes this interpretation, this ability could increase a sense of ownership in those putting up the money and a sense of comfort in those not wanting to deal with government. Recently, the state has started approving leases for instream beneficial uses (e.g., fish life, pollution abatement) in the name of private parties, but stops short of anointing them as bona fide instream water rights—successfully safeguarding the distinction between rights held by the state for the people, and rights held by people for the people.

Another interesting on-the-ground type of question for acquired rights is how the water will be ushered down a stream course. In other words, how far downstream from the location of the old point of diversion will this water, newly baptized as instream, be protected in-channel? If it is only to the next point of diversion of a senior user, that could be a matter of yards on some stream systems. If it is further than that, the question becomes how much further—to the stream's mouth, to the Pacific? To the extent the flow becomes a pulse of water to be protected over some distance, it becomes a management challenge. With a small watermaster corps, few stream gages or diversion meters, and lots of users, keeping that quantity of water from being sipped into nothingness would prove no small matter.

Once an instream water right is established, it is treated like other water rights (though few would guess it, given the reaction from many corners of Oregon): it must yield to senior users; and it is subject to abandonment, forfeiture, and emergency water shortage provisions.

In a lot of rural Oregon, instream water rights are extremely suspect. In part, this might stem from a basic misunderstanding about how they work. Many people believe that instream rights can take water away from existing water right holders. They can't (ORS 537.334). Without knowing that the law provides otherwise, some people may reason if the state establishes some sort of flow minimum, and there's less than that right now, the gap just has to be filled by taking water from those that have it.

In actuality, while instream flows may represent some biological bottom line for fish, they are really the maximum quantity of water than can be protected instream under a given priority date. Just like the quantity of water allowed by any other right. Another factor feeding misunderstanding may be past attempts by some environmental groups to pass a law giving instream rights a really old priority date (such as the date of statehood). This would put instream rights in a superior position for demanding water, but at the cost of most other rights issued by the state. The legislature has never passed such a law.

Another concern some have is that moneyed interests far removed from the community will simply buy out local water rights and convert them to instream uses. If this buy-out retires a lot of farmland, the local impacts in terms of tax base or economic multipliers (such as farm machinery sales) could be serious. It will be interesting to see how instream water right acquisitions, resulting from the local exercise of a free market principle, square with the viability of local communities, traditional champions of private property rights. For now, with rural communities feeling hammered by logging cutbacks or other environmental protections, any tinkering with water rights that appears to benefit the environment causes nervousness.

▣ *Storage*

Oregon distinguishes between diverting water to store it and, once stored, taking it from storage. These are viewed as different activities, each requiring a different permit. Stored water is also distinguished from "natural flow." Natural flow is defined as that quantity of water occurring in a stream given upstream use, but absent any "regulation"—that is, manipulation of streamflow to fill reservoirs (which would keep water back and thus decrease flow for the period of filling) or the release of stored waters (which would augment flows during the period of release.) (Natural flow should not be confused with the quantity of water originally in a stream in its natural state [i.e., before settlement], sometimes called historic flow.)

Unless otherwise specified, when people get a permit, it authorizes them to use (and priority date depending, demand) only natural flow. It would not allow them to require an upstream facility to release stored water. That would defeat the purpose of storing water, which is to keep it for later use. The stored water legally belongs to the person storing it. However, if a reservoir releases more water than is needed by any water rights that can call for it (see "secondary" rights, below), the excess release becomes available for appropriation (ORS 540.045(3)).

Applications to store water must specify the height of the dam, the capacity of the reservoir, and any uses to be made of the stored waters (ORS 537.140(1)(d)). The department must approve in writing the site of and plans for any structure that, in failing, could damage life or property. Plans and specifications are required for any dam that is 10 or more feet high and backs up 3,000,000 or more gallons (ORS 540.350-540.400). The plans may be of a preliminary nature in the application; but no reservoir may be filled before final written approval from the department is obtained (ORS 537.400(4)-(5)). When the department issues a permit for a new storage project to a county, municipality or district, construction must be completed within ten years—a longer time period than for other out-of-stream uses (ORS 537.248).

Storing water, whether for large farm ponds as shown above, or for dams on major river systems, generally requires a water right.(Oregon State Archives, Oregon Dep't. of Agriculture, OAG0475)

Anyone planning to take stored water and apply it to beneficial use must obtain a secondary permit. An application for a secondary permit must identify the source reservoir and show there is an agreement with the reservoir owners to provide enough water for the proposed use (ORS 537.400(1)). The same goes for any instream right that relies on stored water (ORS 537.336(4)). Usually, Oregon's big storage projects are operated by one interest to provide water for others (Chapter 7). For example, Prineville Reservoir in Central Oregon stores water under a permit issued to the federal Bureau of Reclamation. The Bureau, in turn, sells water use contracts to irrigators, who have secondary permits to pump the stored water released into the Crooked River and apply it to their fields.

Another storage-related permit is for "make-up" water—that is, water to maintain a reservoir by replacing water lost to evaporation or seepage. In other words, a permit to store water does not automatically allow the holder to continually dip into the natural flow passing through his or her reservoir to make up for losses (ORS 537.400(2)) without applying for a secondary permit to do so.

Storage fees consist of an application evaluation fee ($300), permit recording fees ($250), and additional charges based on the amount stored ($20 for the first 10 acre-feet and $1 per acre-foot beyond that). Consider a dam that created a respectable-sized lake 500 feet in diameter averaging 10 feet deep. This impoundment would contain about 45 acre-feet of water. Thus, the unending right to store that water costs just over $600.

Another kind of storage that is receiving more attention is underground water storage. This approach, called aquifer storage and recovery (ASR) is described in the groundwater section below.

Compared to some years past, the present is a bust in what is likely the boom and bust cycle of storage development in Oregon. Earlier this century, the development of water resources, including reservoir construction, was paramount in the state. There were a supportive citizenry and fewer regulatory requirements for both big projects (such as the Willamette basin flood control reservoir system) and a host of small projects encouraged by many government agencies (e.g., farm ponds sponsored by the federal Agricultural Stabilization and Conservation Service). Most importantly, there was abundant federal money.

A time of development and money is not a time to sweat the details. And getting state storage permits decades ago probably seemed a very small detail. Lots of dams were built without permits. As water supplies have tightened over the last few years, and more scrutiny has been applied to water management, these and other illegal uses have come to light.

While the state has the authority to drain illegal ponds, bringing down the steel fist of government upon usually unsuspecting landowners is not a welcome prospect, in terms of fairness, public relations, or agency budgets. Accordingly, in 1993 the legislature established a ponds amnesty period that also exempted some off-channel ponds from permit requirements (though requiring a notice to the agency) and allowed owners of others to apply for permits and not be penalized. The response was so great (both in numbers and emotional intensity) that the amnesty was extended and ponds permitting requirements eased. There is now a simplified and cheaper application process for ponds storing less than 9.2 acre-feet or with dams less than 10 feet high, if they do not injure any existing water right and do not pose a significant detrimental impact to fish. ($40 application fee, with $10 for each acre-foot, not to exceed a total cost of $250.) (ORS 537.409)

⊟ *Groundwater*

The steps in the groundwater permitting process are basically the same as those for other permits (ORS 537.515-537.629). However, because groundwater is subject to a different set of physical rules and represents a distinct type of source under Oregon law, there are some differences between groundwater and other permits.

Prior to 1955, permits for groundwater use were only required in eastern Oregon, and then only for groundwater believed to flow in underground channels (Clark 1983). In the Groundwater Act of 1955, the legislature required a permit for the use of any kind of groundwater statewide (except for certain exemptions described in Chapter 2 and for pre-1955 uses that could use water under a registration).

The statutes define groundwater broadly: "any water, except capillary moisture, beneath the land surface or beneath the bed of any stream, lake, reservoir or other body of surface water within the boundaries of this state, whatever may be the geological formation or structure in which such water stands, flows, percolates or otherwise moves" (ORS 537.515).

When it set up the groundwater permitting scheme, the legislature had nearly fifty years of surface water permitting experience to learn from. The emphasis it chose to place on preserving state control of the groundwater resource is telling. The groundwater statutes begin by reciting policies firmly anchoring permits within the context of control:

> ... in order to insure the *preservation of the public welfare,*
> *safety and health* it is necessary that:

Oregon was one of the first states to require a permit for pumping groundwater. To prevent or slow water level declines, the state now restricts new groundwater uses in a certain areas, including a growing number in the Willamette Valley not far from this wellhead near Silverton. (Oregon Department of Agriculture)

(2) Rights to appropriate ground water and priority thereof be acknowledged and protected, *except when, under certain conditions, the public welfare, safety and health require otherwise.*

(3) Beneficial use without waste, *within the capacity of available resources,* be the basis, measure and extent of the right to appropriate groundwater. ...

(7) Reasonably stable ground water levels be determined and *maintained.*

(8) Depletion of ground water supplies below economic levels, impairment of natural quality of ground water by pollution and wasteful practices in connection with ground water be prevented or *controlled* within practicable limits.

(9) Whenever wasteful use of ground water, impairment of ... existing rights to appropriate surface water, declining ground water levels, ... interference among wells, ... overdrawing of ground water supplies or pollution of ground water exists or impends, *controlled use of the ground water concerned be authorized and imposed ... by the commission under the police power of the state ... when ... voluntary ...* action is not taken or is ineffective.

(10) Location, construction, depth, capacity, yield and
other characteristics of and matters in connection with wells
be *controlled* in accordance with the purposes set forth in this
section." (ORS 537.525; emphases added)

The evident care exhibited by the legislature in protecting the groundwater
resource may explain why groundwater applications have to contain more
source information than surface water permits, including the depth to
the water table, if known; the location of each well; the proposed depth,
diameter, and casing method of each well; each well's estimated capacity;
and the horsepower of each pump motor (ORS 537.615).

In addition, as noted in Chapter 4, in deciding whether to issue a
groundwater permit, the state in theory determines, not if the use would
hurt the public interest, but whether it will affirmatively "ensure the
preservation of the public welfare, safety, and health" (ORS 537.621(2)).
While permits for using groundwater may not look very different from any
others, in theory they have a different legal footing. Unlike most surface
water uses, once issued, groundwater permits can still be subject to state
control. The primary means of asserting this control of already-issued
permits is through a critical groundwater area declaration.

The Water Resources Commission is authorized to designate critical
groundwater areas wherever needed to remedy serious groundwater
problems. Designation, which occurs through administrative rules, may be
triggered by excessive water table declines (Figure 19), imminent overdraw
of the groundwater supply, substantial interference among wells, polluted
groundwater and more, as provided in ORS 537.730. A public hearing on
the designation is required. Notice must be mailed to affected groundwater
permittees or certificate holders and all licensed well drillers.

The corrective actions imposed by the commission may include
prohibiting new groundwater permits and limiting how much water can be
withdrawn each day, month, or year (termed "determining the permissible
total withdrawal") from a specific groundwater body. In addition, after a
contested case hearing, the commission may actually restrict the exercise
of existing groundwater rights and permits. Restrictions may include
splitting the permissible total withdrawal among existing water right
holders; establishing preferences for certain types of use (notwithstanding
the priority dates of any associated rights); reducing the amount of a right;
requiring the sealing of polluting wells; or limiting the number of wells
used by a single appropriator.

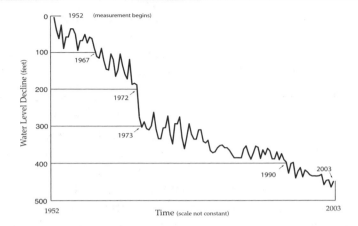

Figure 19: Water level declines in state observation well 445 near Hermiston, Oregon. This graph shows water levels in a well drawing from basalts in the Butter Creek Critical Groundwater Area near Hermiston. Data collected by the Oregon Water Resources Department show about 450 feet of decline over 50 years. (Source: Data from OWRD. Dashed lines indicate missing data)

Critical groundwater areas are bad news. Even in the face of rapidly dropping groundwater levels, few citizens welcome the state singling out their area for an extra layer of regulation. This is especially true when that regulation can get personal by requiring cutbacks in individual water use. Not surprisingly, establishing critical groundwater areas is a laborious, contentious—and therefore expensive—process. For example, the Butter Creek Critical Groundwater Area near Hermiston was first designated in 1976. The decision was taken to the state Court of Appeals and ultimately to the Oregon Supreme Court, which upheld the decision in 1989. Pumping controls went into effect in 1990—fourteen years after the initial decision (Oregon Water Resources Department 1993b). Given this track record, it is not surprising that this most direct means of control provided in the statutes is seldom used. There are six critical groundwater areas in Oregon, which cover a total of about 800 square miles (Figure 20).

Reacting to groundwater problems is unpopular and expensive. That is why the state is turning more and more to solutions that prevent, rather than fix, the problems. One fairly recent preventive mechanism is the groundwater limited area. Groundwater limited areas are locales classified in basin programs for very limited future groundwater uses in order to control existing or prevent worsening groundwater supply problems. They are easily confused with "groundwater management areas" and

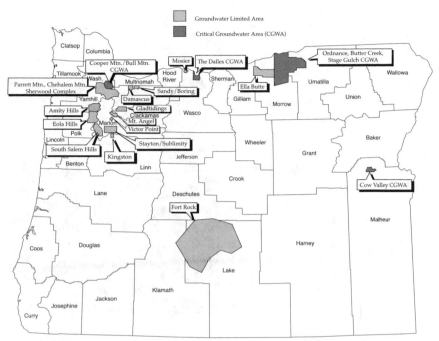

Figure 20: Areas of groundwater supply restriction (does not include areas restricted because of groundwater/surface water connection). (Information from Oregon Water Resources Department)

"areas of groundwater concern"—different designations established by the Oregon Department of Environmental Quality to control groundwater contamination (ORS 468B.175 and .180).

Another preventive approach is the agency's careful review of groundwater applications. Each is checked against existing data to determine if the proposed groundwater use has the "the potential to cause substantial interference" with other wells or surface water users. The hydraulic connection between groundwater and surface water is being more carefully scrutinized than ever before. Groundwater and surface water form a continuum. They are really just water flowing under different time regimes. All groundwater at some point reaches the surface. While Oregon statutes treat them as separate sources, with increased use of the water resource, the interplay between the two becomes important and awfully technical—to be convinced, consult OAR Chapter 690, Division 9.

Consider a stream closed to prevent further over-appropriation. A person proposes to dig a well 10 feet deep into streamside gravels 10 feet from the streambank and pump water out. By the statutory definition, this

Taking a well measurement involves lowering a steel tape (sometimes equipped with electronic sensors) down the bore until it hits water, then reading off the distance. In some areas with hundreds of feet of decline, such as the Butter Creek Critical Groundwater Area shown above, it takes a lot more tape than it used to. (Oregon Water Resources Department)

would be a groundwater use. But clearly, it is taking water belonging to the stream and already allocated to senior users. By law those users must be protected. This groundwater use should be subject to the same regulations as its associated surface water.

Division 9 establishes the basic rule that any well tapping an "unconfined aquifer" (a water-bearing zone more or less open to influence by the atmosphere; e.g., river sands or gravels) within one-quarter mile of a stream is hydraulically connected to the stream. This cut represents the easy call. Where to draw the line in other situations requires a complex analysis of site-specific data such as well depth, distance from surface sources, amount of use, and the geologic character of the surrounding rock. Where the line is drawn can make or break a groundwater application. Often this line becomes the front between factions warring over the proposed use. These factions often hire their own hydrogeological experts to employ science in the art of getting what they want.

The geologic character of the rock surrounding a well is becoming important for another reason: underground storage. Wells are not necessarily a one-way proposition. Under certain circumstances, farmers, cities, or other water users may want to literally "hole away" surplus surface water by injecting it down wells. That surplus would then be pumped up during times of shortage—assuming the geology of the site was tight enough to keep it from escaping. This process is called aquifer storage and recovery (ASR).

ASR can occur through either an existing right or a new right. The legislature has declared that ASR is actually a beneficial use within a beneficial use, inherent in every right that has been or will be issued (ORS 537.531). The Water Resources Department is the sole permitting agency for this activity, which carries some risk of mixing possibly dirty surface water with cleaner groundwater. Therefore, the department must consult with both the Department of Environmental Quality and the Health Division to make sure water quality problems do not arise (ORS 537.534). ASR projects must be undertaken in two phases. The first requires a test program performed under a five-year limited license (a type of temporary authorization; see Chapter 5). Test results on water quality, water storage parameters, water table recovery rates, and preliminary geologic information are then considered by the department in making the final decision whether or not to allow the use (ORS 537.534).

After successful completion of tests, the second phase is obtaining a permit for ongoing storage and recovery of the water. In deciding whether to allow ASR projects, the department evaluates potential impacts of injection on ground water quality and of recovery on existing users of the aquifer, as well as other injurious effects of the project. Water quality issues are addressed through coordination with the Oregon Department of Environmental Quality and the Health Services Division of the Department of Human Services. An ASR limited license or permit may be modified or revoked if harmful impacts occur after the project is in operation (Oregon Water Resources Department. n.d. [d]).

In summary, Oregon law aims high in protecting the groundwater resource and its users. It authorizes a lot of state control over the exercise of groundwater rights. However, that protection is only as good as the availability of groundwater data. Judgments about general groundwater availability, whether or not water tables are declining, impacts of new uses on nearby wells or streams and ultimately the public welfare itself, all hinge on good data. And those data are in short supply (Chapter 1). In addition, when data are sufficient to trigger groundwater controls, the damage has usually already been done and communities are heavily invested in the customary level of (over-) use. And the controls are so unpopular and fiercely resisted that the state no longer considers them a practical management alternative. Until more information is available and/or better controls are provided, managing Oregon's groundwater resource primarily through a permit system is likely to be a hit or miss proposition.

Conditions

In issuing surface water permits, groundwater permits, or instream water right certificates, the Water Resources Department may attach terms, limitations, or conditions to make the use consistent with the public interest (ORS 537.170, .190, .211, .343, .625, .628, .629). Although long allowed, the use of conditions has increased in recent years because of an improved awareness of water limits and the impacts of additional water uses (see Figure 21).

Beyond controlling adverse impacts, the use of conditions offers an institutional allure. Conditions allow an agency to say "yes, but ..." instead of "no way!" And as a collective of living, breathing human beings, the better angel of an agency's nature—given the chance—would rather help people than make them mad. Besides, there is a very real need to meet a growing water demand in Oregon. Without the ability to condition permits, there would be little flexibility in the system. If not pitched in exactly the right way, water use requests would be denied more often than not. Every request would be a long-shot—not a desirable feature in most permitting systems.

When getting a permit becomes the exception rather than the rule, the permitting system is either broken or has served its purpose and turned itself off. Some see the use of conditions as a bureaucratic dodge that merely postpones the inevitable: the close of the allocation period of Oregon's history and the opening of the re-allocation era. They maintain that with a limited resource such as water, permit issuance has to stop sooner or later. Far from being a sign of failure, in their view it is proof the system is preventing over-appropriation and protecting senior users. They would argue other mechanisms such as transfers, storage, or conservation will be available to fill the water demand gap. In their view, it may be better not to lend false hope to permittees or disguise limits by concocting conditions that get lost in the noise of the real world.

But Oregon is growing; not all streams are over-appropriated; the groundwater resource, while not well understood, is a significant supply source; and storage during the high-flow season could supply many new users. Permitting will continue for some time. As long as it does, there will be a need to fine-tune permits through conditions to protect the public interest, including existing users.

The department uses two types of conditions. The first are advisories about existing laws printed in the permit. For example, permits often contain conditions requiring the holder to cease use if told to do so by a watermaster or to use no more water than allowed in the permit. Strictly speaking, these are re-statements of law. They would apply whether or not they were included in the permit. The use is not conditioned on these provisions, but premised on them. However, they are included for both emphasis and education.

The other kind of condition involves operational constraints. These are true conditions, in that they are specific to the permit and, if not met, represent a violation of the law. The Water Resources Department has developed a rather lengthy and evolving menu from which conditions may be selected to meet the needs of a specific permitting context.

For example, most surface water permits currently include some requirement for measuring and reporting water use. For small uses, the condition is really a warning that they may be required to measure in the future; for medium uses, the condition generally requires the permittee to install a measuring device before diverting water; and large uses may be conditioned to require both measuring water use and regularly reporting to the department. Most surface water diversions also have conditions requiring water intakes to have fish screens (built to the specifications of the Oregon Department of Fish and Wildlife) to keep fish out of pumps and ditches. Likewise, some instream water rights are conditioned to exempt future domestic and livestock uses from being cut off to satisfy instream needs. Groundwater permits often include triggers that require well pumping to be reduced or to cease altogether if there is an average water level decline of 3 or more feet per year for five consecutive years; a total water level decline of 15 or more feet; or a decline of 15 or more feet in any older, neighboring well.

While the use of conditions can allow water use with a minimum of adverse impacts, it presents certain problems. First, as new permits are ornamented with unique and site-sensitive requirements, watermasters will probably find it increasingly difficult to keep track of who is supposed to be doing what. Second, as land changes hands and there is no longer first-hand knowledge of all the agreed-upon conditions, the use will gravitate back to the least-troublesome mode of operation. Third, it will take increasing vigilance for the state to do its duty and assure similar conditions are applied consistently to similar uses.

STATE OF OREGON
COUNTY OF KLAMATH
CERTIFICATE OF WATER RIGHT

THIS CERTIFICATE ISSUED TO

NAME   **A**
ADDRESS   **B**
TOWN, OREGON

confirms the right to use the waters of <u>A WELL</u> in the <u>SWAN LAKE BASIN</u> for <u>IRRIGATING 252.2 ACRES.</u>   **C**

This right was perfected under Permit G-9332. The date of priority is MAY 26, 1981. This right is limited to 3.15 CUBIC FEET PER SECOND or its equivalent in case of rotation, measured at the well.   **D**

The well is located as follows:

NW 1/4 SW1/4, SECTION 6, T37S, R10E, W.M.; 1940 FEET NORTH AND 570 FEET EAST FROM SOUTHWEST CORNER OF SECTION 8.   **E**

The amount of water used for irrigation together with the amount secured under any other right existing for the same lands, is limited to a diversion of ONE-EIGHTIETH of one cubic foot per second (or its equivalent) and 3.0 acre-feet for each acre irrigated during the irrigation season of each year. Provided further that in the event of a request for a change in point of appropriation, or repair of this well, the quantity of water allowed herein, together with any other right from this point of appropriation shall not exceed the capacity of this well at the time of perfection of this right.   **F**

The use shall conform to such reasonable rotation system as may be ordered by the proper state officer.

A description of the place of use to which this right is appurtenant is as follows.
    SW1/4 SW 1/4   32.4 ACRES
    SE1/4 SW1/4   32.0 ACRES
            SECTION 20

    NE 1/4 NW1/4   30.6 ACRES   **E**
    NW1/4 NW1/4   31.2 ACRES
    SW1/4 NW1/4   31.2 ACRES
    SE1/4 NW1/4   30.4 ACRES
    NE1/4 SW1/4   32.0 ACRES
    NW1/4 SW1/4   32.4 ACRES
            SECTION 29
TOWNSHIP 37 SOUTH, RANGE 10 EAST, W.M.

The well shall be maintained in accordance with the General Standards for the Construction and Maintenance of Water Wells in Oregon.   **G**

The right to use water for the above purpose is restricted to beneficial use on the lands or place of use described.

WITNESS the signature of the Water Resources Director, affixed July 24, 1992.

Martha O. Pagel   **H**

Recorded in State Record of Water Right Certificate numbered 67564.

Figure 21: How to read a water right

A. *Name and address of person or entity to whom the right was issued. Because water rights "go with the land," this information tends to date quickly, with old-timers' names and obsolete addresses dominating the water right rolls.*

B. *The right's* **source** *and* **beneficial use**: *it's a groundwater right to irrigate 252.2 acres.*

C. *The* **permit** *that allowed the applicant to first take control of water and put it to beneficial use. This document is a water right* **certificate** *because it certifies water was used according to permit terms. If landowners are doing a record check and can only find a permit [usually good for no more than 5 years], they should contact the Oregon Water Resources Department to make sure they're legal.*

D. *The right's core—the* **priority date** *and* **rate** *of use. The holder of this right would have to stop pumping if it interfered with nearby users whose rights pre-date May 26, 1981; or could demand nearby users with newer rights to cease use. The rate of use is limited to 3.15 cubic feet per second — at the wellhead, not at the end of a mile of leaky pipe. The amount of water the water right holder can use is further limited by the "duty," as explained below.*

E. *Precise* **locations** *of the well and the 252.2 irrigated acres by reference to the public land survey grid (Township: north/south locator; Range: east/west locator; Section: one of 36 square-mile boxes making up a Township/Range area; "SW 1/4 NW 1/4," standing for "in the southwest quarter of the northwest quarter," results from quartering the section, then quartering the quarters, which yields sixteen 40 acre subdivisions; "W.M." stands for Willamette Meridian, a statewide baseline.) Public land survey information can be found in deeds and tax assessments. Moving the well location or changing the place of use is illegal without amending the right through a* **transfer**.

F. *Although the water right holder can use up to 3.15 cfs, no more than 756.6 acre-feet of water can be diverted during the irrigation season — i.e., 3 acre-feet for each of the 252.2 acres irrigated. This total is called the "**duty**." In other words, the water user can't pump at 3.15 cfs for the entire irrigation season, for that would generate about 1,140 acre-feet, assuming a 6 month season — nearly 50 percent more than allowed.*

G. *Conditions of use: while these examples re-state water law, other rights' conditions may require environmental protections or measurement.*

H. *The* **certificate number**, *a vital record. It is interesting to note that in this case, the water right process took 11 years from application (as captured in the priority date) to certification (director's signature).*

CHAPTER 5

Other Authorizations

WATER RIGHT CERTIFICATES—AND THEIR IMMATURE FORM, water use permits—are the most common ways the state gives permission to use public water in Oregon. Given Oregon's natural diversity, growing population, and changing economy, it is not surprising that a variety of other water use authorizations have been created to meet changing water use needs. This chapter deals with what are now less common forms of permission, though in the coming years several will not only become more common, but dominant.

Transfers

Over time, people change their minds about how to use the water allowed them under a water right—a new road opens up different ground to farm; a city water intake is washed away and a better replacement location is identified; a new land owner wants to use water not for irrigation, but for a private campground. But water rights are written narrowly and authorize only very specific uses. Legally, these changes cannot just happen. Uncontrolled change, no matter how gradual, can undermine the state's water management system.

Water use "creep" is viewed as a risk to the system and all its users: new activities may drift away from recognized beneficial uses; customary and relied-upon water flow patterns may be disrupted as water gets moved over the landscape; the location of rights is first confused, then lost. State control of changes is required to prevent disorder. That control is achieved through the water right transfer process.

Transfers are orders approving changes in a water right or permit. The changed right or permit keeps the same priority date. A water user must ask the state for permission to change the type of use, point of appropriation, or place of use (ORS 540.520(1)). Uses eligible for transfer include those allowed under adjudication court decrees; water right certificates; and water use permits where the completed use has been surveyed and certification has been requested (ORS 540.505(4)).

A few changes do not require transfers. Water users are allowed to move their points of diversion to follow a naturally changing stream channel, as long as the distance moved is not greater than 500 feet and some other conditions are met (ORS 540.510(5)). In addition, no transfer is required if an irrigation right is used for incidental agricultural uses, stock watering, or other irrigation-related use, as long as it stays within other terms of the right (ORS 540.520 (8)).

The transfer application is straightforward, requiring the applicant's name; a description of current and proposed types of use and places of use; reasons for the requested changes; and evidence that water has been used over the previous five years (ORS 540.520(2)). In addition, the applicant must provide a map of the affected area prepared by a Certified Water Rights Examiner. The state requires applicants to pay fees for transfer examination and filing. Examination fees support the cost of evaluation. Filing fees pay for recording the transfer order, if approved. As for permit applications, transfer fees are set in statute, and fall short of the cost of processing. Transfers cost a total of $350 for each change, plus $175 for each cfs beyond the first (see Table 18). Just as with permit applications, transfer applicants can pay the department to hire temporary staff or to contract with an outside service provider to expedite transfer processing.

The Water Resources Department's transfer process is very similar to its permit application process, with a series of check-points that include proposed and final orders. When the department receives a transfer application, it first checks to see if it is complete and if the water right involved is subject to transfer. If there are problems, the department returns the application and any fees paid to the applicant. If the application is in good shape, the department provides notice to interested parties and invites comment for at least a thirty-day period.

After the close of the comment period, the department issues a preliminary determination of whether the application should be approved or rejected. The department provides a copy to the applicant and gives the applicant at least thirty days to address any issues identified. If the applicant amends the application or provides additional information in support of approval of the application, the department revises the preliminary determination as appropriate.

The department then gives notice of the transfer application and preliminary determination through its weekly notice, by mailing notice to each person who submitted comments, and (except in a few circumstances) by publication in a newspaper in the area of the transfer (the cost to be paid by the applicant).

Any person may file a protest or standing statement if filing occurs within thirty days of the department's notice. If there is a protest, the department holds a hearing and issues a proposed final order to approve or reject the application. If there are no protests or notifications of intent, the department issues a final order consistent with its earlier preliminary determination.

A transfer request must be approved if it meets the basic requirements and if it does not result in an enlargement of the original right (such as using a greater rate or duty of water than currently allowed; increasing the acreage irrigated under a right; failing to keep the original place of use from receiving water from the same source; or diverting more water at the new point of diversion or appropriation than is legally available to that right at the original point of diversion or appropriation), or causes injury to existing water rights (unless the holders of such rights give permission in a written affadavit). The final transfer order sets a time limit for completing the changes requested (OAR Chapter 690, Division 380).

If the original use was allowed under a water right certificate, the department cancels it and issues a new one reflecting the changes. Sometimes only part of a water right certificate is changed. In that case, two new water certificates are issued: one for the changes and one to accurately reflect the unchanged portion, called the "remaining right" (ORS 540.530). If the transfer is to move a point of diversion, the new diversion facility must have a fish screen, if requested by the Oregon Department of Fish and Wildlife (ORS 540.520(4)).

For many years, transfers enabled only permanent change to a water right; because water rights are issued in perpetuity, changes to rights took on that same time horizon. However, many water use changes are not permanent. Starting in 1995, the legislature began allowing some types of temporary transfers. By 2005, it had authorized the department to approve temporary transfers of up to five years for changes in place of use, point of appropriation (if needed to support a changed place of use), and type of use. Any land from which a use is transferred must remain dry for the period of the transfer. However, the time water is not used on these original lands does not count toward forfeiture. The department may revoke temporary transfers at any time if it finds they are causing injury to any existing water right (ORS 540.523).

In another transfer twist, the 1995 legislature broke new ground by allowing a surface water point of diversion to be transferred to a groundwater use (ORS 540.531). Transfer between sources had previously been considered a prohibited practice. In promoting this change, the state was attempting to resolve a couple of problems. First, some cities were finding it difficult to comply with the federal Safe Drinking Water Act which requires extensive treatment of surface water sources. If the surface water source can be transferred to a nearby groundwater source, cities can wind up with both cleaner and cheaper water. Secondly, appropriators in some stream systems use bulldozers to heap up gravel diversion dams in the stream bed each season. Allowing these users to transfer their surface diversions to nearby shallow wells under carefully controlled circumstances can meet the users' needs, lessen impacts on stream habitat, and do away with any risk that fish can get into pumps or ditches (Oregon Water Resources Department 1995a).

There are some other transfer-related processes worth noting briefly. First, in a kind of amnesty for technically illegal water use changes, irrigation and some other districts may petition to transfer places of use to reflect on-the-ground changes that became established over many years. The petition may include an unlimited number of transfers and must be accompanied by thorough documentation (ORS 540.574–540.580). Second, when the government draws down a reservoir (such as on the Columbia to speed salmon migration) and strands an authorized diversion facility, the owner must notify the state but may move the diversion without a transfer. If within sixty days the department finds the new diversion location will injure existing rights, it can disapprove the change (ORS 540.510(6)). Third, water may be applied to new lands under an existing right through Oregon's conserved water program without going through the transfer process (Chapter 6). Fourth, to transfer water out of its basin of origin, special permission is needed (Chapter 11). Fifth, if an individual (not a company or government body) has been using a diversion point for over ten years that is not the one authorized on the water right, the individual may request an abbreviated transfer process to change the diversion point listed in the water right to the actual location. Finally, permanent acquisitions of out-of-stream uses for instream purposes require a transfer, but under OAR Chapter 690, Division 77, temporary instream leasing of out-of-stream rights may proceed without a transfer.

No matter what kind of transfer is requested, by law the department's evaluation cannot consider the public interest—probably because it is presumed that public interest concerns were addressed in approving the original water use. To do it again would be a sort of double jeopardy. However, the extent to which the public interest was considered in issuing past water rights is debatable—most were the slam-dunks of a simpler time. Even if originally considered, many argue what may have been in the public interest decades ago is not necessarily so today. In addition, some believe that because a transfer amounts to a new right equipped with the benefit of an old priority date, in granting this benefit the state should be able to consider the current interests of the public.

Environmental groups naturally tend to support a transfer approach that includes a public interest review. Agricultural and other groups fear this would be disastrous, re-opening already-issued rights and leading to environmental exactions and diminishment of the rights. This debate is anything but academic and the stakes are large. The reason: transfers probably represent one of the most important water sources for the future of Oregon (Oregon Water Resources Department 1995a).

When a concert or a Blazer game or the Pendleton Round-Up is sold out, if you need a ticket, you talk to someone who has one. If you have nose-bleed seats and want a front-row view, you work an exchange. Most surface water supplies in Oregon are just as sold-out as any popular event— only most the people in the seats have life-time passes. Some relief might be found by enlarging the arena (storage) or replacing the deluxe seats with more, slightly smaller versions (conservation). But even these have their limits—and there is always the problem of financing. Increasingly, newcomers wanting water for new uses at different locations will have to draw it from the pool of what already has been allocated. And that means transfers. Of all the areas of water law that will have the greatest impact on twenty-first-century Oregon, transfers are the most important, and will therefore receive the most attention (i.e., fighting) in the coming years.

Limited Licenses

Some water uses just do not need a water right attached in perpetuity to a particular piece of ground. Road construction, crops that need a one-time shot of water (e.g., some vineyards), general construction, and emergencies threatening public health, safety, and welfare illustrate uses of a "short-term or fixed duration." Such uses are eligible for a limited water use license. The

Water Resources Department may issue a limited license for one use for a maximum of five consecutive years. If the department determines that water is available and the proposed use will not impair the public interest, a limited license can be issued with terms and conditions similar to those of a water use permit. Limited licenses can be issued quickly, in as little as three weeks. In an emergency where the water is needed to protect the public health, safety, and welfare, the department may issue a limited license for immediate use (Oregon Water Resources Department 2004a).

Licensees are allowed to use or store surface or groundwater at their own risk. The use gets no priority date and is therefore subordinate to all other authorized users of the same source. In addition, licensees must notify the watermaster when they plan on using water under a limited license and keep a record of the water they use. Furthermore, the department may revoke licenses at any time if a use injures any other water right or minimum perennial streamflow. Lastly, the department may issue limited licenses to place illegal users on a compliance schedule (ORS 537.143). Limited license fees are one of the few water use authorization fees set in rule rather than statute (OAR Chapter 690, Division 340), and are therefore subject to change by the agency. In 2006, the fee was set at $150 for the first, and $15 for each additional, point of diversion.

Registrations

"Registrations," as used in this book, refer to water uses allowed simply by notifying the Water Resources Department. They represent a sign-out sheet approach to providing water for usually very old or very small (again, a relative term) uses. Beyond that, the different types of registrations do not have much in common. Rather, they have popped up through the years to address specific water use situations where the normal permit procedure proved onerous. In other words, in the crossword puzzle of Oregon's water management system, "registration" is frequently the answer to "a twelve-letter word describing an alternative that appears fairer and sounds simpler than permitting, at least to those promoting it." The types of registration are summarized below.

▣ *"Grand-fathered" uses*

People who have asserted that water was already being used on their lands before a permitting system went into effect have been able to register those uses. A registration allows them to continue using water until the state determines the validity of their claims. If the claim is validated, the users receive a water right certificate. Validation occurs through a process called adjudication. In this process, after giving broad notice of its intent to identify all such "old" users of a water body, the department takes testimony and evidence from people, and submits findings to the local circuit court. The court then either affirms or modifies the department's findings and issues a decree establishing the amount, location, and priority of the old rights. Those claiming surface water use prior to 1909 are supposed to have filed their registration by December 31, 1992 (ORS 539.005–539.240). Those claiming groundwater use prior to 1955 are to have filed a registration by May 29, 1961 (ORS 537.585–537.610; ORS 537.635–ORS 537.700). About 70 percent of Oregon has been adjudicated for surface water, mostly east of the Cascades (Figure 22) (Oregon Water Resources Department n.d. [e]). Only one small area (Harbor Bench on the south coast near Brookings) has been adjudicated for groundwater (Lissner 1996). The job of completing adjudication will likely still be going strong well past the centenary of the adoption of the Oregon Water Code in 2009.

⊡ Environmental projects

Certain wetland, stream restoration, and storm-water management projects that require a water use permit qualify for a registration under ORS 537.015–537.032. This type of registration amounts to an expedited preliminary permit. It allows water use to occur prior to a final permitting

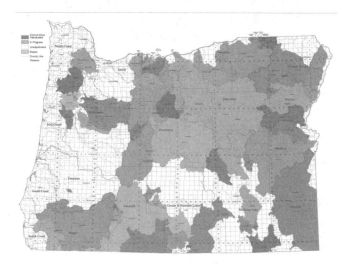

Figure 22: Adjudicated areas in Oregon (dark shading represents completed adjudications, light shading in progress). (Map courtesy of Oregon Water Resources Department)

decision. Before approving a registration, the department must determine whether the proposed project would injure other water rights, harm the public interest, or really result in a benefit. Requested registrations are announced in the department's weekly notice and may be commented on for thirty days. If, based on comments or its own analysis, the department cannot make the required determinations, the registration is denied and no early use may take place: the proposed project must await a decision through the normal water permitting process.

▣ Road projects

It was news, and bad news at that, to a number of county and state road managers that the water they used to build or maintain roads needed to be covered by a water right. They knew that getting a water right for each of potentially hundreds or thousands of diversion points could be expensive, involved, and uncertain. In addition, even if obtained, a right issued in perpetuity for a specific location probably would not serve the mobile and unpredictable needs of managing a road system. Consequently, the legislature created a registration for road-related uses. Under ORS 537.040, a public road agency can register a water use by submitting an application and $300. The application must include a map; identify the maximum amount of water to be used during a calendar year and any 24-hour period; and list the proposed surface water or groundwater sources. Registrations receive no priority date and are subordinate to all other existing or future permits and rights. Use under a registration may be cut off any time the department believes it is adversely affecting the watershed or other water right holders. Registrations, which cannot exceed 50,000 gallons from a single source during any 24-hour period, are good as long as the user submits an annual renewal statement accompanied by $50.

▣ Miscellaneous Registrations

Although it is not referred to as such in the statutes, department staff sometimes call a notice of exempt reservoir a "ponds registration." Under ORS 537.405–537.409, ponds are exempt from regulation if they were built before 1995; store less than 9.2 acre-feet, or have a dam less than 10 feet high; and do not injure existing rights or pose a significant risk to existing fisheries. However, owners of such ponds were required to submit a notice of having an exempt reservoir to the Water Resources Department by January 31, 1997.

Another kind of registration is that required for using "reclaimed" water—water that has been used for municipal purposes, then treated and readied for another use. Most often, the term refers to sewage effluent. In years past, effluent was almost always returned after treatment to mix with the waters of origin and flowed to the next downstream water appropriator. Recently, more and more cities are finding they can kill two birds with one stone by piping the water to other (often paying) users: they can meet increasingly stringent water quality standards by not discharging back into a stream—and they can satisfy water demands of their residents or neighbors, such as golf-course operators or nearby farmers. While this may be good news for the cities and their clients, it may come as a surprise to downstream users who have come to rely on the water.

In an attempt to address the equities of such a situation, the state requires any person using or intending to use reclaimed water to file a registration. The registration form must, among other things identify the source of the reclaimed water and the nature and amount of the proposed use; specify the location of any facilities used to convey the reclaimed water; and indicate whether the reclaimed water is substituting for water used under any permit or right. If the reclaimed water had previously been discharged into a stream for five or more years, and it made up half or more of the stream's flow, the department must notify any affected water right holders. If water right holders show that re-routing the water would impair their ability to satisfy their right, they are entitled to a preference for using the reclaimed water, with one big catch—the water cannot be routed down the stream, but must be piped from the treatment plant, likely at the deprived water right holder's expense (ORS 537.131—537.132).

Reserved Water Rights

Unlike most of the authorizations described elsewhere in this section, reserved water rights represent full-blown rights to use water. However, the process through which they are determined (courts, negotiations, and adjudication rather than the regular application route) and the frequency of their determination (rare) combine to make them an exceptional authorization to use public water.

When the federal government reserves lands for special purposes, it generally also lays claim to enough water to serve those purposes. For example, unless otherwise stipulated, national forests, wildlife refuges, wilderness areas, and wild and scenic rivers (see Chapter 9) would have a call on water with a priority date of the day of their establishment. This legal principle was first put forward by the U.S. Supreme Court in 1908 (in *Winters v. United States*, 207 U.S. 564 (1908)) and is called the Winters doctrine (Oregon Water Resources Department 1995b). The implied reservation of water is called a reserved water right.

Unlike other water rights where the details of use are settled through the permit evaluation process and specified in a water right certificate, reserved rights are generally established first, with the details to be worked out later. "Later" can be a long time coming. For example, many national forests were created in the years bracketing 1900. And many tribal reservations date from the mid- to late-1800s. But, for most of the twentieth century, Oregon spent little effort in nailing down reserved rights.

Arguably, the reserved rights for tribal reservations would be high on the list for determination, since they directly affect both on- and off-reservation economies. However, of the nine federally recognized tribes in Oregon, only one has had its reserved rights addressed (the rights for the Umatilla River basin, including those of the Confederated Tribes of the Umatilla Indian Reservation, were adjudicated in 1916) (Oregon Water Resources Department 1995b). However, in 1987 the legislature established a process for negotiating reserved right determinations of federally recognized Indian tribes (ORS 539.300—539.350). The key word here is "negotiate"; in other states it has been "litigate." Determining the reserved rights of tribal interests in the western United States has been a long, contentious, and expensive court process. Under the Oregon law, the director of the Oregon Water Resources Department is authorized to negotiate water right agreements with all federally recognized Indian tribes in Oregon. The department works with Oregon's nine federally recognized Indian tribes on a government-to-government basis to address water issues of mutual concern. It bases its approach on two principles: 1) the department should endeavor to identify and help protect existing tribal rights to the use of water; and 2) it should forge partnerships with tribes to share responsibility for water and watershed management (Oregon Water Resources Department 2006c).

Thus, the State of Oregon has selected negotiation as the best strategy to work through an often complex and controversial process. It is controversial because reserved water rights usually are far superior to (that is, far older than) most water rights established through state processes—and that can be upsetting to state water right holders who may have an entirely different (and incorrect) understanding of their place in line for water.

It is complex because of the mix of parties involved (state agencies and their counsel, federal trustees and their counsel, tribal members and their counsel, state water rights holders, and the public), past actions (state adjudications that did not include reserved right considerations), and the methods of quantification. These methods usually are predicated on the assumption that irrigated agriculture is the economic engine of Indian reservations. Therefore, the size of the reserved right is keyed to an assessment of how many acres of reservation land may be productively (and practically) irrigated and the quantity of water necessary to serve those lands. However, in a case involving the Klamath tribe (*U.S. v. Adair*, 478 F. Supp. 336 (D. Or. 1979)), the court found that reserved rights also extended to the protection of hunting and fishing rights. The priority date of water rights for irrigation purposes was the date of the establishment of the reservation; that for hunting and fishing purposes dated from time immemorial (Clark 1983).

Reserved water rights (which pre-date many the state has issued) entitle Native Americans to often-unspecified amounts of water, including that needed to satisfy hunting and fishing rights. Celilo Falls, an ancestral fishing site on the Columbia, disappeared in 1957 beneath the waters of Lake Celilo formed when The Dalles Dam began operation. (Oregon State Archives, Oregon Highway Division: Gifford, OHDG211)

The department began reserved right negotiations with the Confederated Tribes of the Warm Springs in 1989 and reached agreement with the tribes in 1997. The Deschutes County Circuit Court ratified the agreement in 2003. The department began working with the Klamath Tribe to determine its reserved rights through the Klamath basin adjudication proceeding, which began in 1975. The claiming period for all parties closed in 1997, and a formal process of alternative dispute resolution has been successfully used in place of litigation to settle most conflicts. The department expects to submit its adjudication findings to the Klamath County Circuit Court in 2008. The department also expects to enter into re-negotiations with the the Confederated Tribes of the Umatilla Reservation. Although reserved rights were determined in 1916, the tribes and the United States have raised questions regarding the legal sufficiency of that settlement (Oregon Water Resources Department 2006c).

Additional Authorizations

Summarized below are some additional authorizations that warrant at least a passing mention.

▣ Supplemental rights

Although long familiar to the department, supplemental rights have only recently appeared clearly and explicitly in statutes. Supplemental rights are water rights issued as a back-up to pre-existing, or primary, rights. The term almost always applies to irrigation uses. Supplemental rights cover the same lands as the primary right (the only instance, other than in municipal rights, where water right "stacking" is allowed) and may be exercised only when the primary right is exhausted during a given season (ORS 540.505(3)). The primary right may give out because of drought, normal seasonal run-off patterns, or for other reasons. An application for a supplemental right undergoes the same kind of review as any other application.

▣ Exchanges

An exchange is a substitution of one source for another in equal or reduced amounts. A person may request the Water Resources Commission for permission to use stored, surface, or groundwater from another source if users of the other source are supplied replacement water. The request can be made if the applicant's water source is not sufficient; it would result in better conservation; or the applicant can develop water for an instream purpose, but cannot convey the water to its point of use. The commission must approve any such request unless other water right holders would be injured, the public interest would be adversely affected, the exchange would be too difficult to administer, or there would not be enough replacement water (ORS 540.533–540.543).

▣ *Drought permits*

After the governor declares a drought (actually, that "a severe, continuing drought exists"), the Water Resources Commission may issue emergency, temporary water use permits to drought-stricken water users (ORS 536.750). Before issuing a drought permit, the department must first find that the use will not injure existing rights or be detrimental to the public interest. If an emergency use does cause injury to existing rights after all, it can be revoked. Each emergency permit must yield to prior permits, rights, and minimum perennial streamflows. If the permit is issued to a state agency or public body, the holder must either submit a water conservation plan first, or be ready to implement a conservation plan when directed by the governor. The term of any permit is decided on a case-by-case basis, but at the latest expires when the drought declaration is canceled (OAR 690-19-0040).

▣ *Use of Conserved Water*

Under Oregon's conserved water program, a water user may obtain approval from the state to use a certain amount of saved water on new lands. This authorization is more fully explained in Chapter 6.

▣ *Hydroelectric Licenses*

Private entities may apply to the state for a license to use water to produce hydroelectric power. Technically, a license is different than a water use permit or a water right. This authorization is described in more detail in Chapter 4.

PART III

Oregon's Waterworks

The primitive conception that water, like air and sunshine, is one of the gifts of Nature which are free to all alike … must give way when countries become densely populated, or when special industries, like agriculture by irrigation, make so large demands on streams that there is not enough water for all. Free water on Manhattan Island is no more a possibility than free forests …

Moisture is necessary to plant growth, and in arid lands this moisture is supplied largely from streams. Hence in such regions the right to use rivers in irrigation is an indispensable requisite to any large creation of wealth in lands. As population increases and civilization advances, there is not only a more extensive but a more intensive use of water. …So extended have the demands for water become … that … great corporations are formed for acquiring water, constructing dams, building storage works, canals, and pipe lines for the conveyance and distribution of water for different purposes … State experiment stations and the Department of Agriculture are studying how economy in the use of water in irrigation may be promoted, and cities find waste in domestic water supplies a serious evil.

There is nothing in farming where rainfall is ample which corresponds to the intensity of feeling which marks the struggle for control of streams in arid lands, or the anxiety which besets irrigators regarding the stability of their water titles. The farmer who remains serene of spirit when he sees his fields burning for lack of water and knows that his loss of crops is due to wasteful use by others is a rare if not impossible character.

Advancing civilization has done more than augment the uses and value of water; it has increased the evils and dangers arising from water. Every reservoir, every diversion dam in a stream, every artificial waterway adds a new element of danger and insecurity to the lives and property below and gives ground for new laws and regulations with respect to the management of water.

Elwood Mead
The Evolution of Property Rights in Water
In *First Biennial Report of the State Engineer to the Governor of Oregon for the Years 1905 - 1906*

CHAPTER 6

Water Rights and Water Wrongs

WHAT HAPPENS "ON THE GROUND" WHEN a citizen goes to exercise a water right? No longer a matter of the theoretical or the promissory, the use becomes real. It takes root in an amazing array of forms—a Hermiston watermelon, a Willamette Valley strawberry, a field of high desert alfalfa, or a robust Umpqua steelhead. The user, whether knowingly or not, enters into a community of water.

This community is characterized by a dramatic give-and-take. After all, members may by right choke off newer users or may be required by older users to watch their own interests wither. A community perhaps not in the sense of Mr. Roger's Neighborhood as much as Homer Simpson's Springfield—where there is as much testiness as tolerance. A community where each member has entered into a contract with the people of the state to use their water beneficially without waste. How that contract is enforced and how users participate in that community are the subjects of this chapter.

Watermasters and Field Enforcement

Enforcement is the act of assuring compliance with Oregon's water laws in general, and the terms of authorized water use, in particular. The state is responsible for enforcement and its primary enforcers are state watermasters, officers appointed by the Water Resources Department to manage water use in their districts.

Districts correspond roughly to the state's major river basins, which means a given watermaster may be expected to cover a bit of territory—such as District 4, the Massachusetts-sized John Day basin. About the only territory not under watermaster jurisdiction is any land within an irrigation district. Once water is diverted by a district, it is under the exclusive control of its directors (ORS 540.270). Oregon employs around twenty watermasters who, with a similar number of county-funded assistants and additional state water resource technicians (such as well inspectors, water right inspectors, and hydrographers), comprise the state's water management field force. The field force is organized around five regions and fifteen satellite offices statewide (Figure 23).

Watermasters' duties include those specified in statute (ORS 540.045), as well as many required or delegated by the director of the Water Resources Department. Their most important duty is distributing water. This means monitoring streamflows or water levels and cutting off junior users to keep water flowing to senior users, when necessary (and it almost always is). In other words, as the traffic cop of Oregon's water system, the watermaster halts some water users when others have the right of way. Most distribution involves surface water, not only because of the predominance of surface water rights, but because it is harder to identify and address interference between wells.

When it is necessary to cut off junior water users, often a figurative white glove and whistle are enough. Letters of reminder, press releases, and, most importantly, personal contact can be very effective means of explaining why a person's use must cease. According to the department, these methods typically produce a 97 percent compliance rate (Oregon Water Resources Department 2005 a). However, the statutes also explicitly authorize a watermaster to distribute water by "regulating, adjusting, and fastening the headgates, valves or other control works." In other words, the state can seize control of a private facility and limit the amount of water

Figure 23: Oregon Water Resources Department field regions and watermaster districts. (Information from Oregon Water Resources Department)

Each summer state watermasters shut off junior users to allow water to flow to senior water right holders. (Oregon Water Resources Department)

pumped or diverted—even to the point of shutting off use altogether. When this happens, the watermaster must attach a signed notice to the water works explaining that it is wholly under control of the watermaster (ORS 540.045(1)). This is called "posting notice." And jumping a fence to do this is no problem, as the director and his or her delegates have power to enter upon any private property (ORS 536.037(1)(e); ORS 537.780(1)(e)).

Shutting off users is one way of "regulating" water use. In this context, regulating water users simply means the state determined it was their turn to stop using water. Making them stop is not punitive. (However, "regulation" can also mean taking action against an illegal use, as explained elsewhere in this chapter.) How do watermasters know when it is time to regulate junior users? First, most have a pretty good feel for the hydraulics of the systems under their watch. Wet years or dry, they are familiar with run-off and water use patterns. They can make an educated guess when flows will drop to the point where junior users need to be cut back. Of course, this is informed by their careful attention to streamflow and water table measurements which they and other field staff gather through the season. It should be noted that when users are cut back, the watermaster can also turn them back on if water conditions improve.

They can also try to get a handle on how water is actually being used by selectively requiring measuring devices along ditches and both above and below reservoirs (ORS 540.310 and 540.330). Ultimately, they receive the most help from the water users themselves, who do not hesitate to blow the whistle when they are not getting enough water. About half the actions taken by watermasters are triggered by such complaints. The other half result from the watermasters' own monitoring. In 2004, a year with fairly normal precipitation and run-off statewide, 294 streams were regulated through nearly ten thousand separate actions (which included not only orders to water users to change what they were doing, but also inspections that found no change was needed) (Oregon Water Resources Department 2005 a).

Regulation of junior rights can reveal a great deal about the pressure on a stream system. It not only suggests the workload of a watermaster, but provides an on-the-ground measure of over-appropriation. For example, if every summer holders of post-1940 water rights get cut off, it probably means the stream became fully appropriated about that time. It also means that water rights issued since then are likely contributing to over-appropriation. Many streams around Oregon are routinely regulated back to the 1800s—even in the best of water years (Figure 24).

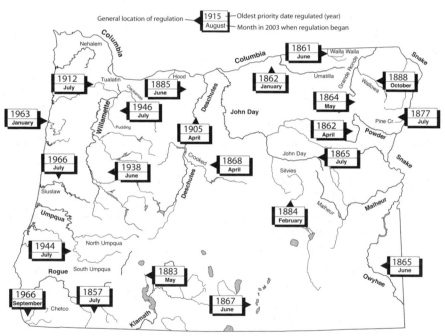

Figure 24: Surface water right regulation 2003. Many streams were already fully appropriated one hundred years ago. Each year shown above represents a dividing line between water rights that were fully satisfied in 2003 and those that were shut off or restricted by watermasters. The map shows that in far eastern and southern Oregon, only the oldest of rights can be assured of getting water. It also shows that regulation (restriction) can occur in almost any season of the year. (Source: Information from Oregon Water Resources Department 2004c)

When watermasters receive a complaint, they investigate. Based on the investigation, they decide on an appropriate action. The first course of action is to find and cut off any illegal uses. The next is frequently to force one or more junior appropriators to stop using water, but this is not always a given. For example, a watermaster is under no obligation to regulate other users to provide enough hydraulic head to push water through a senior appropriator's weed-choked diversion, or to bring water up to the level of a diversion. It is up to the user to maintain diversions and ditches in good operating condition (OAR 690-250-0030). When wells interfere with one another, the outcome vis-à-vis senior and junior users may be unexpected. An interfering junior user will escape the sting of regulation if a senior user can drill deeper and get water. In other words, even though they were there first, senior groundwater users cannot expect to lock out other users simply by claiming any additional use would affect the operation of their typically shallower wells. Sometimes the action is no action. It is not every day that government takes official notice of futility. Yet a watermaster can do just that by declaring that a complaint is a "futile call." When the water freed up by cutting off junior users would evaporate, percolate, or otherwise not get to the senior user in any appreciable amount, the watermaster may disregard the complaint (OAR 690-250-0020).

In addition, users may cooperate during water shortages by agreeing to take turns using the water they are collectively entitled to. Such an agreement is called a "rotation." When written rotation agreements are submitted to watermasters, they must distribute water according to the terms of the agreement (ORS 540.150).

But a watermaster is not a water czar. Water must be distributed in strict accordance with the amounts and priorities established under existing rights. The watermaster is not free to ignore the needs of a senior user—not even to keep water flowing to a city, to a cannery in full operation, or to an in-coming run of threatened salmon.

Additional duties of a watermaster identified by statute include distributing water inside a private system (if asked and compensated) (ORS 540.210); "shepherding" (i.e., protecting from unauthorized use) stored water released down a stream for delivery to intended users (if asked and compensated) (ORS 540.410); filing affidavits of non-use for apparently-abandoned water rights (ORS 540.660); and, last but clearly not the least, "any other duties the … director may require." (ORS 540.045(1)(e))

And the director requires quite a few: stream measurement; gaging station maintenance; precipitation gage monitoring; well inspections; dam safety inspections; attendance at local meetings; reviewing proposed regulations; consulting with other staff on proposed water uses or transfers; participating in instream leasing negotiations; testifying in contested case proceedings; and otherwise acting as a miniature Water Resources Department by being a local contact point for all types of citizen encounters with the system. Watermasters also spend a fair amount of time helping water users resolve conflicts. Of course, they are also expected to keep their eyes peeled for illegal uses (see below).

As both a state officer and a member of the local community, watermasters are under great pressure. They are expected to act impartially on behalf of the people of the state, frequently by personally turning off the water of their neighbors. Sometimes this is to serve other neighbors. Sometimes it is to serve people or interests far removed. As more restrictions are placed on new water uses and more scrutiny is applied to existing uses, watermasters risk increasingly becoming identified with "Salem," the fabled home of armchair bureaucrats who either do not understand, or have it in for, local resource-based economies. Hostility, threats, and even violence directed against these state regulators is not unknown.

Consequently some critics charge that some watermasters have "gone native" and do not provide the enforcement needed to protect state interests. But considering there are twenty watermasters and about the same number of county-funded assistants to oversee approximately 60,000 diversions on 100,000 miles of stream over a 100,000 square-mile state, any enforcement shortcomings should more likely be blamed on a sheer imbalance of numbers rather than misdirected sympathies. Ironically, despite the growth in water rights, water regulations, and general awareness of water resource challenges, the state has fewer watermasters and assistants today than in the early 1980s (Oregon Water Resources Department 2001).

Illegal Water Use

When people have gone to the trouble and expense of working within the system to obtain legal authorization for their water use, they rightfully and understandably object to people using water unlawfully. And they expect the state to protect them from those illegal users. Illegal use takes water away from those who have a right to it. It represents an unquantified and uncontrolled draw on the public's water, risking the health of stream systems, the integrity of groundwater tables, and the livelihoods of many. In short, just like shoplifting, it has a cost to consumers and messes up the bookkeeping. It does not matter whether it results from delinquency, desperation, or diminished capacity, it is illegal and imposes significant costs on the system.

Illegal use includes using water without authorization; in greater amounts than allowed; after having been shut off; or in violation of the terms of a permit or certificate. No person may use water someone else is entitled to or willfully waste water to the detriment of another person. Not surprisingly, neither can anyone interfere with a diversion or use water when they have been told not to. Under these circumstances, a person does not have to be caught red-handed bolt-cutting the lock on their headgate—the use itself is prima facie evidence of guilt (ORS 540.710–ORS 540.730).

What happens when a watermaster finds or is told of an illegal use? The Water Resources Department's internal guidelines on the matter basically say "it depends." This is not surprising given how stretched the agency's enforcement staffing is. It is important to set enforcement priorities, just as police might in deciding whether to ticket a speeding motorist or respond to a bank robbery in progress. In responding to an illegal use, a watermaster considers a number of factors, including whether it is causing direct injury to an existing right; interfering with public health and safety, water quality, or sensitive species; conflicting with local land use plans; or caused by a repeat offender. Watermasters also weigh whether enforcement in a particular circumstance would be a good example to other violators (Oregon Water Resources Department 2001).

The watermaster's response can be formal or informal. Voluntary compliance is preferred whenever possible. Normally this entails personal contacts, phone calls, letters, and other forms of communication, all of which deliver the cease-and-desist message with a dose of public education. However, where this fails, the state may use more persuasive means of obtaining compliance, such as locking down headgates or pumps, leveling fines, filing criminal complaints with the local district attorney, or having the watermaster arrest the violator. Not surprisingly, the last two options, though allowed under ORS 540.050, 540. 060, and 540.990, are rarely used.

Prior to 1989, the state could only pursue criminal convictions that carried the penalties listed in ORS 540.990, penalties that, over the years, have lost any sting they ever had. For example, if someone monkeyed with a diversion or used water in defiance of a watermaster order, they could be fined at least $10 but not more than $250 and/or thrown in jail for not more than 6 months. Because jail was just about out of the question, any fines amounted to a pretty low cost of doing business. And that was only after a potentially expensive (to the state) court fight.

A new enforcement option with more teeth—civil penalties—was authorized by the 1989 Legislature and codified in ORS 536.900–536.935. Under civil penalties, fines are larger and set by administrative rules (not statute), making it easier to maintain their sting through time, rather than having inflation render them a small cost of doing business. The department may fine a person who violates the terms of a permit, certificate, or license; uses water without authorization; disregards a watermaster's order or interferes with waterworks; or violates well maintenance regulations. Civil penalty amounts are set in the department's rules (OAR Chapter 690, Divisions 240 and 260), but no penalty may be greater than $5,000 for each violation (Table 21). However, under most circumstances, the same offense carried into a new day can be treated as a new violation. In deciding the amount of the fine, the department must consider the past history of the person with regard to obeying water law; the financial condition of the violator (including any gain resulting from the illegal use); and the risk posed by the violation to other rights and the public at large.

The department not only gets to keep the money paid as penalties, but can collect the cost of its expenses in dealing with a violator. However, penalties have not represented much of a revenue stream for the department. Any revenue received is peanuts compared to that pocketed by the Department of Environmental Quality, the state's best-known and far more generously staffed environmental enforcer (Table 22). DEQ collects at least $500,000 annually in civil penalties (and some years, many times that amount), thanks in part to rules that allow fines of tens of thousands of dollars for certain violations (OAR Chapter 340, Division 12; Oregon Department of Environmental Quality 2006 a).

Enforcement is not a one-way street, however. Any person who feels they may be injured by the action of a watermaster may approach a local circuit court for an injunction. The court is to issue an injunction only if the watermaster has failed to carry out orders of the Water Resources Commission or of court decrees that determine rights to water use (ORS 540.740).

How much illegal water use is there in Oregon? No one knows. Unauthorized use of water is hard to detect. Determining if land is being illegally irrigated, for example, requires witnessing the irrigation (or deducing the field is green and growing through artificial means), identifying the water source, pinning down the public land survey coordinates of the

Table 21. Civil Penalty Schedule for Water Law Violation

General Violations OAR Chapter 690, Division 260	Penalty Range For First Occurrence[1]		
	Major	Moderate	Minor
Class I: (e.g., using water without permit; violating terms of permit, certificate, license; failing to comply with watermaster order to cease use.)	$1,000	$500	$250
Class II: (e.g., failing to install measuring device when required)	$500	$250	$125
Class III: (e.g., failure to maintain ditch or well.)	$200	$100	$50

[1] Based on risk of substantial harm to other rights or public health and safety Penalties may be increased for repeated violations by the multiple of the repetition; i.e., fines may be doubled for the second violation, tripled for the third, and so on, up to a factor of five.

Well Construction Violations OAR Chapter 690, Division 240	Penalty Range Per Occurrence
Major: (any violation not listed as minor)	$50 to $1,000
Minor: (e.g., failure to properly mark drilling machine, install access port, comply with liner pipe standard, perform yield and drawdown tests)	$25 to $250
Late/Missing Start Card Fee	**Occurrence per calendar year**
First: $150; Second: $250; Third and subequent: $250 plus possible license suspension	

field, consulting water rights records, and tracking down the landowner or leasee to assure the state's records are not in error (which is not easy, as all water right records are not directly tied to county assessor ownership records). Trying to determine the legality of something like a self-supplied industrial plant is even harder, for there is even less to spot from a road, and road-based observations are the rule, given the size of watermaster districts

Table 22. Civil Penalty Comparison: Water Resources Department and Department of Environmental Quality

Year	Water Resources Dept[1]		Dept. of Environmental Quality[2]	
	Civil Penalties (#)	Penalty Assessed	Civil Penalties (#)	Penalty Paid
1993	14	$18,325	277	$777,000
1994	12	$7,775	171	$433,004
1995	11	$36,400	172	$687,147
1996	5	$18,290	199	$561,818
1997	13	$53,191	137	$446,757
1998	8	$20,200	145	$547,183
1999	14	$26,100	138	$635,021
2000	11	$13,500	202	$934,532
2001	18	$44,950	271	$4,117,767
2002	6	$14,575	287	$3,475,783
2003	11	$9,650	254	$1,571,324
2004	5	$2,800	277	$777,000

[1] *Source*: Oregon Water Resources Department. n.d.[i]
[2] *Source*: Oregon Department of Environmental Quality. 2006a

and workloads. Conducting regular diversion inventories may be possible on streams big enough to float a raft or boat, but the effectiveness falls off fast on smaller streams that would entail miles of wading. Finally, detecting excessive water use under existing rights can be almost impossible, because in most cases water use (whether by rate or duty) is neither measured nor recorded (Chapter 7).

However, indications are that illegal use may not be uncommon. First, about one-third of the regulatory actions taken by watermasters in 2004 were against illegal surface water use. Second, in 1992 the department found a number of illegal water uses in the Damascus groundwater limited area and conducted a comprehensive water use inventory. The results: around a hundred unauthorized uses were identified in the 30-square-mile area. Third, the state—long aware that many reservoirs have been built without water use permits—declared a sort of amnesty where landowners could legalize their ponds. Thousands of ponds registrations poured into the Water Resources Department—so many that the legislature continued the amnesty period and eased the water rights laws concerning ponds.

There is probably little deliberate illegal use. Most results from people who simply never were told that (and probably never thought to ask if) a permit was required to use water. By the time they find out, they may have staked their lives and savings on home sites, hay crops, nurseries, orchards, golf courses, RV parks, or practically any other water use under the sun. Illegal use is not solely the province of poorly informed mom-and-pop operators, however. It extends to corporate landholders as well as to state and federal agencies that oversee grazing lands, wildlife refuges, or other lands. The Water Resources Department has the authority to cut them all off cold. Some would term it an obligation. But turning off the spigots would frequently cause severe economic, and sometime ecologic, disruption. Whether from wisdom or weakness, the department prefers to work with people and help them get legal. This approach has taken different forms through the years.

In the past, if an illegal user promised to apply for a water right, it was considered close enough and the water use could continue. But as obtaining a water use permit became more difficult over time, the interaction between a watermaster and a violator became more awkward. The choice boiled down to either looking the other way or cutting them off immediately—neither of which represented a good management solution. In 1995, the legislature authorized the department to use a new enforcement tool: the enforcement order. In certain circumstances, the department can now impose a compliance schedule by issuing a limited license to continue what had been an illegal use. In theory, this gives the violator time to either obtain a permit through the application process or a certificate through the transfer process, or otherwise arrange for an alternative water supply (ORS 537.143).

It should also be mentioned that enforcement is strictly the job of the Water Resources Department. If motivated citizens want to take matters into their own hands and sue illegal users (or the state to compel enforcement), they can forget it. Unlike some federal environmental laws such as the Clean Air, Clean Water, Safe Drinking Water, and Endangered Species acts, there are no provisions in Oregon law allowing citizens to file suit to stop water law violations (Kaufman 1992).

Water Conservation

In theory, every drop of water used under a water right must support some benefit. "Beneficial use shall be the basis, the measure and the limit of all rights to the use of water in this state" (ORS 540.610(1)). Or, more to the point, "the waters within this state belong to the public for use by the people for beneficial purposes without waste" (ORS 536.310(1)). As described in the previous section, the willful waste of water is statutorily prohibited (ORS 540.720). Clearly, conserving water is supposed to be just as important as priority dates, appurtenancy, or any other operating principle of Oregon's water management system. But is it?

Priority dates and places of use are comparatively easy to pin down. But conserving water and avoiding waste are not. There are no clear guidelines, and no dearth of strong opinions. Like pornography, people may not be able to define waste, but feel they know it when they see it. To some people, using scarce Central Oregon water for golf courses instead of alfalfa is a waste; to others, watering some scabby eastern Oregon pasture is wasteful when the water is needed for spring chinook; to still others, giving a stream's remaining water to an instream right instead of a growing city is a waste. Portland's Forecourt Fountain may appear wasteful to a Redmond farmer. The canals of central Oregon racing with Deschutes River water may shout "waste" to a Beaverton fly fisher. Is it wasteful to hose off your driveway in the rain? Waste may be the sound of one side pointing.

Still, given the frequency of its mention in the statutes, the architects of the state's water law must have intended something by waste. Neither is case law silent on the issue. As early as 1915 the state Supreme Court noted, "while the crude and wasteful manner of irrigation formerly in vogue must be replaced by modern, economical methods, the ancient means used for applying water is not a reason for forfeiting the right to a sufficient amount to irrigate the land in a proper manner." Six years later, it reasoned "we have not arrived at the stage of irrigation when farmers can practically lay iron water-pipes, or construct concrete ditches; yet the question that water for irrigation must be used economically and without needless waste is no longer debatable. Public necessity demands such use and conservation of the public waters of the state" (*Foster v. Foster*, 107 Or. 291, 213 P. 895; *In re* Willow Creek, 74 Or 622).

In any event, given the importance of water to Oregon's history, commerce, and general identity, understanding what waste is (both to urban and rural users) seems worth a try. Either to avoid it altogether, or come to terms with it as a necessary evil. For one person's waste is often another's water source. In Oregon, eliminating waste everywhere means drying up wells and wetlands somewhere.

Waste, according to the Water Resources Department, is:

> the continued use of more water than is needed to satisfy the
> specific beneficial uses for which a right was granted. The

need for water shall be based on using the technology and management practices that provide for the efficient use of water considering:

(a) The economic feasibility of use of the technology and management practices by the water user;

(b) The environmental impacts of making modifications;

(c) The available proven technology;

(d) The time needed to make modifications;

(e) Local variations in soil type and weather; and

(f) Relevant water management plans and subbasin conservation plans (OAR 690-400-0010)

This definition strokes the harp-strings of reason. It invokes the principle of beneficial use while soft-pedaling considerations to use in face-to-face dealings with individual water users. It represents a bottoms-up, case-by-case approach harmonizing site characteristics with the user's access to money and technology. Few could argue with it in the abstract. Everyone will argue with it if it is ever applied. But argument may be unavoidable whenever the state attempts to modify human behavior. And surely, wasting water must be considered first and foremost a human behavior.

What is a state to do after declaring waste bad and conservation good? Oregon has begun to explore this relatively unknown territory. Like so many of this state's water management efforts, it requires looking both forwards to future uses and backwards to uses firmly established through already-issued permits and rights.

⊞ *Preventing Waste in Future Uses*

The first and probably easiest step is to build waste prevention into the permitting process for new water uses. It was not until 1995 that the statutes required an explicit analysis of the amount of water requested in an application. The department must conduct an "assessment of the amount of water necessary for the proposed use" (ORS 537.153(3); 537.621(3)). The department's rules (OAR Chapter 690, Division 310) require each applicant to describe why they need the amount of water requested and any measures they propose to prevent waste. Furthermore, in its assessment, the department must specifically consider water use efficiency and the avoidance of waste.

In another permit-based approach to conservation, the department has placed conditions on some new municipal permits by requiring cities to develop conservation plans. These plans must include a schedule for annual water use audits and for installing and maintaining meters; a public education program; an assessment of specific conservation measures, such as leak repair, low water use landscaping, retrofitting inefficient water use fixtures, and conservation-friendly rate structures; a water curtailment plan for water-short periods; and a long-range water supply analysis (OAR Chapter 690, Division 86). In addition, to receive permit extensions, which are highly desired by municipalities, they must first develop a conservation plan (ORS 537.320).

In 1991, the state legislature took another conservation step affecting new users when it passed the "Toilet Bill." The bill required that, effective July 1, 1993, only water-efficient plumbing fixtures be installed in new homes or in remodeling existing homes. Under ORS 447.145, the average amount of water used cannot exceed 1.6 gallons for flush toilets or 2.5 gallons per minute for shower heads and interior faucets.

▣ *Preventing Waste in Existing Uses*

Oregon has taken a somewhat opportunistic approach to promoting conservation in existing uses. For example, in return for allowing irrigation districts a way to use water on lands not described in their original water rights (a process termed "re-mapping"), the legislature required them to submit water conservation plans to the Water Resources Department (ORS 540.572–540.578). Under the department's Division 86 rules, these plans must include an evaluation of water use efficiency opportunities, including quantifying and repairing system losses. In addition, plans must include financing options for future conservation efforts, as well as feasibility analyses of metered pressurized delivery to small parcels; lining canals or using piping instead of open ditches; assisting clients with irrigation scheduling; and rate structures that promote conservation. Lastly, they must establish a schedule for education programs, identifying water curtailment provisions and performing long-range water supply studies.

Another important step was the 1994 conclusion of the Oregon Attorney General's office, which clarified the relationship between forfeiture and conservation (Oregon Department of Justice 1994). The Attorney General's finding (adopted into ORS 540.610 in 1997) states that a user can purposely use less water (i.e., conserve) without eroding the amount of their right. As long as enough water is used to satisfy the purposes listed in it, the right has been completely exercised and is not subject to forfeiture, even if the full

Oregon ranks sixth in the nation for dollars invested in irrigation water conservation efforts—much of it for moving from very old systems to new technologies. In 2004, Farmers Irrigation District near Hood River replaced a leaky, vintage-1900, wooden flume (top photo) with a state-of-the-art pipeline (lower photo) that has zero evaporation loss. (Farmer's Irrigation District)

amount of the right has not been used. This ruling has poked a considerable hole in the West's "use it or lose it" doctrine: if you have a right to 2 cfs and get by with 1 cfs of it for five years running, you have forfeited the other 1 cfs. Naturally, this encourages maximum, often unnecessary, use of water. Oregon's new interpretation (described as "beneficial use capped by rate and duty") allows more thrifty use of water, while preserving a user's right to more should weather or crop selection demand.

But what if the water right holder wants to use some of that "saved" water on new lands? Under a program celebrated nationwide (at least in water management circles), the state will allow it—but if and only if it gets a piece of the action. Oregon's Use of Conserved Water Program (ORS 537.455–ORS 537.500) promotes conservation by allowing water right holders some benefit from conserving water. Rather than losing it, conserving water right holders get to keep a certain portion of water they do not use. The program has three goals: to aggressively promote conservation; to encourage the highest and best use of water by allowing conserved water to be sold or leased; and to encourage local cooperation and coordination in increasing efficiency and improving streamflows. Conserved water is defined as the difference between the amount stated on the right or whatever users are actually able to divert (whichever is smaller) and the bottom-line amount needed to serve the beneficial use under the right.

The Water Resources Commission must first approve any use of conserved water after reviewing information submitted in a conservation proposal. If the proposal is approved, after subtracting any amount needed to cushion effects on other water rights, the conserved water is split into two shares: the water right holder's share and the state's share. The default split is 25 percent for the state and 75 percent for the water right holder. If public funds are used to finance 25 percent or more of the conservation project (and do not have to be paid back), then the state's share increases to that proportion: if the government pays for half of the conservation project, it could claim half of the saved water. However, the water right holder is always guaranteed at least a 25 percent share. If the commission determines that the state's share is needed for instream use, it is converted to an instream water right.. Otherwise, the water becomes available for use by other water right holders in the system. The conserving water right holder may use the water on other lands, or sell or lease rights to the saved water to others.

These relatively new processes may be succeeding. Oregon ranks sixth in the nation for dollars invested in water conservation efforts related to irrigation systems. Plus, it leads the nation in the number of farms transferring water—either by renting or leasing—to environmental uses (Chapter 4) (Oregon Department of Agriculture 2005).

But while conserving water can have significant impacts on local water supplies, it alone will not result in brimming streams or recharged water tables. The dimension of use on a statewide level simply dwarfs the savings from any conservation program, no matter how actively applied. However, an activated approach is essential if Oregon is to secure the benefits of conservative water use, as the law contemplates. So far, the state's approach to conservation can be characterized by what it isn't, as much as by what it is.

First, it is not an exercise in police power. Although the statutes and the department's Division 250 rules identify waste as an illegal practice subject to watermaster control or cut-off, because there are no standards for waste, direct frontal attack is rare.

Second, it is not a measure-your-way-to-efficiency endeavor. A first step toward a comprehensive approach to water conservation might be a universal requirement for water right holders to measure what they use (see Chapter 7). As users gain a better understanding not only of what their rights allow, but what they are taking, a fair amount of self-policing would likely result. And where it did not, the state could sample measurement records and use the information to develop an enforcement strategy.

Third, it is not economic control, as there is no charge for the use of the public's water. Ultimately, waste may be what stops when measurement meets pricing. Even a nominal cost can have a big impact on human behavior: a nickel deposit keeps Oregon's roadsides nearly pop-bottle-free. If some price were attached to using water, many people would track their use and some would experience a nearly irresistible, penny-pinching urge to seek out waste and eliminate it.

Fourth, it is not speedy: the basin-specific efficiency standards and sub-basin conservation plans (called for in the department's 1995 Strategic Plan and in its 1990 Conservation and Efficient Water Use Policy, respectively) never materialized. Plus, despite its start-up fanfare, few have been willing to participate in the Use of Conserved Water Program. As of 2006, the department has received forty-one applications and approved twenty-one. However, the pace of conservation appears to be picking up, especially as cities recognize water supply challenges and seek to extend their permits (which carries a statutory requirement for a conservation plan).

Still, in terms of its century-old water management system, Oregon has figuratively just taken delivery on its water conservation tools. And the difficulty of government getting citizens (by carrot or stick) to do something they have never been required to do before (consciously and consistently conserve) should not be underestimated. But there may be cause for optimism. In the early 2000s, 32 percent of the Oregon irrigators that responded to a survey indicated they had made energy or water conservation improvements in the previous five years. Half had done so to reduce their water requirements. In the year they were surveyed, Oregon farmers spent over $10 million on water conservation improvements—over a quarter of their annual equipment investment in equipment, facilities, land improvement, and computer technology: When asked to identify barriers to making conservation improvements, the most common responses included it just wasn't a priority, difficulty with financing and affordability, and uncertainty about the future availability of water. The farmers also identified neighboring farmers, extension agents, and equipment dealers as those they relied on most for conservation information (U.S. Department of Agriculture 2004 a). This may suggest that education is the key to better use of Oregon's fairly impressive set of conservation tools.

And in recent years, there has been a notable increase in educational efforts. Not only do most major cities (often considered the well-to-do in terms of water user categories) spread the news about conservation through modified rate structures and public promotions, Oregon's irrigation community (land-rich, but frequently cash-poor) is also exploring ways to increase efficiency. Both sectors are forming partnerships with the state, and in particular with one federal agency.

The U.S. Bureau of Reclamation has become increasingly active in promoting water conservation in Oregon. Like many agencies in the late twentieth century, the bureau experienced a mid-life crisis. It has given up the red meat and cigarettes of building big irrigation projects for the arid West, and is seeking renewed life through the bean curd and aerobics of conservation.

Bureau activities have included testing experimental methods and co-sponsoring demonstration projects to increase irrigation water use efficiency, as well as helping local governments identify water supply options, including conservation. The Oregon Water Resources Department has been an important partner (both in terms of dollars and activity) in nearly all of these efforts, as have been numerous other agencies, individuals, local groups, tribes, and irrigation districts.

The bureau has established many conservation partnerships in eastern Oregon, especially in the fisheries-significant John Day, Umatilla, Grande Ronde, and Upper Deschutes basins, where work has included improving canals and other irrigation delivery systems, enhancing instream flows through exchanges and leases, and providing funding to local conservation groups through challenge grants.

The bureau has also worked west of the Cascades to help cities and counties identify options to meet long-term municipal, domestic, and industrial water supply needs, including through increased water conservation. It has also collaborated with local interests to increase instream flows for water quality in the Tualatin, and to develop alternative means of supplying irrigation water, allowing the removal of Savage Rapids Dam on the Rogue (U.S. Bureau of Reclamation 2005).

CHAPTER 7

Water Use: The Big Pitcher

WATER WORKS FOR OREGON EVERY DAY in a thousand ways. It generates power, grows crops, floats boats, feeds fish, splashes tourists, and slakes the thirst of industry, cities, and citizens. How Oregonians use water reflects their identity just as much as how they vote or what they read. But today's water use patterns color the future as much as they frame the past. They are a product of nineteenth-century settlement, twentieth-century industrialization, and recent environmental awareness. Today's users are the custodians of just about all the surface water there is to appropriate; they have the surface supply sewn up. That makes them the custodians also of the state's future, for groundwater supplies are uncertain and new water demands will be met mostly by transforming today's uses. This chapter and the next explore the major sources, types, and extent of today's uses in the belief that knowing our water present is the first step in shaping our water future. But first it may be instructive to look far into the past.

Water and the Original Oregonians

Clearly, appreciation and use of water by Oregon residents did not spring into existence with the arrival of trappers, pioneers, and city folk. The emigrants of the nineteenth century came not to a new land, but a very old one. One where river valleys and headwater canyons had long been mapped in the intimate, ten-thousand-year memory of native peoples. Although their cultures reflected the diversity of Oregon's landscape, they had much in common with each other and, for that matter, with people everywhere: they liked to live by water. In most of Oregon, few villages were more than a walk away from a stream or lake (Drucker 1963; Mack 1990; Clewett and Sundahl 1990).

Rivers provided food: salmon, sturgeon, lampreys, mussels, and more. For many native groups, fishing provided far more animal protein than

Oregon's streams were the supermarkets of the original Oregonians where they found abundant supplies of favorite foods: salmon, sturgeon, mussels and lampreys. (Oregon State Archives, Oregon Highway Division, OHDM005)

hunting (Drucker 1963). Perhaps the closest native people came to using machinery was the fence-like weirs and fish traps they built on smaller streams. Of course, native Oregonians also used water for domestic purposes, such as cooking (e.g., the Umpqua's preparation of acorn meal or the Chetco's fish head and tail soup), health (e.g., sweat lodges), and, of course, drinking (some coastal people kept cedar water buckets in their dwellings). Rivers were also their roads: both historic migration and localized trade took advantage of river corridors. Oregon natives perfected a variety of water craft, ranging from crude log rafts to sleek whitewater dugouts. Heated water was used by some to soften and shape the interior of canoes (Mack 1990; Clewett and Sundahl 1990; LaLande 1990; Schonchin 1990; Drucker 1963).

Obviously, given their way of life, the natives' water demand (as we might define it today) was small. And their technology matched that demand. They lived in acceptance of what rivers offered; there was simply no demand (and therefore no means were developed) to squeeze a stream from its banks to serve agriculture, cities, or industry. This approach to water fostered a native outlook that was, and remains today, markedly different from that of modern American society, which values the technological and utilitarian. While the achievements of that society are in many ways quite remarkable, they have taken a toll on the health of Oregon's water resources.

A contemporary Oregon Native Americans is quoted in a regional study as saying, "Whites show up to a river, and they have to put a dam on it. They can't leave anything alone. It doesn't matter what it is, they've got to change it." The study notes the strongest views were on the damage to rivers, with a near-universal complaint that local rivers have been destroyed. In one instance, an older man expressed his frustration, saying, "I don't want to look at it any more. It's not my river. I'm mad about that." The people interviewed for the study blamed low streamflows and poor water quality on clear-cutting, grazing, damming, and the upstream withdrawal of water for irrigation. Frequently, they stressed the need for more sensible irrigation practices that left some water instream (Liberman 1990).

Ironically, today the potential for irrigation is a primary factor in determining the amount of water legally allocated to tribes under reserved water rights (see Chapter 5). However, traditional instream uses, both on- and off-reservation appear to be receiving greater attention in water rights negotiations with Oregon's Native American interests (Oregon Water Resources Department 1995b). Interestingly, it might also be argued that Oregon is on a winding evolutionary path to a way of looking at its water resources that is more compatible with that of native people, as evidenced by the attempts now underway to forestall species extinction, especially salmon and steelhead; the burgeoning popularity of broad-based local watershed councils (a modern echo of a tribe-per-basin, the population pattern of antiquity?); and a growing interest in Oregon's tools for achieving sustainable water use through conservation. The trick will be whether Oregonians, living today on silicon and petroleum, can fashion our water future from the lessons of obsidian and cedar. And that future turns in no small way on the particulars of the present.

Today's Big Picture

When it comes to describing Oregon's water use, we must settle for an artist's sketch rather than a photograph. There is little direct information available to provide a photographic level of detail. Most information is inferred from surrogates like irrigated acreage, population figures, or industrial surveys, all tied to some theoretical water need statistic. The reason: water use measurement is not, nor has it ever been required, for the vast majority of water users.

⊡ *Measuring Water Use*

There is no state water meter clocking water use. Not year by year, let alone moment by moment. Although watermasters have become proficient in counting sprinkler heads, estimating pump horsepower, or assessing other indicators, there is no universal requirement in Oregon that water users watch what they use. Rather, the state has four relatively limited measuring authorities at its disposal; they are limited in that they apply to only small groups of users at a time.

In certain specific instances, the Water Resources Department may require individuals to install measuring devices along ditches (ORS 540.310) or on streams feeding reservoirs (ORS 540.330), but this requirement is applied sparingly. In addition, since 1987 the state has had the authority if "necessary" to declare Serious Water Management Problem Areas in locales experiencing severe water difficulties and require water use measurement (ORS 540.435). None have ever been declared. Third, when issuing water use permits the Water Resources Department may impose conditions requiring metering, but these conditions are a relatively new practice and reserved for larger uses only. The most common measuring condition only alerts permittees to the future possibility that they might be asked to measure.

The fourth measurement authority is the strongest and most broadly applied. Under ORS 537.099, any government agency that holds a water right must annually report the amount, period, and purpose of its water use to the Water Resources Department. Government agencies include all state and federal agencies, local governments, irrigation districts, and water control districts. Interestingly, the requirement also includes the Water

Unlike this canal (shown with gaging station on the left), few water diversions in Oregon are measured; there is no universal requirement for people diverting the public's water to keep track of what they use. (Oregon State Archives, Oregon Dept. of Agriculture, OAG0052)

Resources Department, which must report to itself on the instream water rights it holds for the people of the state (see Instream Water Use, below). Thus, each year city, fish hatchery, school, cemetery, county park, and many other agency personnel fill out water use reports. OAR Chapter 690, Division 85 requires all measurements (except for small uses) to meet an 85 percent accuracy standard. The rules also provide for reporting waivers and extensions for a variety of reasons. Nearly twelve thousand entities are required to report annually (Brown 1996).

⊞ *Statewide Water Use Summary*

It should be acknowledged at the outset that estimates of water use often vary. Surveys performed with different assumptions or different methods, or during different decades, can yield different results. The figures in this chapter are intended as rough indicators of use, rather than precise measures.

A 1955 legislative working group noted: "There is little relationship between the acreage of land holding permits or rights to water for irrigation and the amount of land actually being irrigated" (Water Resources Committee 1955). This is true of all water right records. A lot of water right deadwood has built up since passage of Oregon's 1909 Water Code, and consequently Oregon's water records need a lot of pruning. Rights that have been abandoned on the ground often have a life of their own on paper. Holders of rights that are being exercised may be using way more or considerably less than the amount of record. Thus, statistics generated from water rights records need to be used cautiously, especially when being compared with data from other sources. But these records can be used to approximate the magnitude and proportions of water use.

Table 23 displays an estimate of Oregon's annual water withdrawals— that is, water which is diverted or pumped out of or away from its source; the figures do not include instream or hydroelectric uses (addressed in separate sections), which make use of water in-channel. Further, the figures represent uptake, not consumption. For example, water sprayed onto an irrigated field may run back into a stream and to other downstream uses.

All told, nearly 8 million acre-feet are moved through pumps or ditches in the state (Hutson et al. 2004). This amount is about 12 percent of the 66 million acre-feet carried by Oregon's streams, alone (see Chapter 1). Clearly, out-of-stream (or -aquifer) uses partake of only a modest portion of the state's total annual water production. However, it would be a mistake to

Table 23. Oregon Annual Water Withdrawals in 2000.					
	Irrigation	Public Supply	Indus-trial	Domestic	Thermo-electric	Total
Total amount withdrawn (acre-feet)	6,815,680	634,486	218,595	85,420	17,151	7,768,530
% of total water withdrawn	87.7	8.2	2.8	1.1	0.2	
Surface Water						
% of total surface water withdrawn	89.1	7.5	3.1	0.1	0.2	
% of total water withdrawn	76.3	6.5	2.6	0.1	0.2	
Surface water withdrawal as % of total category withdrawal	87.0	79.0	93.8	10.5	83.7	
Groundwater						
% of total groundwater withdrawn	79.8	11.9	1.2	6.9	0.2	
% of total water withdrawn	11.4	1.7	0.2	1.0	-	
Groundwater withdrawal as % of total category withdrawal	13.0	20.8	6.2	89.6	16.1	

Source: Data from Hutson et al. 2000. Total amount converted from million gallons per day. (Data on livestock, mining, aquaculture not collected)

view the remaining water as unused, much less as wasting to the sea. Non-consumptive uses such as hydroelectric power, instream fish and wildlife uses, recreation, and channel-maintaining floods all depend on the ebb and flow of the water that is not directly taken up and used. In addition, water uptake peaks during the lowest-flow months, thus amplifying its effect on stream and aquifer systems.

By far the largest water withdrawal category is irrigation, representing nearly 90 percent of total withdrawals. Statewide, the predominant source is surface water, which, on the whole, supplies over 85 percent of water uses (except for domestic use, which relies on groundwater). Around 75 percent of water withdrawals occur in eastern Oregon, the driest part of the state. Here, 97 percent of the water used is for irrigation, compared to about 60 percent west of the Cascades (Hutson et al. 2004).

Major Sources

As noted in Chapter 2, water use is defined primarily by source and purpose. This section deals with Oregon's three major sources: surface water (e.g., streams), storage (water behind reservoirs), and groundwater (water from wells). The next section covers the uses to which pumped or diverted water is put.

⊞ Surface

The Oregon Water Resources Department has record of over seventy thousand individual out-of-channel diversions from springs or streams (not counting "diversions," where dams hold water back for storage or hydroelectric production) (Oregon Water Resources Department. 2006 d). They range in size from the minuscule (less than 0.0005 cfs—about one-quarter gallon per minute) to the riverine (well over 500 cfs—the flow of a respectable river, such as the Trask or Umatilla rivers in May). However, about half the diversions are allowed less than 0.1 cfs (about 45 gallons per minute; see Figure 25). Most of the surface water use occurs in eastern Oregon (Table 24). Because of its thousands of livestock ponds, the Bureau

Half the surface water diversions in Oregon are for less than 0.1 cubic feet per second—about 45 gallons a minute. (Oregon State Archives, Oregon Highway Division, OHD2186)

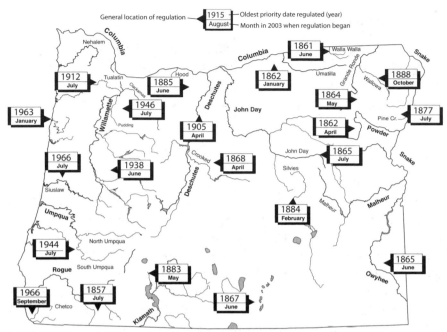

Figure 25: Surface water rights by size of diversion (excluding hydroelectric and instream rights). Oregon's water right profile is dominated by relatively small users, a fact which may intensify management challenges in the future. (Source: Data from Oregon Water Resources Department)

Table 24. Surface Water Withdrawals by Drainage		
USGS Principal Drainage	State Basin Equivalents	Estimated Surface Water Withdrawals (*million gallons a day*)
Middle Columbia	Deschutes, John Day, Umatilla	1,620
Middle Snake	Owyhee, Malheur R., Powder	1,580
Klamath	Klamath	722
Closed	Goose/Summer, Malheur lakes	607
Willamette	Willamette	466
Coastal	North (part), Mid, South; Rogue & Umpqua	426
Lower Snake	Grande Ronde	195
Lower Columbia	North Coast (Columbia R. tributaries), Sandy	163

Source: Data from Broad and Nebert 1990

of Land Management probably holds more individual rights than any other single entity. In terms of volume (cumulative direct flow volume, that is, not reservoir volume), however, the single biggest player is most likely the State of Oregon with its hundreds of instream rights (see below), not to mention rights for parks, wildlife refuges, and fish hatcheries.

⊞ *Groundwater*

Groundwater is a precious resource and is becoming more precious every day. Oregonians are hooked on groundwater. They use almost one billion (not million, billion) gallons of it each day—over a year that amount is on par with the Metolius, Siletz, or Tualatin rivers (Hutson et al. 2004). Seventy percent of Oregonians (including over 90 percent of rural residents) rely on groundwater as their primary or secondary drinking water source (Oregon Department of Environmental Quality 2003a). There are also an estimated two hundred thousand domestic wells in use in Oregon (Oregon Water Resources Department n.d. [c]). Eastern Oregonians rely on groundwater for their drinking more than western Oregonians (Task Force on Drinking Water Construction Funding and Regionalization 1991). The two biggest uses of groundwater are irrigation (at 80 percent) and public water supply (about 12 percent) (Hutson et al. 2004).

Although Oregonians have depended to some degree on wells for much of the state's recorded history, the development of groundwater as a significant resource did not get underway until after World War II. This development was driven by population growth, expanding agricultural markets, and technological advances, including improved pumping equipment and cheap, readily available electrical power (Water Resources

Table 25. Aquifer water usage (million gallons per day; Year 2000 estimate)				
Aquifer	Public Supply	Industrial	Irrigation	Totals
Basin and Range basin-fill aquifers	3.91	0.00	171.31	175.22
Pacific Northwest basin-fill aquifers	7.9	2.52	40.56	50.98
Snake River Plain basin-fill aquifers	2.6	0.00	25.35	27.95
Willamette Lowland basin-fill aquifers	53.31	3.6	236.62	293.53
Pacific Northwest volcanic-rock aquifers	31.76	0.91	119.22	151.89
Columbia Plateau basaltic-rock aquifers	18.98	5.06	146.58	170.62
Subtotal	118.46	12.09	739.64	870.19
Percent of Grand Total	14	1	85	

Source: U.S. Geological Survey n.d.[a]

Committee 1955). Groundwater development has also been spurred by periodic droughts; often, when people's surface water sources dry up, they drill wells, which are less subject to the immediate impacts of climate. To date, the Oregon Water Resources Department has issued around sixteen thousand groundwater rights (Oregon Water Resources Department 2006d). In addition, as federal drinking water standards grow more rigorous (favoring cleaner groundwater sources) and surface water sources become fully (if not over-) appropriated, use of Oregon's groundwater resource will certainly increase.

The U.S. Geological Survey estimates that, in 2000, approximately 63 percent of the total groundwater used in Oregon comes from basin fill and alluvial aquifers—relatively near-surface, loose material such as sands and gravels. These aquifers provide about 64 percent of all the groundwater pumped for irrigation, and nearly 60 percent of that pumped for public supply. The Columbia River basalts, thick beds of dark lava with water-holding fractures and between-bed zones, contribute about 19 percent of supply. The rest comes from other volcanic formations. Of all the water pumped out of aquifers, around 85 percent is used for irrigation (Table 25) (U.S. Geological Survey n.d.[a]).

⊡ Storage

Building reservoirs would appear to be a reflex reaction to Oregon's runoff patterns. Whether for augmenting supply or controlling runoff, reservoirs are as old as settlement. In addition, some dams were built beginning in the late 1800s to supply electricity to growing cities. However, one of the earliest reasons for storing water was to float logs to downstream mills. "Splash" dams were built on streams too small to float logs on their own. When enough water and logs were backed up, the dam was blown and the logs flushed to downstream mills. By 1900 Oregon had more than one hundred sixty splash dams in coastal basins and the Columbia basin (Williamson et al. 1995). The dams frequently blocked salmon runs and the flash floods of water, mud, and logs wiped out streambed and riparian habitat. Consequently Oregon outlawed splash dams in 1958 (ORS 541.455).

Major dam building picked up around 1900 and continued for the next sixty years. The main sponsor of these big dams was the federal government. For the huge Columbia River dams, the primary beneficiaries were agriculture (both for irrigation and the inland waterway created allowing cheap transport of products) and Northwest citizens (through some of the cheapest electricity in the nation). For Oregon's other large dams, the primary beneficiaries were irrigated agriculture and flood-prone communities. Today, there are nearly fifteen thousand reservoirs recorded in Oregon's water rights records (Oregon Water Resources Department 2006d) Interestingly, the big dams that store water in Oregon and the rest of the Northwest capture only a portion of the total runoff and do so a season at a time. Reservoirs in the Columbia River basin retain only about 30 percent of the region's annual runoff, compared to the Colorado River basin, where reservoirs can store four times the region's annual precipitation (Service 2004). The Northwest's water-storing capacity is essentially built on the bet that reservoirs will have no trouble refilling each year—a bet that climate change may very well take water managers up on.

The twin pillars of Oregon's storage infrastructure are the U.S. Army Corps of Engineers and the U.S. Bureau of Reclamation.

In the Flood Control Act of 1936, Congress declared flood control on navigable waters or their tributaries to be in the general public interest and therefore properly within the jurisdiction of the federal government. The U.S. Army Corps of Engineers is primarily responsible for carrying out the provisions of the Act, in cooperation with states and local interests. Corps reservoirs are generally operated to serve more purposes than Bureau of Reclamation projects, including flood control, hydroelectric power generation, navigation, irrigation, municipal and industrial water supply,

Owyhee Dam impounds over 1 million acre-feet, making it second only to Crater Lake in terms of Oregon's largest-volume water bodies. Finished in 1932, the 417 foot high structure was the architectural forerunner of Hoover Dam (U.S. Bureau of Reclamation)

water quality, fisheries, and recreation. Use of water stored for irrigation purposes in Corps projects is usually managed by the bureau.

In Oregon, Corps projects are almost entirely west of the Cascades (Table 26). The really big projects are in the Willamette basin, where thirteen dams store over 2.3 million acre-feet. These include the dams most evident to people as they travel through the Cascade passes, such as Detroit, Green Peter, and Lookout Point. The Corps also stores over 500,000 acre-feet in two Rogue River basin projects. Willow Creek Dam, in north-central Oregon, stores about 6,000 acre-feet. Additionally, the Corps operates the mainstem Columbia River dams where the John Day Dam impounds 534,000 acre-feet for storage purposes. The other Columbia River dams in Oregon (Bonneville, The Dalles, and McNary) are "run-of-river," not storage, operations (U.S. Army Corps of Engineers 1996a). This means they back up just enough water to route through turbines to generate hydro-power.

The National Reclamation Act of 1902 gave the federal government (through the Bureau of Reclamation and its predecessors) the authority to finance large irrigation projects in the western United States. Oregon's Klamath project was the first to have lands reserved by the Secretary of Interior for project construction. As early as 1940, the U.S. Bureau of Reclamation had built eight dams in the state impounding over 1.7 million acre-feet—already about 70 percent of the agency's current total storage capacity in Oregon (U.S. Bureau of Reclamation 1981).

The bureau stores about 2.5 million acre-feet of water in Oregon (Table 27). This includes Owyhee Reservoir in southeast Oregon—with 1.1 million acre-feet, the state's largest reservoir and its second-largest standing water body (by volume). For most projects, the bureau is responsible for project construction and issuing contracts to irrigation districts and individuals for use of its stored water. All but one of its twenty-four storage dams in Oregon (McKay Dam on the Umatilla Project) are operated by water-user organizations (see Water Providers section, below). In 2005, the bureau's twelve reservoir projects in Oregon supplied water to nearly 400,000 acres (about 20 percent of Oregon's irrigated land) that produced a gross crop value of $200 million. In addition, roughly 3.1 million visitor-days were spent at recreation sites on the bureau's Oregon projects (U.S. Bureau of Reclamation 2005).

These projects are a good deal for Oregon's irrigators, who under reclamation law often receive a hefty financial boost from the federal government. Since the beginning of the reclamation program in 1902, the federal government has subsidized the repayment of irrigators' share of the construction costs of water projects. By law, these construction costs are repaid without interest. In addition, irrigators generally have forty or more years to make their payments—often after a start-up grace period (which can last up to ten years) during which they receive water but are not billed for repayment. Third, irrigators' repayment obligations are often downsized to fit their "ability to pay" and the cost shifted to other project beneficiaries (especially hydroelectric producers). The cost of this financial assistance can add up. For example, the federal subsidy associated with Oregon's Tualatin project was $30.6 million out of a $31.5 million irrigation component—or 97 percent of the cost (General Accounting Office 1996). In contrast, interest is charged to municipal and industrial users of project water, which results in much higher costs for those uses (Wahl 1989).

Oregon also has a few big, non-federal reservoirs. While they can impound a large volume of water, they are operated primarily to generate hydro-power, not as multi-purpose projects. They include the Pelton-Round Butte dam complex on the Deschutes River run by Portland General Electric and Idaho Power Company's dams on the Snake River. These facilities are described in more detail in the Hydroelectric Use section, below.

How a storage project is operated depends on its purposes. Water supply reservoirs (such as Portland's Bull Run dams) store large volumes to assure predictable annual water quantity and to outlast most droughts. Storage at irrigation reservoirs (such as Ochoco Reservoir) is timed to create a maximum supply at the beginning of the growing season and a safe holdover amount afterwards. Flood-control reservoirs (e.g., the Willamette basin's Detroit or Lookout Point reservoirs) maintain small permanent pools, but can hold back large run-off bursts from storms or melting snow, and thus prevent downstream flooding. These reservoirs are generally emptied as quickly as possible after high runoff to refresh their storage capacities. Hydroelectric reservoirs (such as the Clackamas River's North Fork Reservoir) store and release water on a schedule that meets regional energy demand. Re-regulation reservoirs below hydroelectric dams (such as Big Cliff Dam below Detroit) stabilize rivers by damping flow fluctuations caused by daily power generation releases (U.S. Army Corps of Engineers 1996a).

As in all other states, the purposes of Oregon's federal reservoirs were spelled out when Congress authorized them (if not in the authorizing law, then in reports to Congress supporting authorization). At times, project operations may have to be adjusted to comply with new federal laws such

Dexter (foreground) and Lookout Point (background) Reservoirs are two of the 13 flood control reservoirs operated in Oregon by the U.S. Army Corps of Engineers. (Oregon State Archives, Oregon Water Resources Dept., OWR0075)

as the Clean Water Act or Endangered Species Act (U.S. Army Corps of Engineers 1992). However, in Oregon most project purposes were locked in based on how the world looked in 1950 or before. For example, all but one of the Willamette system reservoirs were authorized by 1950— and five of these (representing over half of system storage capacity) by 1938 (U.S. Army Corps of Engineers 1989). To use water for other (and perhaps by today's standards, more pressing) purposes is difficult when it is not outright impossible. For example, Congress authorized the bureau's Prineville Reservoir for irrigation and flood control. Only. That water is off-limits to fish, cities, or recreationists or other uses that may be dying to get it. But they just can't—at least not in any straightforward fashion.

In some projects, releases are sometimes made under the technical guise of an authorized purpose, when in actuality they serve another. For example, it is unlikely summer water quality flows in the Willamette River (5,000 cfs at Albany and 6,500 cfs at Salem) could be met without upstream storage releases. The releases are a regular occurrence, but technically are for navigation rather than the unauthorized but critical maintenance of the Willamette's water quality (Oregon Water Resources Department 1994).

Table 26. Corps of Engineers Reservoir Projects

Reservoir [dam ht., ft.][1]	Stream Location[2]	Vol.[2] (acre-ft)	Area[2] (acres)	Authorized Purposes[3]								
				FC	FW	H	I	N	R	Rr	WQ	WS
Applegate [242]	Applegate R.	82,200	988	✓	✓	✓	✓		✓			
Big Cliff [141]	N. Santiam R.	7,000[4]	146[4]			✓				✓	✓	
Blue River [312]	Blue R.	85,000	935	✓	✓				✓			
Cottage Grove [103]	Coast Frk. Willamette R.	33,500	1,139	✓	✓		✓	✓	✓		✓	
Cougar [519]	S. Frk. McKenzie R.	219,300	1,280	✓	✓	✓	✓	✓	✓		✓	
Detroit [450]	N. Frk. Santiam R.	455,000	3,580	✓	✓	✓	✓	✓	✓		✓	
Dexter [117]	M. Frk. Willamette R.	27,500	1,025			✓			✓		✓	
Dorena [145]	Row R.	77,600	1,840	✓	✓		✓	✓	✓		✓	
Fall Creek [205]	Fall Cr.	125,000	1,860	✓	✓		✓	✓	✓	✓	✓	
Fern Ridge [49]	Long Tom R.	101,200	9,360	✓	✓		✓	✓	✓		✓	
Foster [126]	South Santiam R.	61,000	1,220	✓	✓	✓	✓	✓	✓	✓	✓	
Green Peter [378]	Middle Santiam R.	430,000	3,720	✓	✓	✓	✓	✓	✓		✓	
Hills Creek [341]	M. Frk. Willamette R.	356,000	2,735	✓	✓	✓	✓	✓	✓		✓	
Lookout Point [276]	M. Frk. Willamette R.	453,000	4,360	✓	✓	✓	✓	✓	✓		✓	
Lost Creek [345]	Rogue R.	465,000	3,428	✓	✓	✓	✓		✓		✓	✓
Willow Creek [165]	Willow Cr.	14,100[4]	269[4]	✓			✓		✓		✓	✓

1 Oregon Water Resources Department 1996b
2 Johnson et al. 1985, except where indicated.
3 U.S. Army Corps of Engineers 1992; FC = Flood control; FW = Fish/Wildlife; H = Hydroelectric; I = Irrigation; R = Recreation; Rr = Re-regulation; WQ = Water quality; WS = Water supply
4 U.S. Army Corps of Engineers 1996a

While dams have a justifiably bad environmental reputation, the reservoirs they create are frequently utilitarian, and occasionally beautiful. Detroit Dam cut off salmon runs and today is one of the state's most popular and scenic recreation areas. (U.S. Army Corps of Engineers. Image File: Det0457.jpg. Date: 01JUL1985. Photographer: Bob Heims)

Updating a storage project's authorization literally takes an Act of Congress. It is a very involved, time-consuming process rife with potential conflict. Agricultural interests may often feel that most projects were built for them and that by paying decades of federal water-delivery fees, they have essentially paid off the government's project costs. Further, they can resent what they see as the late-arrival of heretofore dam-hating urban interests, seeking to horn in on the stored water and possibly disrupt traditional water use practices. Municipalities and fisheries interests, on the other hand, view the projects as property of all the taxpayers, and bridle at being excluded from, or made to pay through the nose for, public water stored at public expense. Resolving these different points of view will be a twenty-first-century challenge.

In the 1990s, the Corps, the State of Oregon, and multiple local interests began to look into reauthorizing the Willamette Basin Project. Early work indicated that a net economic benefit would result from modifying operations to increase minimum flows, fill reservoirs earlier and drain them later, and change the order of reservoir draw-downs. However, the reauthorization process was placed on hold until federal fish and wildlife agencies promulgated requirements to address Willamette basin salmon and steelhead needs under the Endangered Species Act.

In summary, dams and reservoirs have fashioned the Oregon we know today. We drink from them (Bull Run), draw cheap power from them (Bonneville), have flood-proofed communities because of them (Willow Creek Dam), sail on them (Fern Ridge), fish in them (Detroit), irrigate from them (Wickiup), and photograph them (Trillium Lake). While they have supplied tremendous benefits—including contributing to the fabled clean-up of the Willamette River in the 1950s and 1960s by doubling its flow during the summer (Oregon Department of Environmental Quality 2006b)—big dams are expensive to build and have left an indelible environmental legacy.

The Columbia River dams have proven disastrous for the region's salmon runs. In some cases, the impact of tributary dams has been even more damaging. Chinook salmon are now extinct above the Hells Canyon Dam complex on the Snake River, above Pelton and Round Butte dams on the Deschutes River, and above upper basin dams in the Willamette, Umpqua, Rogue, Umatilla and Walla Walla rivers (Williamson et al. 1995). Whether for urban water supply (such as Portland's Bull Run dams, which blocked all access to over 18 miles of coho and steelhead habitat) or for agricultural and power production (such as the upper Deschutes dam complex which cut off anadromous fish runs to three major river systems), this regrettable legacy has weighed heavily in the discussions of future storage options for the state (Oregon Department of Fish and Wildlife 1995). But dam changes are afoot in Oregon. In a historic turnabout, plans are being developed to reintroduce salmon above Willamette and Deschutes basin dams, and a number of old dams are scheduled for removal, such as Marmot Dam on the Sandy and Savage Rapids Dam on the Rogue.

Table 27. Bureau of Reclamation Projects in Oregon (data from U.S. Bureau of Reclamation, 1981)

Project name: Arnold (water rec'd. from
  Crane Prairie Res.)
  Stream location: Deschutes River
  Irrigation acres served: 4,292
  Canals: 11 miles
  Laterals: 25 miles
  Irrigation District: Arnold I.D.

Project name: Baker
  Dam/reservoir [height]: Thief Valley
    [73 ft]
  Stream location: Powder River
  Volume: 17,600 acre-feet
  Area: 740 acres
  Irrigation acres served: 7,281
  Authorized uses: irrigation
  Irrigation District: Lower Powder River

  Dam/reservoir [height]: Mason [173 ft]
  Stream location: Powder River
  Volume: 95,500 acre-feet
  Area: 2,235 acres
  Irrigation acres served: 18,532
  Authorized uses: irrigation, flood control,
    fish and wildlife
  Irrigation District: Baker Valley

Project name: Burnt River
  Dam/reservoir [height]: Unity [82 ft]
  Stream location: Burnt River
  Volume: 25,800 acre-feet
  Area: 926 acres
  Irrigation acres served: 15,616
  Authorized uses: irrigation
  Irrigation District: Burnt River

Project name: Crescent
  Dam/reservoir [height]: Crescent Lake
    [40 ft]
  Stream location: Crescent Creek
  Volume: 86,900 acre-feet
  Area: 4,050 acres
  Irrigation acres served: 8,000
  Authorized uses: irrigation
  Irrigation District: Tumalo

Note: Irrigation acres served number does
not include supplemental acreage.

Project name: Crooked River
  Dam/reservoir [height]: Arthur C.
    Bowman [245 ft]
  Stream location: Crooked River
  Volume: 154,700 acre-feet
  Area: 3,030 acres
  Irrigation acres served: 19,070
  Canals: 49 miles
  Laterals: 48 miles
  Authorized uses: irrigation, flood control
  Irrigation District: Ochoco

  Dam/reservoir [height]: Ochoco
    [125 ft]
  Stream location: Ochoco Creek
  Volume: 48,000 acre-feet
  Area: 1,100 acres
  Authorized uses: irrigation
  Irrigation District: Ochoco

Project name: Deschutes
  Dam/reservoir [height]: Wickiup
    [100 ft]
  Stream location: Deschutes River
  Volume: 200,000 acre-feet
  Area: 11,170 acres
  Irrigation acres served: 50,000
  Canals: 66 miles
  Laterals: 231 miles
  Authorized uses: irrigation
  Irrigation District: North Unit

  Dam/reservoir [height]: Crane Prairie
    [36 ft]
  Stream location: Deschutes River
  Volume: 55,300 acre-feet
  Area: 4,940 acres
  Authorized uses: irrigation
  Irrigation District: Central Oregon

  Dam/reservoir [height]: Haystack
    [105 ft]
  Stream location: Deschutes River/
    Haystack Creek
  Volume: 5,635 acre-feet
  Area: 233 acres
  Authorized uses: irrigation
  Irrigation District: North Unit

Table 26 (continued)

Project name: Grants Pass
  Dam/reservoir [height]: Savage Rapids
    [39 ft]
  Stream location: Rogue River
  Volume: 0 acre-feet (diversion dam)
  Irrigation acres served: 10,081
  Canals: 67 miles
  Laterals: 40 miles
  Irrigation District: Grants Pass

Project name: Klamath (*storage dams only*)
  Dam/reservoir [height]: Link River
    [22 ft]
  Stream location: Upper Klamath Lake/
    Link River
  Volume: 873,000 acre-feet
  Area: 90,900 acres
  Irrigation acres served: 223,693
  Canals: 185 miles
  Laterals: 532 miles
  Authorized uses: irrigation, power
  Operator: Pacific Power & Light

  Dam/reservoir [height]: Clear Lake
    Reservoir [42 ft]
  Stream location: (drains Tule Lake)
  Volume: 527,000 acre-feet
  Area: 25,800 acres
  Authorized uses: irrigation
  Operator: Bureau of Reclamation

  Dam/reservoir [height]: Gerber [85 ft]
  Stream location: (drains Tule Lake)
  Volume: 94,000 acre-feet
  Area: 3,800 acres
  Authorized uses: irrigation
  Operator: Bureau of Reclamation; Tule
    Lake and Langell Valley Districts
    operate several diversion dams

Project name: Owyhee
  Dam/reservoir [height]: Owyhee [417 ft]
  Stream location: Owyhee River
  Volume: 1,120,000 acre-feet
  Area: 12,742 acres
  Irrigation acres served: 105,249
  Canals: 172 miles
  Laterals: 543 miles
  Authorized uses: irrigation, flood control
  Irrigation District: North and South
    Boards of Control

Project name: Rogue (information excludes
    diversion dams)
  Dam/reservoir [height]: Fourmile Lake
    [25 ft]
  Volume: 15,600 acre-feet
  Area: 960 acres
  Irrigation acres served: 34,180
  Canals: 250 miles
  Laterals: 197 miles
  Authorized uses: irrigation
  Irrigation District: Medford

  Dam/reservoir [height]: Fish Lake
    [49 ft]
  Volume: 7,956 acre-feet
  Area: 415 acres
  Authorized uses: irrigation
  Irrigation District: Medford
  Dam/reservoir [height]: Hyatt [53 ft]
  Stream location: Keene Creek
  Volume: 16,200 acre-feet
  Area: 880 acres
  Authorized uses: irrigation, power, flood
    control
  Irrigation District: Talent

  Dam/reservoir [height]: Emigrant
    [204 ft]
  Stream location: Emigrant Creek
  Volume: 40,500 acre-feet
  Area: 806 acres
  Authorized uses: irrigation, flood control,
    fish and wildlife
  Irrigation District: Talent

  Dam/reservoir [height]: Howard Prairie
    [100 ft]
  Stream location: Beaver Creek
  Volume: 62,100 acre-feet
  Area: 1,900 acres
  Authorized uses: irrigation, power, flood
    control
  Irrigation District: Talent

  Dam/reservoir [height]: Keene Creek
    [78 ft]
  Stream location: Keene Creek
  Volume: 340 acre-feet
  Area: 15 acres
  Authorized uses: irrigation, power, flood
    control
  Irrigation District: Talent

Table 26 (continued)

Project name: Rogue (continued)
Dam/reservoir [height]: Agate [86 ft]
Stream location: Dry Creek
Volume: 4,780 acre-feet
Area: 216 acres
Authorized uses: irrigation, fish and
  wildlife
Irrigation District: Rogue River Valley

Project name: The Dalles (pump project)
Dam/reservoir: (three holding reservoirs)
Stream location: Columbia River
Irrigation acres served: 5,655
Canals: 1.2 miles (buried pipeline)
Laterals: 45 miles (buried pipeline)
Authorized uses: irrigation
Irrigation District: The Dalles

Project name: Tualatin
Dam/reservoir [height]: Scoggins [151 ft]
Stream location: Scoggins Creek
Volume: 59,910 acre-feet
Area: 1,132 acres
Irrigation acres served: 17,000 (1977
  estimate of "ultimate" potential)
Laterals: 88 miles
Authorized uses: irrigation, water quality,
  municipal/industrial, recreation, fish
  and wildlife, flood control
Operator: Bureau of Reclamation; pump
  plants by Tualatin Valley

Project name: Umatilla
Dam/reservoir [height]: McKay [165 ft]
Stream location: McKay Creek
Volume: 73,800 acre-feet
Area: 1,290 acres
Irrigation acres served: 20,916 (includes
  "individual storage contractors)
Canals: 137 miles
Laterals: 106 miles
Authorized uses: irrigation
Operator: Bureau of Reclamation;
  Stanfield and Westland operate own
  canals

Dam/reservoir [height]: Cold Springs
  [100 ft]
Stream location: Umatilla River through
  feed canal
Volume: 50,000 acre-feet
Area: 1,550 acres
Authorized uses: irrigation
Irrigation District: Hermiston and West
  Extension operate diversion dams,
  canals

Project name: Vale
Dam/reservoir [height]: Agency Valley
  [110 ft]
Stream location: North Fork Malheur
  River
Volume: 60,000 acre-feet
Area: 1,900 acres
Irrigation acres served: 34,993
Canals: 86 miles
Laterals: 279 miles
Authorized uses: irrigation, flood control
Irrigation District: Vale

Dam/reservoir [height]: Warm Springs
  [106 ft]
Stream location: Middle Fork Malheur
  River
Volume: 192,400 acre-feet
Area: 4,600 acres
Authorized uses: irrigation, flood control
Irrigation District: Vale

Dam/reservoir [height]: Bully Creek
  [121 ft]
Stream location: Bully Creek
Volume: 31,600 acre-feet
Area: 985 acres
Authorized uses: irrigation, flood control,
  fish and wildlife
Irrigation District: Vale

Project name: Wapinitia
Dam/reservoir [height]: Wasco [59 ft]
Stream location: Clear Creek
Volume: 13,060 acre-feet
Area: 557 acres
Irrigation acres served: 2,108
Canals: 38 miles
Laterals: 63 miles
Authorized uses: irrigation
Irrigation District: Juniper Flat

CHAPTER 8

Water at Work

The previous chapter summarized Oregon's water use by pattern and source. This chapter presents a more detailed look at how Oregonians put water to work, and how they get it to come out the ends of pipes—whether at home or in fields.

Major Types of Use

⊞ Irrigation

Oregon's irrigation history began with European settlement. The first instance of irrigation was probably for gardening at early missions in the 1830s and '40s. However, the first official record of irrigation is for an 1852 diversion from Wagner Creek in Jackson County by one Jacob Wagner. By 1909, nearly 1.2 million acres had already been reported as having been irrigated—about 95 percent in eastern Oregon (Water Resources Committee 1955). However, this figure is somewhat at odds with other estimates of actively irrigated lands at the turn of the twentieth century. Like so many records that result from a long claim-staking process (i.e., water rights adjudication), it is probable that irrigated lands were either over-reported or were actually developed at one time, but converted to other land uses or abandoned. The growth of irrigated agriculture was rapid through the 1950s, but has leveled off since that time (Figure 26). According to the 2002 Census of Agriculture, Oregon now irrigates about 1.9 million acres.

Irrigation is Oregon's largest water consumptive use. It makes up about 88 percent of Oregon's total water withdrawals. Irrigation accounts for over ten times much water as the next largest use, public supply (Hutson et al. 2004). In 2003, Oregon was third in the nation in the number of farms and ranches that irrigated, and ninth in the amount of acreage irrigated. Nearly

Irrigation, the largest consumptive water use in Oregon, has created a sound agricultural economy, changing our landscape in the process. Here the irrigated circles formed by center pivot systems are in close proximity to their water source, the Columbia River near Hermiston. (Oregon Department of Agriculture)

eighteen thousand farms or ranches (45 percent of all farms and ranches in Oregon) irrigate in some fashion. Eighty-five percent of the value of all Oregon crops comes from farms that irrigate. Farms that apply irrigation to every acre of their operation account for 55 percent of the value of all Oregon's crops. The number of irrigating farms is about evenly split between western and eastern Oregon, but 80 percent of Oregon's irrigated acreage lies east of the Cascades. About one-third of the state's irrigated lands are served, not by on-farm diversions or pumps, but through deliveries of stored water by districts or other organizations (Oregon Department of Agriculture 2005). About 65 percent of all land irrigated was from surface water; 35 percent from groundwater. About 40 percent of irrigated lands are served through gravity-flow systems (ditches, canals, flood irrigation), and 60 percent through sprinklers. Of gravity-flow irrigation, over 70 percent occurs through unlined surface ditches (U.S. Department of Agriculture 2004a).

Of the eighteen thousand Oregon farms that irrigate, nearly three-quarters water less than 50 acres, and they account for only 8 percent of the total acreage irrigated. On the other end of the scale, 2 percent of farms irrigate 1,000 acres or more, and they account for about 40 percent of the total acreage irrigated (Table 28) (U.S. Department of Agriculture 2004b). Klamath County has the most irrigated land in the state, while Jackson County has the greatest number of irrigated farms. Of the top ten irrigation counties, all but Marion are east of the Cascades (Table 29).

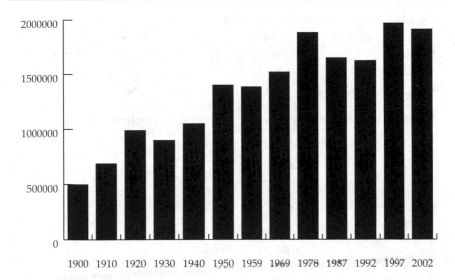

Figure 26: Irrigated acreage by year. (Source: 1900-1950: Water Resources Committee 1955; 1959-1992; U.S. Bureau of Census 1992; 1997 and 2002, U.S. Department of Agriculture 2004b)

The most prevalent forms of sprinkler irrigation in Oregon are center pivot, sideroll, and hand-move systems, which have application efficiencies ranging from 65 to 90 percent. Flood irrigation serves 70 percent of the lands irrigated by gravity systems, with furrow irrigation providing water to about 25 percent (Tables 30 and 31) (U.S. Department of Agriculture 2004a; U.S. Bureau of Reclamation n.d.).

Nearly 70 percent of Oregon's irrigated lands are in alfalfa, hay, or pasture (Table 32). Thus, it could be said that what Oregon's water really grows most is cows. All told, in 2002 the market value of agricultural products sold from farms with irrigated land was over $2.4 billion. (That is roughly 80 percent of the total market value of all agricultural products sold in Oregon.) For non-irrigated farms, the market value was about one-third

Table 28. Irrigated Acreage by Size Classes

	Acres irrigated on farm						
	1-9	10-49	50-99	100-219	220-499	500-999	1,000+
Number of farms	7,626	5,350	1,467	1,243	1,206	544	340
Acres irrigated in size class	27,679	117,126	102,121	169,054	369,559	374,137	747,951

Source: U.S. Department of Agriculture 2004b

of that. It is interesting that this economic yield could be even greater with firmer water supplies. According to the U.S. Department of Agriculture's 2003 Farm and Ranch Irrigation Survey, over four hundred thousand of Oregon's irrigated acres suffered diminished yields from surface water shortages alone (U.S. Department of Agriculture 2004a).

Table 29. Oregon Irrigation by County (ranked by irrigated acreage)

Location	Irrigated acreage	No. of irrigated farms
Oregon	1,907,627	17,776
Klamath	242,153	916
Malheur	223,263	1,120
Lake	194,320	315
Harney	133,008	301
Baker	127,077	552
Umatilla	121,909	1,011
Marion	100,415	1,315
Morrow	94,798	197
Crook	77,861	501
Union	64,901	409
Jefferson	56,954	311
Jackson	49,887	1,428
Deschutes	44,436	1,425
Grant	41,854	251
Wallowa	40,323	351
Wasco	31,874	284
Linn	30,503	590
Clackamas	26,927	1,151
Washington	25,182	689
Yamhill	24,907	543
Lane	22,001	791
Benton	20,655	287
Hood River	19,596	521
Douglas	15,602	622
Wheeler	13,524	85
Polk	12,984	247
Coos	10,848	294
Josephine	9,415	520
Multnomah	7,780	262
Gilliam	6,830	27
Tillamook	6,066	93
Columbia	2,818	136
Curry	2,804	82
Sherman	1,949	24
Lincoln	1,361	92
Clatsop	842	33

Source: U.S. Department of Agriculture 2004b

Table 30. Irrigation System Use in Oregon by Acres Irrigated

Types	acres
Sprinkler	
Center Pivot	367,743
Sideroll	270,704
Hand Move	224,704
Big Gun	80,786
Solid Set	72,236
Linear	32,038
Subtotal	1,048,211
Gravity flow	
Flood	
controlled	255,218
uncontrolled	242,728
Furrows	170,319
Other	16,999
Subtotal	685,264
Total	1,731,660*

* There is a slight difference in total irrigated acreage between the 2002 Census of Agriculture and the 2003 Farm and Ranch Survey.

Source: U.S. Department of Agriculture 2004a

Table 31. Irrigation Method Efficiency

Irrigation Method	Efficiency* (Percent)
Center Pivot	75 - 90
Linear Move	75 - 90
Row Crop Drip	85 - 90
Drip	85 - 90
Microspray	85 - 90
Bubbler: Low Head	80 - 90
Furrow: Modern	60 - 90
Solid Set	70 - 85
Hand Move	65 - 85
Side-Roll	65 - 85
Traveling Gun	60 - 75
Furrow	50 - 80

*Values represent range obtainable with correct design and management. Actual efficiencies dependent on local conditions (e.g., temperature, humidity, wind).

Source: U.S. Bureau of Reclamation n.d.

Table 32. Irrigated Crop Summary

Crop	Acreage	% of Total Irrigated Acreage
Alfalfa and other hay	741,000	42.8
Pasture grasses	417,779	24.1
Wheat	143,046	8.3
Corn for silage	52,991	3.1
Vegetables	45,032	2.6
Potatoes	44,788	2.6
Orchards and vineyards	37,600	2.2
Berry production	31,021	1.8
Other (e.g., nursery production)	142,558	8.2

Source: U.S. Department of Agriculture 2004a

▣ Municipal and Domestic Use

This category includes public supply systems (such as those for cities) that distribute water to customers (including households and industries), as well as self-supplied domestic use (homes with pumps). All told, an estimated 720,000 acre-feet of water is withdrawn annually for these uses. Taken together, they represent the second-largest water withdrawal category in the state. Statewide, an average 70 percent of public and domestic supplies are drawn from surface water. In eastern Oregon, nearly 60 percent of public supply comes from groundwater, while in western Oregon, surface water provides about 86 percent of public supply. Nearly seven hundred

Keeping Oregon's rapidly growing population supplied with clean water is no easy matter: on the average, each Oregonian uses over 200 gallons per day. Oregon's population is expected to double in the next 50 years. (Portland Water Bureau)

thousand Oregonians rely on self-supplied domestic water sources, with about four hundred thousand of these pumping their water from their own domestic wells (Hutson et al. 2004; U.S. Geological Survey n.d.[b]; Oregon Department of Human Services 2003).

In 1995, Oregon had an estimated per capita public and domestic use of 134 gallons per day—the lowest of the Northwest states (Washington's was 144, and Idaho's 197, gpd) (Houston et al. 2003). Well-designed water systems allow for up to 100 gallons per day for each person in a household. Although this level of service might be supplied by a source providing 6 gallons per minute, 10 is a better bet. The modern-day mansions (that seem to be becoming more rule than exception in today's suburbia) might need up to 18 gallons per minute, assuming they have six bedrooms and three bathrooms.

A study of how water was actually used in twelve hundred North American homes conducted in the late 1990s found that toilet flushing, clothes washing, and bathing account for nearly two-thirds of average indoor water use. Faucet use (aggregated for all purposes, e.g., for drinking, cooking, and handwashing) was the fourth-largest indoor category. Dishwashing and bathing were the smallest uses of water in households (Dziegielewski 2000). Table 33 displays indoor water uses measured in the study.

Providing Oregon's growing population with clean, affordable water falls mostly to urban water providers, as cities are where the newly born or -arrived will usually dwell. Urban providers are either the cities themselves or districts formed to serve city residents. For example, the Portland metropolitan area is served not only by the Portland Water Bureau, but

Table 33. Household Water Uses (Indoor), ranked high to low				
Use	Average Frequency[1]	Average Volume[2]	Average Daily Use[3]	% of Total Daily Use
Toilet flushing	5.05	3.7	18.5	26.7
Clothes washing	0.37	40.6	15.0	21.7
Showering	0.7	16.6	11.6	16.8
Faucet use	17.6	0.6	10.9	15.7
Leaks	0.46	20.7	9.5	13.7
Other domestic	-	-	1.6	2.3
Bathing	0.05	23.8	1.2	1.7
Dishwashing	0.1	10.0	1.0	1.4
Total indoor use			69.3	100

[1] Events/person/day. [2] Gallons/event. [3] Gallons/person/day
Source: Dziegielewski 2000

Table 34. Drinking Water Sources for Oregon's Larger Cities (with greater than 3,500 connections)

City	Connections	Population Served	Sources[1]
Albany	13,000	39,000	South Santiam River
Ashland	9,784	20,430	Ashland Cr., Talent Irrigation Ditch
Astoria	3,872	9,813	Bear Cr., Cedar Cr., Middle Lake
Baker City	4,400	9,950	Mill Creek, Goodrich Cr. Intake, Elk Cr. Intake, Marble Springs Reservoir, Salmon Creek
Beaverton	15,140	62,000	Tualatin River, Sorrento ASR well
Bend	19,100	45,500	Bridge Cr., wells
Canby	4,126	14,000	Molalla River, springs
Central Point	4,313	12,230	Big Butte Springs, through Medford Water Commission
Coos Bay-North Bend	11,000	37,000	Pony Cr.; wells
Corvallis	14,000	52,000	Willamette River; Rock Cr.; Griffith Cr.
Cottage Grove	3,631	8,910	Layng Cr.; Prather Cr.; well
Dallas	4,378	13,270	Rickreall Cr.
Eugene	57,450	165,000	McKenzie River
Forest Grove	4,000	16,275	Clear Cr., Roaring Cr., Deep Cr., Thomas Cr., Smith Cr.
Grants Pass	9,455	26,000	Rogue River
Gresham	15,770	64,224	(from Portland Water Bureau)
Hermiston	4,400	14,120	Columbia River; wells
Hillsboro	20,150	79,340	Tualatin River
Keizer	9,900	30,500	wells
Klamath Falls	15,077	38,000	wells
La Grande	4,800	12,500	wells; Beaver Cr.
Lake Oswego	11,000	35,750	Clackamas River; well
Lebanon	4,600	13,000	Santiam Canal (South Santiam)
Lincoln City	5,360	14,013	Schooner Cr.; Rock Cr.
McMinnville	11,718	27,600	Haskins Reservoir; McGuire Reservoir
Medford	26,255	83,454	Big Butte Springs; Rogue River
Milwaukie	6,500	20,050	Wells; Portland Water Bureau; Clackamas Water District; Oak Lodge Water District
Newberg	5,300	19,750	Wells; springs
Newport	4,500	9,650	Big Creek; Siletz River
Ontario	3,500	11,140	Snake River; wells
Oregon City	8,471	26,217	Clackamas River; well
Pendleton	5,700	16,830	Wells; Umatilla River
Portland	163,900	482,500	Bull Run River; wells
Redmond	7,400	17,450	Wells

City	Connections	Population Served	Sources[1]
Rockwood PUD	13,500	53,900	(through Portland Water Bureau)
Roseburg	10,266	26,660	North Umpqua River
Salem	51,800	170,000	Santiam River; wells
Sherwood	5,063	12,230	Wells; Tualatin River
Silverton	3,800	6,700	Abiqua Cr.; Silver Cr.
Springfield	19,021	56,000	Wells; Willamette River
St Helens	3,800	10,500	Wells
The Dalles	4,400	12,250	Mill Cr.; wells
Tigard	16,000	47,000	Through Lake Oswego Municipal Water, Portland Water Bur., Tualatin Valley Water Dist., City of Beaverton; wells
Troutdale	4,300	14,300	Wells
Tualatin	6,046	25,464	Through Portland Water Bureau
Washington County Communities (through Tualatin Valley Water District)	47,400	158,000	Through Portland Water Bureau, Hillsboro Joint Water Commission; wells
West Linn	7,763	24,000	Through Oregon City/South Fork Water Board, City of Lake Oswego
Wilsonville	3,965	15,880	Wells; Willamette River
Woodburn	5,553	20,000	Wells

[1] (inc. permanent, seasonal & emergency sources)
Source: Oregon Department of Human Services n.d.

by over sixty other water districts and companies (Portland Water Bureau 1992). As described above, public suppliers obtain their water from both surface and groundwater sources, or often a combination of the two. It is common for cities to have access to several sources, for either primary or back-up supplies. Alternatively, many times cities buy water wholesale from other entities.

In these times of burgeoning water demand, increasing attention is being paid to protecting the quality and quantity of municipal supplies. Although it may not be widely recognized, there is probably no more direct connection between nature and city dwellers than the water they drink. What they have to do to it before they can drink it can speak volumes about their city and the watershed they count on. The names of their city's water source is becoming just as important as their ZIP code or computer passwords. Table 34 lists the water supply sources for a number of Oregon cities.

⊞ *Hydroelectric Uses*

Upon their arrival in the Oregon country, settlers almost immediately put water to work grinding grain or sawing logs. The first power-driven saw mill was built near Champoeg in the northern Willamette Valley in 1836. By the dawn of the twentieth century, hydroelectricity was being generated in Oregon. In fact, the first long-distance, commercial power transmission is reputed to have occurred between Oregon City and Portland in 1899 (Water Resources Committee 1955). Today, hydropower provides about 44 percent of the state's electric supply (Oregon Department of Energy 2005).

The Oregon Water Resources Department has authorized 166 hydroelectric facilities built by private utilities, irrigation districts, and cities. This total excludes federal facilities, which are not subject to state permitting requirements (Oregon Water Resources Department 1993b). A relative handful of these federal projects, however, account for two-thirds of Oregon's 5,000-plus megawatt hydroelectric power potential. By itself, Oregon's share of the 3,400-megawatt generating capacity of just four Columbia River dams (Bonneville, The Dalles, John Day, and McNary) represents over 34 percent of the state's total. These dams are among the biggest hydropower projects in the nation. The John Day Dam is ranked third and The Dalles sixth in U.S. power production. The federal government also maintains a 2,100- and a 49-megawatt generating capacity in the Willamette and Rogue river basins, respectively (U.S. Army Corps of Engineers. n.d.).

Oregon's largest non-federal hydro-electric facilities are on the Snake and Deschutes rivers (Table 35). Idaho Power Company's three Snake River dams (Brownlee, Hells Canyon, and Oxbow) have a combined

Pelton Dam on the Deschutes River, which generates enough electricity to power nearly forty-five thousand homes, is operated by Portland General Electric, a private utility, under licenses from both the state and federal governments. (Portland General Electric)

Table 35. Oregon's Major Non-Federal Hydroelectric Projects (ranked by capacity)

Licensee	Project Name	River	Kilowatt Capacity	Fed. Exp. Date[4]	State Exp. Date[5]
Idaho Power Co.	Brownlee	Snake	585,400[1]	7/1/2005	12/31/2010
Idaho Power Co.	Hells Canyon	Snake	391,000[1]	7/1/2005	12/31/2017
Portland General Electric	Round Butte	Deschutes	338,000[2]	05/31/2055	12/31/2010
Idaho Power Co.	Oxbow	Snake	190,000[1]	7/31/2005	12/31/2011
Pacificorp	Lemolo & others	North Umpqua	188,660[4]	10/31/2038	12/31/2002
Pacificorp	John Boyle	Klamath	151,000[4]	2/28/2006	12/31/2006
Portland General Electric	Clackamas project: North Fork/Faraday/ Oak Grove	Clackamas	136,600[4]	8/31/2006	8/31/2006
Eugene Water & Electric Board	Carmen	McKenzie	108,000[3]	11/30/2008	--
Portland General Electric	Pelton	Deschutes	108,000[2]	05/31/2055	12/31/2011
Portland General Electric	Sullivan	Willamette	16,000[2]	12/31/2004	—

Sources:
[1] Idaho Power Company websites accessed March 16, 2006: http://www.idahopower.com/ riversrec/relicensing/hellscanyon/brownlee/brownleefacts.asp; http://www.idahopower. com/riversrec/relicensing/hellscanyon/oxbow/oxbowfacts.asp; http://www.idahopower. com/riversrec/relicensing/hellscanyon/hellscanyon/hellscanyonfacts.asp
[2] Portland General Electric websites accessed March 16, 2006: http://www. portlandgeneral.com/about_pge/news/peltonroundbutte/factsheet.asp; http://www. portlandgeneral.com/about_pge/news/sullivan_relicensing.asp
[3] EWEB website accessed March 16, 2006: http://www.eweb.org/news/carmensmith/about. htm
[4] Federal Energy Regulatory Commission. 2005. Outstanding Licenses as of 07/08/2005; Downloaded from http://www.ferc.gov/industries/hydropower/gen-info/licenses.xls; February 20, 2006
[5] Oregon Water Resources Department. n.d.[f] Hydroelectric Information database. Accessed February 20, 2006 from http://www.wrd.state.or.us/OWRD/SW/hydro_info.

licensed generating capacity of about 1,200 megawatts. Portland General Electric operates Pelton and Round Butte dams on the Deschutes; they have a licensed generating capacity of 108 and 338 megawatts, respectively. The Eugene Water and Electric Board's Carmen facility on the McKenzie River is also noteworthy, having a 108-megawatt capacity (Oregon Water Resources Department n.d.[f]; Portland General Electric n.d. [b]; Eugene Water and Electric Board n.d., Idaho Power Company n.d.).

It takes an estimated 539,000,000 acre-feet of water to generate the 40,800 million kilowatt-hours produced in Oregon in 1990 (Solley, Pierce, and Perlman 1993). This amount reflects not only the full blast of the Columbia, but the multiple use of water by turbines as it descends a drainage. Most hydropower production occurs in the Willamette, Rogue, Deschutes, and Snake-Owyhee drainages.

▣ Instream Uses

Instream water uses have a presence and importance that go beyond their official measure. Adding up the flow rates allowed under instream water rights does not come close to representing actual use. For water flowing in streams is always and everywhere used to transport the sediment and nutrient load of watersheds, to support the web of life from bank to bank and from headwaters to mouth, or to be wondered at or waded in by generations of Oregonians. To suggest these uses exist only where the state licenses them would be ridiculous. So too might fitting the pulse and shine of Oregon streams to the flatness of a bookkeeper's ledger. But Oregon knows from experience the results of not originally accounting for instream flows in its water management system (Chapter 4). If instream water rights do not measure the true extent of instream water use, they at least interject instream considerations into the water use debate and establish streamflow protection priorities.

Of roughly fifty thousand surface water rights in Oregon, about fifteen hundred are for instream uses, which total over 80 million acre-feet. About 70 percent of these rights are in Western Oregon and account for roughly 78 percent of the total instream volume. On the eastside, the Deschutes and Klamath basins have the most instream water rights by volume (with approximately eight and two million acre-feet, respectively) (Oregon Water Resources Department 2006e). In terms of water allocations, then, use levels allowed under instream rights appear to dwarf other uses—including irrigation—by a factor of twelve. But that standing is very deceiving, given the very recent priority dates of most instream rights. Like other water rights, instream rights can only leverage whatever water is left after pre-existing users take their share—and in Oregon, those pre-existing users have about a hundred-year head start. The situation might be compared to a doughnut shop where the instream family of use, arriving late 5 minutes before closing, has been given a coupon good for 20 dozen, but a number of 99—and the sign at the counter says "Now serving number 15." Their order, compared to most others in line, is probably one of the biggest. However, the likelihood of getting their order filled is another matter.

But how much of the instream "order" is currently being filled? It is difficult to say. Instream use suffers from the same lack of measurement as most other use categories. The Water Resources Department, which holds instream water rights in trust for the people of the state, is required to report annual use under those rights. However, water flowing through a wide-open stream is harder to measure than water flowing through a pump or headgate. In addition, the state, as the largest water right-holding entity in Oregon, is faced with a vexing management challenge in tracking all instream right stream segments.

Instream right satisfaction relies to a large part on weather—rainy years or years with good snowpacks mean more instream water rights get water. The Oregon Water Resources Department has been taking an increasingly proactive approach to its management of instream water rights. Watermasters identify four or five actions annually to pursue in the watersheds identified in the Oregon Plan for Salmon and Watersheds (Chapter 11) as high priorities for restoration. Usually these entail establishing and updating inventories of major diversions, more closely monitoring streamflows, and providing assistance for the instream water rights leasing program (Oregon Water Resources Department 2005a) Despite these increased and targeted efforts, getting to the point where instream rights are satisfied most years is very slow going.

The department in cooperation with the Oregon Progress Board maintains an instream water right benchmark—Oregon Benchmark 79, the percentage of key streams meeting minimum flow rights (Chapter 11 has a more detailed description). The percentage of these key streams where minimum flow requirements are met twelve months per year has varied from 22 to 76 percent, with an average of 45 percent. The state has a long-term goal of 40 percent, which still means that, on average, 60 percent of the key streams can expect to fall short of instream objectives every year. Without a bigger measuring budget (at one point the department estimated

Table 36. Types of Public Water Systems

System Type	Service Features	No. in State	Population Served	Examples
Community	15-plus connections; or 25-plus people, used year-round.	870	2,975,581	Cities, water districts, rural subdivisions
Non-community				
Non-transient	Non-residential, but serving at least 25 of the same people over 6 months per year.	340	73,929	Factories, schools
Transient	Transient population of at least 25 people	1,420	224,468	Parks, campgrounds, restaurants
State-regulated	4 - 14 connections; or serves 10 - 24 people.	927	15,844	Small mobile home parks, subdivisions, rural residential
Total		3,557	3,289,822	

Source: Oregon Department of Human Services 2006; OAR Chapter 33, Div. 61

it would cost about $2 million for installation and nearly $1 million a year to gage flows for all instream rights [Oregon Water Resources Department 1995c]) or development of cost-effective alternatives (e.g., new measuring technologies, increased interagency cooperation, recruitment of technically trained volunteers), the degree to which Oregonians' instream rights are being met will remain largely a matter of conjecture.

Water Providers

The sources tapped and uses made of Oregon's water resources are only a part of the state's water management tapestry. How the water gets to those who need it is also important. For, in most instances, Oregonians depend on others to get their water for them. Unlike California with its State Water Project (twenty-eight dams and reservoirs, twenty-five pumping and generating plants, and nearly 660 miles of aqueducts serving 23 million people and 755,000 acres of farmland) the State of Oregon has no water infrastructure (California Department of Water Resources 2005; California Department of Water Resources. n.d. [a]). There is no equivalent

Table 37. Public Water Systems by County

	# of public water systems	# of systems with groundwater sources	# of systems with surface water sources	Total population served
Baker	33	31	2	14,422
Benton	84	79	5	64,047
Clackamas	360	333	27	325,581
Clatsop	37	20	17	40,622
Columbia	97	87	10	33,783
Coos	74	52	22	54,797
Crook	54	54	0	13,644
Curry	69	61	8	19,635
Deschutes	189	188	1	120,597
Douglas	159	119	40	95,237
Gilliam	6	6	0	1,574
Grant	28	28	0	5,363
Harney	36	36	0	5,920
Hood River	21	20	1	22,181
Jackson	281	264	17	167,554
Jefferson	37	35	2	25,952
Josephine	249	236	13	54,011
Klamath	187	186	1	65,689
Lake	41	41	0	7,106
Lane	343	318	25	308,753
Lincoln	83	55	28	55,369
Linn	209	198	11	104,220
Malheur	36	35	1	18,694
Marion	252	241	11	303,047
Morrow	25	24	1	10,105
Multnomah	63	53	10	646,263
Polk	26	20	6	40,850
Sherman	15	15	0	3,355
Tillamook	87	60	27	30,185
Umatilla	103	100	3	60,017
Union	29	29	0	20,809
Wallowa	23	21	2	5,118
Wasco	53	52	1	22,809
Washington	111	94	17	449,279
Wheeler	7	7	0	1,030
Yamhill	99	83	16	73,214

Source: Oregon Department of Human Services n.d.

Drinking water systems come in all sizes, such as those for Salem (top) and Mitchell (lower), but all require maintenance especially to replace aging components, not to mention meeting growth needs. A state task force recently estimated water rate increases of up to 400 percent may be needed in some locales to meet drinking water needs. (Top photo: City of Salem; lower photo: Oregon Health Division)

of the state highway system for moving water. Instead, people get water from cities, water associations, irrigation districts, and ditch companies organized and operated as water providers under Oregon law. Generally, these organizations can be divided into two camps: those that provide drinking water and those that transport agricultural water. However, as Oregon grows, agricultural suppliers may be called upon, and profit by, serving urban or suburban uses.

⊞ *Drinking Water Providers*

In 2006, there were 3,557 water systems in Oregon, serving 90 percent of Oregon's population (Oregon Department of Human Services 2006).

These systems are subject to water quality standards enforced by the Oregon Health Division (part of the Oregon Department of Human Services). Under Oregon's Drinking Water Quality Act (ORS Chapter 448), OHS's Office of Public Health Systems administers both state and federal drinking-water laws through its Drinking Water Program. Regulations governing Oregon's public water systems are in OAR Chapter 333, Division 61.

OHD regulates certain public water systems that provide water through pipes (such as cities) or other means (such as water delivery trucks). OHD regulates systems that pipe water to the public for human consumption if they have more than three service connections, or supply water to a public or commercial establishment that operates at least sixty days per year, and that is used by ten or more individuals per day. OHD oversees public water systems that provide water through constructed conveyances other than pipes to at least fifteen service connections or regularly serve at least twenty-five individuals daily at least sixty days of the year. Public systems are subdivided into categories based on service type (Table 36).

Clackamas County has the most water systems (360) in the state, followed by Lane, Jackson, Marion, and Josephine counties, each having around 250 or more (Table 37). Eastern Oregonians are especially dependent on groundwater for their drinking supplies: eleven of the eighteen counties east of the Cascades serve effectively 100 percent of their system populations with wells (Oregon Department of Human Services n.d.) The largest public supplier is the City of Portland, which serves drinking water to approximately 790,000 Oregonians, over one-fifth of the population of Oregon (Portland Water Bureau 2006).

In a pattern that has important ramifications for current and future drinking-water supplies, a few big systems serve most of the state's population. Hundreds of other small, frequently struggling, systems are spread across the state to serve everyone else. About 87 percent of the public water systems in Oregon serve five hundred or fewer people each (Oregon Department of Human Services 2003). Over half of Oregon's community water systems serve just two hundred or fewer people—in aggregate, less than 2 percent of the total population. Just 5 percent of the community water systems provide water for roughly 75 percent of the total community system population. The fifty largest public water systems (each serving more than ten thousand people) supply water for 60 percent of Oregon's population. About half of Oregonians citizens rely solely on groundwater (mostly small systems) and 30 percent on surface water (mostly large systems). Another 20 percent rely on a mix of the two (Oregon Department of Human Services n.d.; Oregon Department of Environmental Quality 2004a).

As Oregon grows at rates above the national average and as new drinking-water quality regulations take effect, water systems around the state are feeling the pinch. Many were built decades ago in simpler times characterized by fewer people, fewer regulations, and lesser expectations of service. Statewide, Oregon is faced with the need to invest heavily in treatment, monitoring, and system operation. A 2001 report by the Environmental Protection Agency estimated that Oregon needed over $2.7 billion to meet its twenty-year drinking-water infrastructure needs. Roughly half of that amount was for transmission and distribution systems, and over 20 percent for treatment facilities (U.S. Environmental Protection Agency 2001).

Thus, assuring clean, dependable, and affordable supplies of drinking water is becoming a lot more costly and complex. Some providers will not be able to handle such a burden by themselves and may have to enter into cooperative agreements, or merge, with other suppliers in their region. Many will seek to raise their water rates. Public funds will continue to play a critical role, though their availability, especially at the federal level, is limited.

Whether they are classed by the state as community or non-community servers, drinking-water providers can organize under several authorities established in Oregon's statutes. Cities are granted the authority to own and operate water supply facilities by ORS 225.020. They are authorized to supply water for nearly any purpose to customers inside or outside city limits; levy assessments against property to pay for service; and purchase, own, appropriate, and condemn land, rights of way, water, or water rights. Domestic water supply districts may be organized by communities to serve their residents. As municipal corporations, these districts have the powers of eminent domain and may levy taxes, issue both general obligation and revenue bonds, and appropriate real property and water rights (ORS Chapter 264). In addition, water authorities may be formed to integrate water supply services to rapidly growing incorporated and unincorporated areas (ORS 450.600–450.989).

Getting water to crops can require expensive investments in infrastructure, such as the canal systems of south central (top photo) or the pipelines of eastern (lower photo) Oregon. (Top photo: U.S. Bureau of Reclamation. Photograph Number: C12-200-10007. Date Taken : 2004-10-6. "Aerial view of the Link River Diversion Dam located just west of Klamath Falls in Oregon." Lower photo: Oregon State Archives, Oregon Highway Division, OHD4847)

The Oregon Public Utility Commission oversees rate structures of investor-owned water utilities. The PUC regulates water rates if the utility serves five hundred or more customers or its average monthly residential rate exceeds $24 and 20 percent or more of its customers petition the commission requesting regulation of rates. In 2004, approximately eighty water utilities were subject to PUC jurisdiction (Oregon Public Utility Commission n.d.).

▣ *Agricultural Water Providers*

In the decades bracketing the beginning of the twentieth century, Oregonians' near-instinctual drive (we are, after all, residents of the Beaver State) to collect, stop, drain, re-route, and deliver water prompted the creation of a variety of legal authorities. Oregon enacted laws to allow formation of state-recognized districts for irrigation, water control, drainage, water improvement, and diking. Initially, these functions were the province of private enterprise (such as the still-existing ditch companies). However, few undertakings were profitable, and the resources and powers of government were called upon to stabilize and promote land development (Oregon Water Resources Department 1987).

Today, Oregon has about two hundred fifty agricultural water providers. Although each represents a different mix of purposes and authorities (including providing water for domestic and municipal purposes), their most important role is delivering water for irrigation (Table 38). By far the largest players in the agricultural water provision game are irrigation districts, which by some estimates serve half of Oregon's irrigated acreage (Lee 1996).

The remaining acreage is split fairly evenly among ditch companies and water control, water improvement, and drainage districts. Ditch companies, some of the state's oldest providers, are concentrated in northeast Oregon (especially Baker County), where a plethora of companies serves smaller acreages. Water improvement districts are well distributed across eastern Oregon, whereas water control districts seem to be creatures largely of the Willamette Valley. Drainage and diking districts occur along the coast, the lower Columbia River, and the wetlands of Klamath and Malheur counties.

Table 38. Types of Agricultural Water Providers

Provider (Year Authorized)	Purpose	Powers
Irrigation Districts (ORS Chapter 545) 1895	To provide for construction of works for irrigation or to provide for betterment, operation or maintenance of works already constructed.	Acquire domestic or municipal water works; place lien on crops; construct drainage works.
Ditch Companies (ORS Chapter 541) Formed under 1891 Act of the Laws of Oregon	Construct and maintain ditches to deliver water from private ditch companies to all persons whose lands lie adjacent to or within reach of the line of the ditch.	Appropriate and divert water; condemn land for ditch rights of way.
Water Control Districts (ORS Chapter 553) 1947	To acquire, purchase, construct, improve, operate and maintain drainage, irrigation, flood and surface water control works and to improve the agricultural and other uses of lands.	Appropriate and acquire water rights; sell, lease and deliver water for irrigation and other purpose inside and outside district; levy taxes; manage facilities for secondary purposes of domestic, municipal, irrigation, recreation, wildlife, fish life and water quality enhancement.
Drainage Districts (ORS Chapter 547) 1915	To reclaim swamp, overflowed lands or irrigated lands.	Locate irrigation works; file for water appropriations to irrigate district lands; construct, operate and maintain irrigation works condemn lands.

CHAPTER 9

Surface Water Safeguards

THERE ARE MANY USES OF SURFACE water that go beyond pumping it out of a channel. The state's water use permitting system often brushes up against, but does not control, activities such as waste disposal, navigation, streambed protection, recreation, and habitat protection. This chapter briefly summarizes Oregon's major programs addressing these activities, especially as they relate to Oregon's water rights system.

Protecting Surface Water Quality

Oregon's Department of Environmental Quality is charged, no less, with maintaining and improving Oregon's environment. A big piece of this Herculean task is protecting and restoring the state's public waters. Most of DEQ's water management mission and authority is found in ORS Chapter 468b. The agency oversees a wide variety of programs and regulates a host of activities that affect water, ranging from the obvious (piping waste into creeks) to the rather exquisitely phrased "deposit of vehicles and accessories into state waters" (which, according to the rules, can be a *good* thing when used to control erosion or create artificial reefs [OAR Chapter 340, Division 46]).

On the whole, however, the department's major role is establishing water quality standards to protect the beneficial uses (such as drinking-water supplies, fish habitat, or recreation) it designates for each stream. These water quality-related beneficial uses (which match many but not all of the Water Resources Department's water supply-related beneficial uses) are compiled basin by basin in OAR Chapter 340, Division 41. DEQ assures compliance with the standards through monitoring, education, and enforcement programs.

DEQ has maintained a surface-water quality monitoring network since 1976. The department focuses on a list of "rivers of special interest" whose

Oregonians have a tradition of demanding clean water. An aroused public in the 1930s helped create the state Sanitary Authority, the predecessor of today's Department of Environmental Quality. (Oregon Historical Society, CN0022-253)

260 segments add up to nearly 7,000 miles. About half of these river miles (roughly 4 percent of Oregon's total river miles) are routinely monitored to provide background water quality data on oxygen content, solids, pH, temperature, bacteria, dissolved gases, and chemical substances—and account for about 90 percent of the state's "piped" pollution (Oregon Department of Environmental Quality n.d. [a]).

▣ *Water Quality Limited Streams*

Section 303(d) of the federal Clean Water Act (33 USC Section 26) requires each state to identify streams and lakes that do not meet water quality standards. Oregon's list identifies about a thousand water quality limited streams running over 13,000 miles in total length that fail at least one water quality standard (Figure 27). The standards most often violated are for temperature, bacteria, and dissolved oxygen (Oregon Department of Environmental Quality. 2003b). DEQ attempts to achieve the quality standards by establishing Total Maximum Daily Loads (TMDLs) on high-priority water quality limited streams (Table 39). TMDLs set the maximum amount of pollutants that may enter a stream. This pollution load is divided ("allocated") among different sources such as: industrial and sewage

Table 39. EPA-Approved Total Maximum Daily Load Allocations (listed by approval date)

Water body (*Basin/# of TMDL segments*)	Water Quality Concern Addressed	TMDL Parameters	EPA Approval Date
Columbia & Willamette rivers (8)	Fish consumption, wildlife	Dioxin	2/25/1991
Bear Creek (Rogue/3)	Algae, DO, pH	Ammonia, BOD, Phosphorus	12/8/1992
Clear Lake (Mid Coast/1)	Protection of water supply	Phosphorus	12/8/1992
Garrison Lake (South Coast/1)	Algae, aquatic weeds	Phosphorus	12/8/1992
Tualatin River (Willamette/12)	DO	Ammonia	12/8/1992
Yamhill River (Willamette/3)	Algae, pH	Phosphorus	12/8/1992
Pudding River (Willamette/16)	DO	Ammonia, BOD	10/18/1993
Tualatin River (Willamette/12)	Algae, pH	Phosphorus	1/27/1994
Rickreall Creek (Willamette/1)	DO	BOD	4/18/1994
Coast Fork Willamette River (Willamette/2)	Algae, DO, pH	Ammonia, Phosphorus	5/17/1996
Coquille R (South Coast/3)	DO	BOD	7/3/1996
Columbia Slough (Willamette/10)	Water contact, DO, pH, algae, fish consumption	Chlorophyll a, dissolved oxygen, pH, phosphorus, bacteria, DDE/DDT, PCBs, Pb (lead), Dieldrin and 2,3,7,8 TCDD (dioxin)	11/25/1998
Upper Sucker Creek (Rogue/1)	Temperature	Temperature	5/4/1999
Upper Grande Ronde Subbasin (Grande Ronde/73)	Temperature, pH, algae, DO, sedimentation	Temperature, sediment, nitrogen, phosphorus	5/3/2000
Upper South Fork Coquille River (South Coast/4)	Temperature	Temperature	3/23/2001
Umatilla Basin (Umatilla/45)	Temperature, pH, sedimentation, turbidity, aquatic weeds, algae	Temperature, pH, sedimentation, turbidity, aquatic weeds, algae	5/9/2001
Tillamook Watershed (North Coast/40)	Temperature, bacteria	Temperature, bacteria	7/31/2001

Water body (*Basin/# of TMDL segments*)	Water Quality Concern Addressed	TMDL Parameters	EPA Approval Date
Tualatin Subbasin (Willamette/101)	Temperature, bacteria, DO, algae, pH	Temperature, bacteria, DO, settleable volatile solids, ammonia, chlorophyll a, pH, phosphorus	8/7/2001
Little River Watershed (North Umpqua/16)	Temperature, pH, sedimentation	Temperature, pH, sediment	1/29/2002
Western Hood Subbasin (Hood/7)	Temperature	Temperature	1/30/2002
Nestucca Bay Watershed (North Coast/6)	Temperature, bacteria, sediment	Temperature, bacteria, sediment	5/13/2002
Lower Sucker Creek Watershed (Illinois/3)	Temperature	Temperature	5/30/2002
Lobster Creek Watershed (Rogue/3)	Temperature	Temperature	6/13/2002
Upper Klamath Lake Drainage (Klamath/32)	Temperature, pH, DO, chlorophyll a	Temperature, pH, DO, chlorophyll a	8/7/2002
Lower Columbia River (Lower Columbia/7)	Total Dissolved Gas	Total Dissolved Gas	11/18/2002
North Coast Subbasins (North Coast/56)	Temperature, bacteria	Temperature, bacteria	8/20/2003
Alvord Lake Subbasin (Malheur Lake/7)	Temperature, dissolved oxygen	Temperature, dissolved oxygen	2/11/2004
Applegate Subbasin (Rogue/17)	Temperature, sediment	Temperature, sediment	2/11/2004
Snake River-Hells Canyon Reach (Snake River/15)	Temperature, total dissolved gas, DDT, DDE, DDD, Dieldrin	Temperature, total dissolved gas, DDT, DDE, DDD, Dieldrin	3/1/2004
Snake River-Hells Canyon Reach (Snake River/5)	Phosphorus, sediment, dissolved oxygen	Temperature, sediment, dissolved oxygen	9/9/2004
Sandy Basin (Sandy/8)	Temperature, bacteria	Temperature, bacteria	4/14/2005
Walla Walla Subbasin (Walla Walla/4)	Temperature	Temperature	9/29/2005

Source: Oregon Department of Environmental Quality n.d.[b]

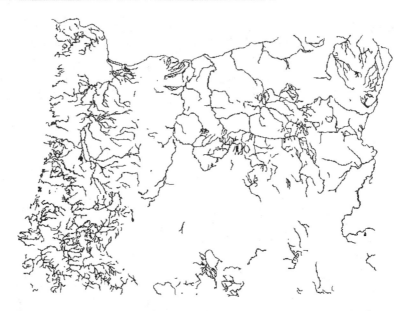

*Figure 27: Water quality limited streams. (Source: Oregon Department of
Environmental Quality)*

discharges; runoff from farms, forests, and cities; and natural processes
that contribute organic matter or soil nutrients (Oregon Department
of Environmental Quality 1996a and 1996b). Given that TMDLs are all
initially exceeded (otherwise the stream would be meeting the standards),
the share of the pollution load becomes a reduction target for those
contributing to a particular source.

As of 2006, DEQ has established TMDLs and allocations for at least some
streams in all of Oregon's major river basins, except the mid-coast, John Day,
Malheur, Owyhee, Willamette, and Goose and Summer Lakes basins (Table
39). Work continues in or is scheduled for all basins, with completion of
all TMDLs and allocations set for 2010. Rather than attempting to establish
TMDLs stream-by-stream in isolation from each other, in 2000 DEQ began
using a broader, more watershed-based approach. Streams are addressed
either in subbasin groups, or in the case of the Willamette, an entire basin
at a time. This should not only speed the process, but also allows for a
greater recognition of basin inter-relationships (Oregon Department of
Environmental Quality 2005a).

Often, pollutant concentrations are affected by low streamflows.
Accordingly, DEQ has applied for instream water rights on fifteen "TMDL"
stream segments (Oregon Water Resources Department 1996a).

▣ *Discharge Permits*

One of the state's strongest lines of defense against water pollution is its permit program. No pollutants can be purposely routed to and discharged into Oregon waters without a DEQ-issued permit. The two major types of permits are for the federal National Pollutant Discharge Elimination System (NPDES) and the state's Water Pollution Control Facility (WPCF) program.

NPDES permits control wastewater discharges to surface waters, such as those from industrial facilities or municipal sewage-treatment plants. The NPDES program is federally mandated and administered by the state. WPCF permitting is a state program regulating all wastewater disposal methods that result in no direct discharge to, but which may affect, surface waters, such as irrigation, evapotranspiration lagoons, industrial seepage pits, and certain subsurface sewage-Oregon disposal systems (Oregon Department of Environmental Quality 1997a).

Permits both limit and legalize water pollution. They grant state permission to individual industrial facilities or sewage-treatment plants to use public waters as a waste disposal system. Permits limit the amount of pollution based on a stream's ability to take it. Permittees must monitor their discharges and report the results to DEQ. If discharges violate water quality standards, permittees must either work to change their operations or face fines. DEQ administers over four thousand water quality permits including 2,645 NPDES permits and 1,450 WPCF permits. Of these, about one thousand are for sewage-treatment systems and approximately two hundred for industrial dischargers (Oregon Department of Environmental Quality 2005b; Oregon Department of Environmental Quality 2003a).

DEQ regulates roughly 1,000 municipal sewage treatment plants in Oregon, such as Salem's which discharges to the Willamette River. (City of Salem)

DEQ offers financial assistance to communities attempting to improve water quality through projects addressing wastewater treatment, stormwater runoff, and estuary management. Assistance is in the form of low-cost loans through the State Revolving Fund program. Loans may finance such projects as treatment plant improvements to stop sewage overflows into streams or creation of artificial wetlands to help treat wastewater. The department also provides low-cost loans to communities where the state is requiring homeowners to connect to new sewer systems (Oregon Department of Environmental Quality 1996c).

▣ *Nonpoint Source Pollution*

The pollution that pours from pipes into streams is as significant as it is visible. But it is also tightly controlled. The biggest problem for Oregon's streams and lakes comes not from pipes, however, but from the pollution washed in from the landscape at large (Oregon Department of Environmental Quality 1996c). This nonpoint pollution is a shared contribution from everybody that lives or works in a watershed. It comes from the oily driveways and pesticide-drenched backyards of city folk; from the over-worked septic tanks of hobby farmers; from the fertilizers, feedlots, and furrows of real farm folk; and from the torn-up ground and road-cuts of developers and loggers.

The goal of DEQ's Non-point Source (NPS) Program is to prevent and eliminate nonpoint pollution in all waterbodies in the state. The department is emphasizing ten programs that focus on watershed protection, voluntary stewardship, and partnerships between all watershed interests. The programs include: defining conditions needed to support beneficial uses; stream and watershed assessment; coordinated watershed planning; education

Today, Oregon's water pollution problems come less from "point" sources than from "non-point" causes such as poor land management practices in our watersheds. (Oregon Dept. of Environmental Quality)

and demonstration projects; and technical and financial assistance. DEQ arranges for and administers grant funds available through Section 319 of the Water Quality Act of 1987. Oregon's share of these funds, critical to NPS Program success, have grown steadily from roughly $500,000 in 1990 to $2 million in 2005 (Oregon Department of Environmental Quality 2005c; Oregon Department of Environmental Quality 2006c).

⊞ *Columbia and Willamette River Studies*

Through the 1980s, concerns grew about the health of the lower Columbia River. In 1990 Washington and Oregon joined to gather additional information to assess its condition more accurately. The Lower Columbia River Bi-State Water Quality Program's principal purposes were to identify water quality problems; determine if beneficial uses were impaired; and find solutions to identified problems. The six-year public/private partnership was jointly administered by the Washington Department of Ecology and the Oregon Department of Environmental Quality, assisted by a twenty-member Bi-State Steering Committee.

The committee's study of high-priority issues addressed fish and wildlife management, human health protection, and water quality improvement measures. The committee recommended better regulation and monitoring of nonpoint source pollutants (including PCBs, organochlorine pesticides, dioxins and furans, and metals) improved land use controls, and the promotion of citizen-based river stewardship throughout the Lower Columbia basin. The committee also recommended greater caution in the consumption of fish and increased study of the levels and causes of contaminants in both sediment and animal tissues. Most importantly, it urged the states of Oregon and Washington to phase out the discharge (both point and nonpoint) of all known bioaccumulative toxic pollutants by 2010 (Oregon Department of Environmental Quality 1997b). The committee also concluded that the river's future was uncertain.

Based on these findings, in 1995 the Lower Columbia River Estuary was designated as one of twenty-eight estuaries in the National Estuary Program. The Lower Columbia River Estuary Partnership—a two-state, public-private, non-profit organization—oversees implementation of an estuary management plan for the lower 146 miles of the Columbia River. LCREP uses a watershed approach that involves the close cooperation of twenty-eight cities, nine counties, and the states of Oregon and Washington. The management plan identifies forty-three actions addressing three major goals: on-the-ground improvements for habitat and land use; heightened education and information and government coordination; and, reduction of toxic and conventional pollutants (Lower Columbia River Estuary Partnership. n.d.).

In 1991, the DEQ began a study of the water quality of the Willamette River and its tributaries. The study, cooperatively funded by the public and private sectors, focused on pollution types and sources and their effect on the river's health and the basin's ecosystem. The study assessed dissolved oxygen levels, toxics, point and nonpoint sources, bacteria, and ecological condition. Early findings suggested that most pollution enters the river from nonpoint sources, especially agricultural activity. The study also found evidence that the biological health of the mainstem of the river is good above Eugene, fair between Eugene and Wilsonville, and poor from Wilsonville through Portland. In addition, the number of fish species present at each of nineteen sample sites appeared to decrease the further downstream the samples were taken (Oregon Department of Environmental Quality 1997c).

In 1996, Governor John Kitzhaber appointed a twelve-member Willamette Basin Task Force to recommend actions needed to restore the river to all its beneficial uses. The task force made over a hundred recommendations, including the formation of a standing body to restore and protect the Willamette River and its watershed. In response, in 1998 the Governor established the Willamette Restoration Initiative to prioritize, supplement, and implement the task force recommendations. The Initiative developed the Willamette Restoration Strategy (the "Willamette Chapter" of the Oregon Plan for Salmon and Watersheds), focusing on twenty-seven critical actions needed to address water quality, water supply, fish and wildlife habitat, and institutional needs. The Initiative grew into a private, non-profit corporation called the Willamette Partnership, which focuses on developing an ecosystem marketplace to speed achievement of Willamette basin conservation goals (Willamette Restoration Initiative 2001;Willamette Partnership 2005).

◨ *Other Surface Water Quality Programs*

In addition to DEQ's core programs on water quality, a few others have been developing of late in the state departments of Agriculture and Forestry. Uncomfortable with the notion of being regulated from Portland sky-rises by latté-lapping environmental bureaucrats, Oregon's influential farming and timber interests got behind legislation placing some water quality regulation in the hands of more harvest-based natural resource agencies.

The Oregon Department of Agriculture oversees the Confined Animal Feeding Operation program which, crudely put, seeks to keep crap out of creeks. The program requires certain livestock owners (mostly dairies) to obtain an operating permit and manage animal wastes carefully (ORS 468B.200). ODA also is responsible for agricultural water quality management plans. Where DEQ identifies a water quality limited stream and designates TMDLs, ODA must develop and carry out a plan to prevent and control water pollution caused by agricultural activities in the affected area. The department may require landowners to follow specific agricultural and cropping practices, but it may not prohibit a practice without a "scientific" basis. ODA can charge affected landowners up to $200 per year to help defray the costs of the program (ORS 568.900—568.933).

The Oregon Department of Forestry, through the authority of the Oregon Forest Practices Act, establishes best management practices "to insure that to the maximum extent practicable nonpoint source discharges from forest operations" do not violate water quality standards (ORS 527.765). As long as forest operators (e.g., loggers, roadbuilders, skidder drivers) follow the practices, they cannot be charged with violating water quality standards— no matter the actual impact of activities on the water body.

Fish Passage and Screening

The mechanics of diverting water often pose risks to fish. They may have trouble migrating over or around dams, and they can be sucked or suckered into pumps and ditches. Oregon has addressed these problems statutorily and, more recently fiscally, with an active fish passage and screening program.

▣ *Passage*

Oregon's fish passage laws are not the result of any twenty-first-century Earth Week rallies—they have been on the books a long time. According to the Oregon Department of Fish and Wildlife, when the Oregon Territory was established in 1848, its new constitution prohibited people from blocking salmon streams and specifically addressed the construction of fish passage facilities. In addition, the first game laws passed in 1872 required fishways over dams. Over time, Oregon statutes came to require passage at all artificial obstructions, although the state's fish and game authorities

Getting fish over or around instream structures can be as difficult as it is legally necessary. The enormous "push up" diversion dam in the top photo clearly poses problems for migrating fish. However, even expensive fish passage facilities (below) don't always do the trick—this fishway with its poor approach and rocky entrance is hardly welcoming to fish on the move. (Both photos: Oregon Department of Fish and Wildlife)

waived these requirements under a variety of circumstances. However, in 1996, the Attorney General determined that ODFW did not actually have the authority to waive fish passage requirements. In response, the legislature established such an authority, but within the context of emphasizing a state policy to require passage while providing water users with technical and financial assistance (Oregon Department of Fish and Wildlife n.d. [a]).

Under current law, anyone owning or operating an artificial obstruction (like a storage- or diversion-dam) on a stream inhabited, or historically inhabited, by native migratory fish must be sure the obstruction allows passage of those fish. "Native migratory fish" includes species in addition to salmonids (the poster fish of passage problems) such as sturgeon, lamprey, and suckers. Beginning in 2001, those responsible for artificial obstructions are required to take action to provide fish passage upon certain triggering events, such as building a new obstruction, replacing or abandoning an existing obstruction, or a fundamental change in permit status (e.g., new water right, renewed hydroelectric license). ODFW must approve passage plans, and may also issue exemptions and waivers (ORS 509.580 -910; OAR Chapter 635, Division 412). ODFW oversees a fish passage program to inventory obstructions, inspect fishways, establish facility design criteria, and assist project owners in planning and building fish passage facilities, some of which may qualify for ODFW grant funds or for tax credits (50 percent of net certified project costs up to $5,000). ODFW places a premium on a cooperative approach with owners of obstructions and has never resorted to formal action to force an owner to provide passage.

⊡ Screening

If water diversions are unscreened, fish can get sucked into or against pump intakes, stranded in canals, or spread over fields as finned fertilizer. It is likely that hundreds of thousands of fish are killed each year as a result of unscreened diversions. According to some estimates, around 80 percent of water diversions in the Pacific Northwest are unscreened and can pose a major risk to fish (U.S. Fish & Wildlife Service n.d.). Of the estimated seventy-three thousand surface water diversions in Oregon, about twelve hundreed have fish screens. ODFW puts the number of diversions still needing screens at about twenty-three hundred, mostly in eastern Oregon. It is estimated that more than 25 percent of these affect threatened, endangered, or sensitive fish species (Oregon Department of Fish and Wildlife 2006).

Generally, any person diverting water from a fish-bearing stream at a rate of 30 or more cfs must install, operate, and maintain a fish screening or by-pass device determined necessary by ODFW (ORS 498.311; 509.615). Any person diverting water at a lesser rate from a fish-bearing stream may be required by ODFW to install screens or by-passes under some circumstance (ORS 498.306).

A fish screen can be far more than just some wire mesh slapped over a diversion. Screening diversion ditches requires a device that stops fish, not flow. The most common device is a rotating drum-screen that allows stream-borne debris to pass, but turns fish aside. Water flowing in a ditch turns a paddle wheel that turns the cylindrical drum screen. The rotating drum conveys sticks, grass, and other flotsam over its top and on down the ditch without plugging it up, while the constantly turning drum face keeps fish out. Like any other machines, these need to be carefully designed, sized to the fit the job, and constantly maintained.

The state's screening program began in 1947. Today, it has both a state and federal funding component. Federal funds support approximately six

Fish screens (top photo) are clever devices using water power to turn a drum which deflects fish, but keeps water flowing in irrigation ditches. The dead steelhead in the ditch below shows what can happen without—or with poorly maintained—screens. (Top photo: Oregon Water Resources Department; lower photo: Oregon Department of Fish and Wildlife)

hundred fish screens in the Deschutes, Umatilla, John Day, and Grande Ronde basins. The funds are provided through the Bonneville Power Administration and the National Marine Fisheries Service, mostly in response to the Northwest Power Planning Council's Fish and Wildlife Program. There is also federally funded screening in the Rogue basin. ODFW operates "screen shops" (where screens are built) in Central Point, Corvallis, John Day, and Enterprise (Oregon Department of Fish and Wildlife n.d.[b]).

State funds support a cost-sharing program for water diversions of any size. Under this program, the state foots the bill for up to 60 percent or $10,000 of the cost of screen installation and is responsible for major repair and maintenance at diversions less than 30 cfs. At diversions of 30 cfs or greater, the water user is responsible for all maintenance. The program targets at least 75 diversions per year for screening. Every two years ODFW must notify diversion operators (the top two hundred fifty on a regularly updated priority listing of thirty-five hundred unscreened diversions) of screening needs and cost-sharing opportunities. The funding sources include the state lottery, a twenty-five-cent surcharge on all sport fishing licenses, and the Fisheries Restoration and Irrigation Mitigation Act of 2000.

FRIMA (Public Law 106–502) created a voluntary fish passage partnership program administered by the Department of the Interior in Oregon, Washington, Idaho, and western Montana. The law requires a local partner cost-share of at least 35 percent, though in practice local partners pony up nearly 60 percent. From 2002 through 2005, ten fish screening and six fishway projects were installed under the program in Oregon. FRIMA provided nearly $3 million and local partners over $3.4 million to complete these projects.

Threatened and Endangered Species

Programs to protect threatened or endangered species can affect state water management, though impacts may be felt more in the future than they have been in the past. In Oregon, species protection falls mostly under the authority of the federal government. The federal Endangered Species Act, first passed by Congress in 1973 (16 USC, Section 1531), gives the Secretary of Commerce, acting through the National Oceanic and Atmospheric Administration's Fisheries Service (formerly the National Marine Fisheries Service), jurisdiction over commercial and ocean recreational fish species.

The Secretary of Interior, through the U. S. Fish and Wildlife Service, has jurisdiction over all other species (except for certain insect pests, to which the ESA does not apply). In terms of Oregon's fish species, that means NOAA Fisheries is responsible for salmon and steelhead, while USFWS is responsible for bull trout, Klamath basin suckers, and the like.

The Oregon Department of Fish and Wildlife is responsible for the state's endangered species program (ORS 496.172–496.192; OAR Chapter 635, Division 100). In that it is Oregon's policy (ORS 496.012) to "prevent the serious depletion of any indigenous species," like the federal government, the state also maintains a list of threatened or endangered species. The state program works in concert with the federal program, but takes a back seat to it in a practical sense. The state's authority for protecting the threatened and endangered species on its list extends only to state actions and to activities on state lands. The federal program, as described below, simply has more punch.

The agencies are responsible for identifying threatened or endangered species. An endangered species is one in danger of extinction throughout

Some of Oregon's most threatened fish populations are its unique stocks of desert fish. The Lahontan cutthroat trout (top photo) and the Warner sucker (lower photo) are protected by both the federal and state governments. (both photos: Oregon Department of Fish and Wildlife)

all or a significant portion of its range. A threatened species is one likely to become endangered within the foreseeable future. Although separate, the state and federal lists are very similar, especially for the "charismatic fauna" of trout, salmon, and mammals. All told, as of 2004, thirty-five species were considered threatened or endangered by the state. The federal government includes an additional sixteen Oregon species not listed by the state. Ten fish species appear on both lists (Table 40). Other water-dependent creatures may end up on the lists over time. For example, the state has identified 139 sensitive species (i.e., likely to become threatened or endangered), including thirty-one fish and thirty-three amphibians and reptilian species in all sensitive categories (critical, vulnerable, naturally rare, and undetermined). When the agencies list a species as threatened or endangered, either on their own motion or in response to petitions, they also must designate habitat critical for species survival (Oregon Department of Fish and Wildlife 2004; Oregon Department of Fish and Wildlife 1997).

Because water both qualifies as habitat and supports habitat throughout Oregon, the ESA figures prominently as a factor shaping the state's water future. Section 7 of the ESA requires each federal agency to insure that its actions do not jeopardize threatened or endangered species. This requirement might kill a new hydro project needing a federal license or dash any hope of federal funding for a storage reservoir. in addition, Section 9 prohibits any person from "taking" an endangered species. "Take" means to "harass, harm, pursue, hunt, shoot, wound, kill, trap, capture, or collect, or to attempt to engage in any such conduct." Harm has long been understood by the federal agencies to include adversely affecting habitat, a view largely upheld by the U.S. Supreme Court in 1995. In *Babbit v. Sweet Home Chapter of Communities for a Great Oregon* (115 S. Ct. 2407, 132 L. Ed. 2d597 (1995)), the court found habitat modification could be considered a form of harm and thus be a taking if it could be linked to the actual injury or death of a member of a listed species.

The extent to which endangered species put Oregon water rights at risk, or vice versa, is not well known. For one thing, the question of what the ESA means in terms of the exercise of water use permits and water rights in Oregon has not been definitively answered—which is to say, it has not been tested in court. However, analysis of the Act and court decisions leaves little doubt that in the event of a conflict with the ESA, state-issued water rights must yield. The State of Oregon's most visible water management response to the ESA in terms of future permits and rights has probably

Table 40. At-Risk Water-Dependent Species

Fish listed on either state or federal threatened or endangered species list

Common Name	Scientific Name	State Status	Federal Status
Snake River Chinook Salmon (Spring/Summer)	*Oncorhynchus tshawytscha*	T	T
Snake River Chinook Salmon (Fall)	*Oncorhynchus tshawytscha*	T	T
Lower Columbia River Coho Salmon	*Oncorhynchus kisutch*	E	
Columbia River Chum Salmon	*Oncorhynchus keta*	S	T
Oregon Coast Coho Salmon	*Oncorhynchus kisutch*	S	T
Southern Oregon Coho Salmon	*Oncorhynchus kisutch*	S	T
Upper Willamette River Steelhead	*Oncorhynchus mykiss irideus*	S	T
Lower Columbia River Steelhead	*Oncorhynchus mykiss irideus*	S	T
Middle Columbia River Steelhead	*Oncorhynchus mykiss gairdneri*	S	T
Snake River Steelhead	*Oncorhynchus mykiss gairdneri*	S	T
Snake River Sockeye Salmon	*Oncorhynchus nerka*	-	E
Upper Columbia River Spring Chinook Salmon	*Oncorhynchus tshawytscha*	-	E
Lower Columbia River Chinook Salmon	*Oncorhynchus tshawytscha*	S	T
Upper Willamette River Chinook Salmon	*Oncorhynchus tshawytscha*	-	T
Bull Trout	*Salvelinus confluentus*	S	T
Lahontan Cutthroat Trout	*Oncorhynchus clarki henshawi*	T	T
Oregon Chub	*Oregonichthys crameri*	S	E
Hutton Spring Tui Chub	*Gila bicolor* ssp.	T	T
Borax Lake Chub	*Gila boraxobius*	E	E
Warner Sucker	*Catostomus warnerensis*	T	T
Lost River Sucker	*Deltistes luxatus*	E	E
Shortnose Sucker	*Chasmistes brevirostris*	E	E
Bosket Spring Speckled Dace	*Rhinichthys osculus* ssp.	T	T

T = "Threatened"; E = "Endangered"; S = "Sensitive"

been the Water Resources Department's adoption of special public interest standards for processing new water use permit requests (OAR Chapter 690, Division 33; see Chapter 3). However, the weight of ESA falls not just prospectively, but on existing water users as well.

In 1992, a federal district court held that a fish screen problem with a California irrigation district's diversion represented a take under the ESA (788 F. Supp. 1126 (N.D. Cal, 10 March 1992)). The court ordered the district to cease pumping during fish migration, noting that an action does not have to be the only, or even over-riding, cause of harm to a listed

Table 40 continued. Aquatic Species Listed as Sensitive-Critical by State of Oregon *(excluding those listed opposite)*

Common Name	Scientific Name
Fish	
Goose Lake Lamprey	*Lampetra tridentata* ssp.
Sheldon Tui Chub	*Gila bicolor eurysoma*
Summer Basin Tui Chub	*Gila bicolor* ssp.
Goose Lake Sucker	*Catostomus occidentalis lacusanerinus*
Coastal Cutthroat Trout	*Oncorhynchus clarki clarki*
Coastal Steelhead	*Oncorhynchus mykiss* ssp.
Malheur Mottled Sculpin	*Cottus bairdi* ssp.
Amphibians	
Columbia Seep Salamander	*Rhyacotriton kezeri*
Northern Leopard Frog	*Rana pipiens*
Oregon Spotted Frog	*Rana pretiosa*

Source: Oregon Department of Fish and Wildlife 2004; Oregon Department of Fish and Wildlife 1997

species—if it contributes to the harm, it may be prohibited (Achterman 1992). In a 1997 case involving the State of Massachusetts, a federal court held that a state agency may be liable for an ESA take for authorizing private actions (i.e., through state permits) that could harm a listed species (*Strahan v Cox*, 127 F 3rd 155 (1st Cir 1997)) (Pagel 2002).

However, in 2001, a court found for the first time that irrigators may be entitled to compensation when federal project water deliveries are withheld for ESA purposes (*Tulare Lake Basin Water Storage District v. United States*, 49 Fed. Cl. 313 (2001)). These California irrigators had state-issued rights for water delivered through contracts with the Bureau of Reclamation. This decision, which hinged on specific language peculiar to these contracts, was later called into question by a court dealing with Oregon's Klamath basin water crisis.

The Klamath basin is perhaps the best example of what can happen when water management goes awry. It is victim to a perfect storm of over-allocation, overdue adjudication, deferred endangered species protections, and inter-state and inter-sovereign complexity. In short, it is a very Western place where too many promises were made for too long; where people in and outside the basin who depend on Klamath water for livelihoods or life are learning that the price of empty promises is economic hardship, uncertainty, and the rage that rises as tradition withers. (Much of the information that follows comes from the University of Colorado Center for Science and Technology Policy Research's Klamath Basin Project Web site.)

The Klamath was one of the first federal reclamation projects. Dams were built in the early 1900s to provide a stable water source and land was drained and made available for farming. Little or no attention was paid to the needs of tribal interests or fish species. For decades, the project achieved its relatively simple purposes that fit perfectly simpler times. But times changed. The Klamath tribe, despite being officially terminated in the 1950s (and re-established in the 1980s), never lost its legal right to (or its passionate interest in) water. Starting in 1979, court decisions into the early 2000s affirmed that the tribe was entitled to water with a priority date of "time immemorial" to support its fishing and hunting rights and traditions.

Adjudication began in 1975 and may actually be completed by 2008. During this generation-long process, the Klamath tribe asserted its water right interests, as did ultimately the United States (for the Forest Service, National Park Service, U.S. Fish and Wildlife Service, Bureau of Land Management, and others—though it initially sued the state because it did not want to be subject to a state adjudication) and about five thousand other claimants, as well. As the adjudication dragged on, the stakes rose. Who had water right seniority was no longer a pro-forma exercise in applied prior appropriation theory. Those with some of the oldest claims to water had decidedly different ideas about how the water should be used than those who had developed a 100-year habit of using it out-of-stream. The tribes were primarily interested in fisheries and the federal government was beginning to change from boosting farmirrigation to supporting instream uses for fish.

In 1988, the Lost River and shortnose suckers were federally listed as endangered species. In the drought years of 1992 and 1994, the Bureau of Reclamation, which operates the Klamath Project, had to cut back on water deliveries (for the first time ever) to maintain reservoir levels needed by these fish—triggering a law suit that wound up in the U.S. Supreme Court.

Irrigators sued the federal government, claiming that the bureau's decision violated part of the ESA that required consideration of economic impacts. They also held more generally that the actions violated the federal Administrative Procedure Act. In the ensuing court battles, the government held that the ESA allowed citizens to sue only on behalf of species, not to challenge protective actions—a position the U.S. Supreme

Court declared erroneous in a March 1997 decision. In *Bennet et al. v. Spear et al.*, the Court found there was no basis in the ESA or other federal law to confine to environmentalists alone the right to sue over endangered species provisions, essentially opening the door to more lawsuits from those feeling aggrieved by species protections.

The Klamath River once was the nation's third-most productive salmon river (Los Angeles *Times* 2006). But in 1997, coho salmon were listed as a threatened species. This added another level of complexity. Not only did lake levels have to be maintained, but downstream flows for coho, as well. Litigation by water users and environmental groups continued. In 2000, the Ninth Circuit ruled that the bureau was legally obligated "to meet the requirements of the ESA, requirements that override the water rights of the irrigators" (15 F. Supp. 2d 990, 996 (D. Or. 1998), 204 F.3d 1206 (9 Cir. 2000)).

Events reached a climax in 2001 with one of the most severe droughts of recent decades. Shortly after a drought was officially declared, the U.S. Fish and Wildlife Service and NOAA Fisheries Service released new findings that even higher lake levels and greater streamflows were needed to protect suckers and salmon than previously thought. Subsequently, a district court judge ruled the Klamath Project was in violation of the Endangered Species Act and could not deliver irrigating water. No deliveries of water from the Klamath Project to contract water irrigators occurred during the 2001 growing season, and crops withered (University of Colorado n.d.).

The local community was outraged, as were farming interests nationwide. Protests were held, including a "bucket brigade," where, in an act of civil disobedience, hundreds of people symbolically dipped and passed buckets of water out of the Klamath River. The national news networks covered the story. White House staff became involved and questions were raised about the sufficiency of the science behind the decisions. And, of course, lawsuits were filed. But in 2001, a federal district court ruled that Klamath Project water rights were subservient to the government's ESA requirements, and established the priorities for managing water as: 1) species listed under the ESA; 2) tribal trust obligations; 3) delivering irrigation water under contract; and 4) providing water for national wildlife refuges in the Klamath basin (145 F.Supp2d 1192 (D.C. Or. 2001).

A National Research Council assessment of the science leading to the decision to halt irrigation deliveries found insufficient support for a direct relationship between higher lake and river levels and fish survival. Accordingly, the federal government allowed full water deliveries in 2002, reducing the flow of the Klamath River (University of Colorado. n.d.). That fall, at least thirty-five thousand and perhaps as many as seventy thousand adult salmon died in the lower Klamath River—the first major adult salmon mortality event ever recorded for the Klamath. A report by California Department of Fish and Game attributed the loss to a disease outbreak compounded by a large run of salmon that may have been concentrated by the abnormally low flows (California Department of Fish and Game 2004). In 2006, the likelihood of low flows out of the KIamath and the potential for a similar fish-kill led to severe restrictions on off-shore salmon fishing over a 700-mile-long area off the Oregon and Northern California coasts to protect returning Klamath River chinook salmon.

In an attempt to meet water needs for the ESA-listed species, in 2001 the bureau began a water bank program, "through which willing buyers and sellers ... provide additional water supplies for fish and wildlife purposes and to enhance the tribal trust resources." The bureau funded two types of activities to supply the water bank. The first was paying farmers to idle irrigated lands. The second was for groundwater substitution, where pumped groundwater was used for irrigation that normally would have used surface water. While this helped with the water situation for fish and provided some relief for irrigators, the program also had a downside. Water bank activities resulted in about an eight-fold increase in groundwater pumping in the Klamath Valley and Tule Lake subbasins, which caused severe well interference at some locations, seasonal water table drops of 10 to 20 feet where pumping was most intense, and year-to-year declines of 2 to 8 feet over extensive areas around large pumping centers (U.S. Geological Survey 2005).

Adding insult to injury from an irrigation perspective, in 2005 PacifiCorp gave notice to state officials that, under its new license, it planned to raise its electric rates for power from its Klamath River hydro-projects. Back in the early 1900s, the government had required the company's predecessor to set a very low rate structure as a condition of its license to build the dams. In 2006, the Oregon Public Utility Commission approved an approximately ten-fold rate increase to take effect over a seven-year period. Because irrigators depend on electricity to run surface water pumps and groundwater wells, and often operate on slim profit margins, this rate increase represents yet another challenge to continuing operation (Klamath Basin Crisis 2005; Oregon Public Utility Commission 2006).

In 2005, the federal claims court held that irrigators could not claim that a property rights taking had occurred when irrigation was shut off in 2001.

Rather, any claim of injury needed to be addressed under contract law; and the court observed that many problems would need to be overcome by the plaintiffs to be successful. Further, the court distinguished its decision from the 2001 "Tulare Lake" case, where under similar circumstances, irrigators had prevailed, noting "the decision appears to be wrong on some counts [and] incomplete in others" (*Klamath Basin Irrigation Dist. V. United States*, 67 Fed. Cl. 504 (2005)).

In summary, the Klamath seems to offer a water management drama of operatic proportion. Not having enough water has split salt-of-the-earth workers into two camps, pitting upper basin farmers against coastal salmon fishermen. The once-suppressed interests of Native Americans and fish have rebounded, finding a volume and venue sufficient to reshape the water contours of the entire West. Conservative, law-and-order types found themselves engaging in acts of civil disobedience to protest the change. The White House intervened in events. And an arcane hydroelectric process emerged as a critical determinant of the basin's future economy. In short, there's no telling what will happen when push finally comes to shove in water conflicts.

Oregon's experience with ESA in terms of water management is not confined to the Klamath. The management of the Columbia and Snake river dams for salmon recovery has received much attention, with issues ranging from whether to breach any of the dams, to spilling water to increase flows for migrating juvenile salmon (at the cost of lost power generation and revenues), to hazing sea lions that have come to view fish ladders as buffet tables.

In addition, the Willamette basin flood-control dams operated by the U.S. Army Corps of Engineers have been found by NOAA Fisheries and the U.S. Fish and Wildlife Service to likely adversely affect a number of species federally listed as threatened or endangered: spring chinook salmon, steelhead trout, bull trout, and Oregon chub (a minnow). As of 2006, these agencies are continuing to develop requirements the Corps will have to meet to avoid jeopardy to the listed fish. Likely measures include providing new or improved fish passage facilities (many of the dams cut off fish migration altogether), temperature control facilities, and adjusting the storage and release schedule to better serve fish needs (U.S. Army Corps of Engineers 2000). The latter will be felt by more than fish, especially during dry years. In the drought year of 2001, the Corps essentially drained Detroit Reservoir in the summer to meet flows required for fish and water quality. Boat ramps were left high and dry and tourism plummeted. That the reservoir could simply go away, that there was no guaranteed water level or right to have the water there, came as a great shock to businesses and tourists, alike (*Oregonian* 2001).

Other examples of Oregon's brushes with the ESA over water management include a negotiated settlement between NOAA Fisheries and the Grants Pass Irrigation District over its diversion facility, Savage Rapids Dam on the Rogue River. NOAA Fisheries had filed suit against the district for take of listed coho salmon resulting from inadequate fish passage and fish screening at the dam. The district ultimately agreed to switch to pumps to divert water and remove the dam—an $18 million proposition that was authorized (but not provided) in the fiscal year 2005 federal budget (Pagel 2002; Mail Tribune 2005). In 2000, the U.S. Fish and Wildlife Service notified several irrigation districts in the upper Walla Walla River basin in Oregon that their customary diversion practices dewatered the river and harmed bull trout, a federally listed species. The districts reached a negotiated agreement with the Service and other interested parties to avoid immediate enforcement action by sponsoring near-term stream flow enhancement efforts and agreeing to develop a long-term conservation plan (Pagel 2002).

Streambed and Wetlands Management

▣ Streambed Ownership

If the public owns all the water from all sources of supply (ORS 537.110), who owns Oregon's watercourses, the channels through which the water flows? For many waterways in Oregon, the answer is the same: the public. When Oregon was admitted to the Union in 1859, like other states before and after, it was granted title to all lands underlying navigable waterways and tidal waters (the legal doctrines defining such title are referred to as "navigability" and "tidality," respectively). Whereas tidal waters are fairly easily identified, the same has not been true of navigable waters. Navigability has been defined through federal court cases to include Oregon streams that in their natural condition were actually and commonly used for, or could have been used for, transporting goods or people. The term navigability may conjure up images of river-boat captains negotiating tricky sandbars, but it means both more and less than that. For example, the courts have held that Native American canoes, fishermen's drift boats, ferries, and even log drives are sufficient evidence of navigability (Oregon Division of State Lands 1996a).

Determining streambed ownership through navigability determinations often depends on historical analyses of travel and commerce. The photograph of these hunters canoeing a Coast Range stream in the 1800s would offer convincing evidence that this particular stream was indeed navigable. (Oregon Historical Society, L.G. Davidson, OrHi4371)

Thus, a stream is navigable in the same sort of way a mountain is climbable: if someone is known to have done it, it is; if no one knows for sure, it still may be. Deciding who owns a streambed is the responsibility of Oregon's Department of State Lands. This state agency conducts the research necessary to determine ownership. However, it can only do so if directed by the State Land Board (the Governor, the Secretary of State and the State Treasurer), which must first find there is a broad public interest at stake. The department must begin any study with public notice, then draft a report and hold public hearings. If the Land Board accepts a department recommendation to assert an ownership claim, it must provide public notice and contact each affected landowner. The board's assertion, considered an order in other than a contested case, is appealable to local circuit courts (ORS 274.400–274.412).

The state has compiled navigability information for three hundred river segments (Oregon Division of State Lands 1995a) but has formally asserted ownership claims on only twelve rivers (Oregon Department of State Lands n.d.[a]) (Figure 28). In addition, the department manages lands underlying the Territorial Sea (which extends 3 miles offshore), tidal lands of many rivers along the coast (including the Nehalem, Alsea, Coos, Siuslaw, Umpqua, Rogue, and Chetco) and a number of lakes, such as Devils, Klamath, Siltcoos, Tenmile, Upper Klamath, and Woahink.

In all, the department manages over 800,000 acres of navigable waterway and tidal holdings. Ultimately, only courts can decide streambed ownership

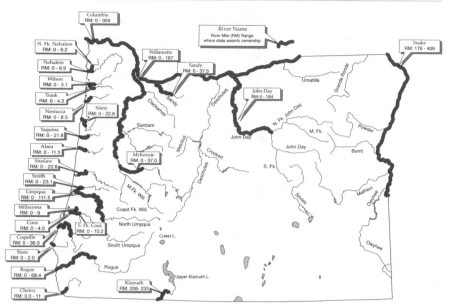

Figure 28: River beds and banks claimed by state (not all waterways where state asserts ownership are shown; other rivers included for locational reference only). (Source: Information from Oregon Division of State Lands 1996b; Oregon Department of State Lands. n.d.[a].)

questions. If the courts say the state is the owner, public title extends from the bed, up the banks, and all the way to the normal high-water line. Lands defined by this boundary are referred to as being "submerged and submersible." The determination of state ownership has several important impacts.

First, it means that the public has a right to use (e.g., walk around on, picnic on top of, fish from) the beds and banks. Even if a stream is not navigable and the bed is privately owned, the courts have held the public still enjoys a "floatage easement" that allows people to float over and make "reasonable, incidental use" (which the courts did not define) of the bed.

Second, it may mean that some landowners need to take a second look at their property deeds and tax records. The deed may say that private ownership extends to the middle of the stream, but if the court has decided the waterway is navigable, it means the state has been the one, true owner all along and the deed is likely mistaken. If taxes have been paid on the streambed, the landowner may be able to recover them (Oregon Division of State Lands 1996a). It should be noted that on some rivers (Willamette, Coquille, Coos, and Umpqua), past legislatures permitted state title to be conveyed to adjacent landowners, and therefore private ownership may include submerged and submersible lands in some locations (Oregon Division of State Lands 1996b).

Third, use of state property normally comes at a price, and use of streambeds is no exception. Under Oregon's Constitution, the Land Board must manage state-owned lands for "the greatest benefit for the people of the state, consistent with the conservation of this resource under sound techniques of land management" (Oregon Constitution, Article VIII, Section 5). Thus, while it must be balanced with the greater public interest, obtaining revenues from these lands is an important state objective— especially since all moneys are earmarked for Oregon's schools. The state generally requires people using its submerged or submersible lands to obtain a lease or easement (ORS 274.040–274.994). Activities or structures requiring a lease include marinas and moorages, log rafts, sand and gravel operations, hydroelectric facilities, houseboats, fish processing, and ship yards. Some uses do not require leases, such as wharves; navigation aids placed by public agencies; and various structures, piers, docks/floats owned, operated by or for a government agency; and non-commercial, private-use docks, floats, floating recreational cabins, and water sport structures. In addition, the owner of land abutting state-owned submersible land receives a preference in any lease offered by the state. The department currently has about four hundred waterway leases on over thirty waterways (Oregon Division of State Lands 1995a; Oregon Department of State Lands. n.d.[b]).

▣ Fill and Removal

The Department of State Lands also has been given responsibility for regulating material put into or removed from state waters. Oregon's Removal-Fill Law (ORS 196.795–196.990) was enacted in 1967. It requires a state permit for removing, filling, or altering more than 50 cubic yards of material within the beds and banks of most of the waters of the state. These waters include natural waterways, perennial and intermittent streams, wetlands and the Territorial Sea (ORS 196.800). There are a variety of permit exemptions, including emergency work, provided notice is given the state (ORS 196.810); dams or diversions allowed under a water use permit or water right (ORS 196.905); a variety of customary agricultural activities (ORS 196.905); or forestland activities regulated under the state's Forest Practices Act (ORS 196.905). However, the permitting threshold drops from 50 cubic yards to zero on state scenic waterways (except for recreational prospecting) (ORS 390.835) and streams that provide critical salmon spawning habitat (ORS 196. 810). State scenic waterways are more fully described below. Critical salmon waters are primarily in the coastal drainages and the Clackamas, Hood, Grande Ronde, and Imnaha river basins.

▣ *Wetlands*

As noted above, "waters of the state" subject to Oregon's fill-and-removal law include wetlands. Oregon has roughly 1.4 million acres of wetlands today, compared to about 2.4 million acres when it became a state in 1859 (Oregon Division of State Lands 1995b). Wetlands are especially important for Oregon's biodiversity. Thirty percent of Oregon's terrestrial vertebrates (164 species) depend on freshwater marshes, and Oregon's estuaries are used by more than one hundred bird species and thirty-five species of fish and shellfish. Oregon's wetlands also support many rare plants, as well as sensitive, threatened, and endangered species such as coho salmon, spotted frogs, western pond turtle, and sandhill crane. Wetlands also help regulate water flows, reduce flooding, and improve water quality (Oregon Habitat Joint Venture 2004).

Oregon's policy is to "Promote the protection, conservation and best use of wetland resources, their functions and values through the integration and close coordination of state-wide planning goals, local comprehensive plans and state and federal regulatory programs" (ORS 196.672). Furthermore, the legislature has stated that "Wetland management is a matter of this state's concern since benefits and impacts related to wetland resources can be international, national, regional and state wide in scope" (ORS 196.668).

The Department of State Lands is responsible for Oregon's wetland programs. It maintains a comprehensive wetlands inventory and establishes the state's uniform system of wetland identification. Oregon also encourages the adoption of wetlands conservation plans by local governments. These plans inventory local wetlands, assess their values, establish protective policies, specify mitigation measures, and provide for monitoring. Wetland conservation plans are implemented through policies and ordinances in state-mandated local comprehensive land use plans (Metropolitan Service District 1991).

Perhaps befitting a state whose residents are known as "webfoots," Oregon has over 1.4 million acres of wetlands, ranging from coastal bogs (right), to the expanses of eastern Oregon desert marshes (below). Still, it is estimated that the state has lost 1 million wetland acres since statehood in 1859. (top photo: Oregon State Archives, Oregon Highway Division, OHD2506; lower photo : © Photographer: Robert Brown Agency: Dreamstime.com)

The department works closely with local governments and landowners to reduce impacts of development, with a goal of no net loss of wetlands. Mitigation is required as a condition of any state permit that allows people to place fill or excavate in a wetland. In this case, "mitigation" means creating, restoring, or enhancing wetlands to replace or compensate for the wetland area and functions lost through the development (Oregon Department of State Lands 1999). One important and relatively new way of providing mitigation is through wetlands mitigation banks.

A mitigation bank is a wetland site created or restored by a public or private entity to establish wetland value credits. The credits represent biological values translated into dollars. If approved by the department, instead of trying to make up for wetland impacts elsewhere on their property, people may literally buy mitigation by purchasing credits from the bank. The first mitigation bank was established in Astoria during the 1985 biennium with a federal grant. Legislation passed in 1995 allowed private mitigation banks to be established under rules adopted by the Land Board (Oregon Legislative Fiscal Office n.d.[a]). As of 2006, there were eighteen wetland mitigation banks in the state (Oregon Department of State Lands n.d.[c]).

Removal and fill activities are also regulated by the federal government, primarily under the authority of Section 404 of the Clean Water Act. This section requires authorization from the U.S. Army Corps of Engineers for discharging material into all waters of the United States, including wetlands. The federal Environmental Protection Agency also has authority under federal water pollution laws to approve or deny any "404" permit. The department has worked closely with the Corps in developing a joint permitting authority.

In 1995, the legislature directed the department to streamline the removal-fill permit process by eliminating duplication between state and federal fill and removal programs. After extensive coordination between state and federal partners, the department and the U.S. Army Corps of Engineers initiated the Statewide Programmatic General Permit process as a two-year pilot program in 2006. The program is intended to make the state and federal wetland and waterway removal-fill permit process a "one-stop shop" for those applying for a permit. In addition to the state's removal-fill law and the federal Clean Water Act (Sections 404 and 401), at least eight other major state and federal statutes and programs are integrated in the program, including such institutional behemoths as Oregon's scenic waterway law, the federal Oregon Coastal Management

Program, the federal Rivers and Harbors Act, and elements of the National Environmental Policy Act. Under the program, the department is the lead agency and could authorize activities for up to 0.5 acre of wetland fill and/or up to 1,000 cubic yards of fill or removal in all waters (Oregon Department of State Lands. n.d.[d]).

River Corridor Protection

There are a number of stream protection approaches in Oregon, including the Willamette River Greenway program (ORS 390.310–390.368), riparian area protections under the Oregon Forest Practices Act (ORS Chapter 527), federal land management plans (including the bi-state effort to manage the 300,000-acre Columbia Gorge National Scenic Area (16 USC, Sec. 544)), and an increasing number of local initiatives growing out of Oregon's watershed management activities (Chapter 11). However, two that stand out in terms of water management impacts are the fraternal twins of river corridor protection: Oregon's scenic waterway program and the federal government's wild and scenic river program.

Oregon has over 1,200 miles of rivers protected under state and federal scenic programs. Most designated rivers, like the Rogue River shown near Prospect above, are assigned complementary state and federal protection. (Oregon State Archives, Oregon Highway Division, OHD0877)

▣ *State Scenic Waterways*

The Oregon Parks and Recreation Department is responsible for the state's scenic waterway program, established by a 2-to-1 vote of the people in 1970 to protect free-flowing rivers. The Scenic Waterways Act targets rivers, lakes, and adjacent lands that "possess outstanding scenic, fish, wildlife, geological, botanical, historic, archeologic, and outdoor recreation values of present and future benefit to the public" (ORS 390.815). The Act prohibits dams, reservoirs, other impoundments, placer mining (though recreational prospecting is allowed), and most new diversion facilities in order to maintain the free-flow of water in quantities necessary for recreation, fish, and wildlife, which are declared the highest and best uses in scenic waterways (ORS 390.835). (The superiority of these three uses has had significant impacts on water use permitting, as explained in Chapter 3.) Scenic water designation, however, does not affect the use of existing water rights, allow public use of private property without landowner permission, or mandate the removal of existing developments.

A scenic waterway includes the stream and all land and tributaries within one-quarter mile of its banks (ORS 390.805). There may be no filling or

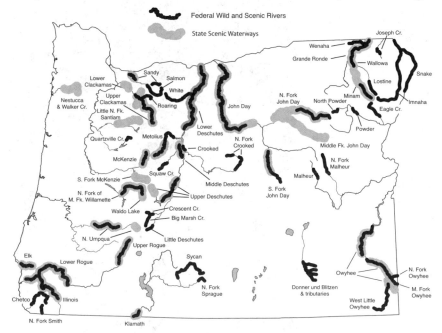

Figure 29: Scenic waters designations. (Source: Information from Oregon Parks and Recreation Department)

altering of streambeds, or removal of streambed material without approval from the Department of State Lands (ORS 390.835). New development or changes to existing land uses must be reviewed by OPRD before taking place. Each scenic waterway has specific management rules (OAR 736-40-0041 through -0078) assigning up to six management classifications. These allow land use intensities that essentially range from primitive to practically suburban. OPRD must be notified of proposed new development projects or changes in existing uses, which are then either approved or denied. If the proposal is denied, OPRD opens negotiations with the landowner. If these are not successful, the agency can make an offer on the property. Otherwise, the proposed use may begin within one year of the notification.

Scenic waterways may be established in three ways: adoption by the governor; designation by the legislature; or vote of the public on an initiative. By far the most common method has been by initiative—Oregon's scenic waterway system nearly doubled in size with the passage of Ballot Measure 7 in 1988. There are now nineteen river segments and one lake covering nearly 1,100 shoreline miles. Figure 29 displays scenic waterway locations (Oregon Parks and Recreation Department. n.d.).

▣ *Federal Wild and Scenic Rivers*

The federal wild and scenic river program parallels the state's program in many ways. The goals of both programs are similar, but the mechanics of protection differ. The Wild and Scenic Rivers Act was passed in 1968 as Public Law 90-542 to protect rivers that possess "outstandingly remarkable scenic, recreational, geologic, fish and wildlife, historic, cultural or other similar values." There are two ways rivers can be added to the Wild and Scenic River System: they can be studied by a federal agency and designated by Congress, or recommended by a state and designated by the Secretary of the Interior. All but two of Oregon's wild and scenic rivers have been designated by Congress. Passage of the 1988 Omnibus Rivers bill made Oregon tops in the number of designated rivers in any state and third in the number of miles protected (Root 1989). Portions of the Klamath River were added to the system by Secretary of Interior Bruce Babbitt in 1995 and the Wallowa River in 1996, in response to earlier requests by Governor Barbara Roberts.

Instead of a rigid corridor boundary (such as the state's one-quarter mile buffer), the Act originally capped designated lands at an average of 320 acres per mile on both sides of the river. The river corridor, which may expand

and contract along its length, is managed by either the U.S. Department of Agriculture or the U.S. Department of the Interior, depending on which agency manages the most land in the corridor. In Oregon, the management of the federally designated rivers falls almost exclusively to the Forest Service and the Bureau of Land Management. If corridor lands are primarily privately owned (a rarity in Oregon), management may occur through coordination with local zoning, jurisdictional transfer between agencies, land donations, or condemnation. In Oregon, there is also considerable reliance on the state scenic waterway program for dually designated rivers—most scenic waterways are also federal wild and scenic rivers for all or a portion of their length (Figure 29).

The major protection given to rivers arises from Section 7 of the Act, which prohibits the Federal Energy Regulatory Commission from licensing any dam, water conduit, reservoir, power house, transmission line, or other project work that directly affects any designated river. Furthermore, the Act precludes any federal agency from constructing or assisting any water resources project that would directly and adversely affect designated rivers.

Just as the effects of the state's Scenic Waterway Act on water allocation decisions were not fully appreciated when waterways were first designated, there may be also be unanticipated impacts from wild and scenic designations. The reason: each designation carries with it an implied federal reservation for the amount of unappropriated water needed to protect the river. The Wild and Scenic Rivers Act includes specific water right language that twists and turns but finally backs into an assertion of federal interest in streamflows:

> Designation of any stream or portion thereof as a national wild, scenic or recreational river area shall not be construed as a reservation of the waters of such streams for purposes *other than those specified in this chapter, or in quantities greater than necessary to accomplish these purposes.* ...
>
> The jurisdiction of the States over waters of any stream included in a national wild, scenic or recreational river area shall be unaffected by this chapter *to the extent that such jurisdiction may be exercised without impairing the purposes of this chapter or its administration.*
>
> (16 U.S.C., Section 1284 (b)–(d), emphases added)

That federal water rights are created along with the designation of a wild and scenic river (unless specifically waived) is backed up not only by the nature of the emphasized qualifications above, but also by a clear legislative history. These rights are for water not yet allocated under state water rights and therefore subject to senior rights. The amount of the right would vary according to levels of previous appropriation and the values for which the river was designated (Baldwin 1990).

For many wild and scenic rivers, however, a federal water claim of this sort would be mooted either by their remote headwater locations (where there would be no upstream water users for the federal government to demand water from) or by the over-appropriated nature of the streams at the time of designation (where senior users already have superior rights to most or all of the flow). Nevertheless, it represents something of a blank check drawn against Oregon waters that warrants quantification to settle the books.

The U.S. Forest Service and the Bureau of Land Management filed reserved right claims in the Klamath Basin Adjudication for instream flows to support recreation, boating, scenic enjoyment, and fishlife pursuant to the Wild and Scenic River Act's designation of rivers in the Klamath basin. The U.S. Forest Service also filed for instream flows to support purposes under the Wilderness Act, including recreational, scenic, scientific, educational, conservation, and historic uses, as well as instream flows related uses under the Multiple Use Sustained Yield Act (Oregon Water Resources Department n.d.[g]).

Recreation

No treatment of Oregon's streams would be complete without some mention of Oregonians' instinctive urge to play in the water. While this urge to play does not usually have a direct affect on water supply management (except the potential for instream rights to protect recreational stream flows and the pressure applied to reservoir managers to provide for both slack- and fast-water recreation; see Chapter 11), it certainly makes for a close relationship between the people and their waters. State law declares it a matter of public interest to increase outdoor recreation opportunities by developing waterways and water facilities and providing access to public waters having recreational values (ORS 390.010). The Oregon Parks and Recreation Department is the primary agency in carrying out this policy.

In response, Oregon has designated nearly 1,100 miles of state scenic waterways (see River Corridor Protection, above) and established a system of parks and protections along its largest river through the Willamette River Greenway (ORS 390.310—ORS 390.368). In addition, a significant number of Oregon's 231 state parks are located along streams and lakes.

According to a 2001 survey conducted by OPRD, Oregonians spent over nineteen million user days each year engaging in outdoor water recreation (e.g., fishing, canoeing, boating). In 2003, the Oregon Department of Fish and Wildlife sold over 460,000 annual adult resident and non-resident fishing licenses. ODFW estimates that in 2001, angling generated over $364 million in economic value and supported nearly thirteen thousand jobs (Oregon Department of Fish and Wildlife. n.d.[c]; Oregon Department of Fish and Wildlife. n.d.[d]). Between 1987 and 2002, fishing activity increased 44 percent, while non-motorized boating (including ocean recreation) increased 138 percent. One of the eight "top statewide outdoor recreation issues" identified by OPRD is the need for more water-based recreation resources and facilities. There is a growing need for increased access for motorized and non-motorized recreational water activities, particularly in rapidly growing areas of the state. Needed resources include boating and angling opportunities near residential areas, boat ramps and camping facilities for motorized boating, and non-motorized boating

Oregonians have long enjoyed a close relationship with their waters. The Deschutes River has been a treasured trout stream for generations. (Photo by Richard Frank Bastasch)

facilities for canoes, kayaks and drift boats. (Oregon Parks and Recreation Department 2003)

There are two hundred thousand registered motor and sailboats in Oregon and an estimated five hundred thousand canoes, rafts, kayaks, and drift boats. The Oregon State Marine Board is charged with regulating the state's recreational boaters, as well as licensing its guides (ORS 830.110). In addition to titling and registering Oregon's motorized recreation vessels, the board sets boating regulations; trains and contracts with county sheriffs and the State Police for boating safety and law enforcement; sponsors boating-education courses and water-safety programs; and awards grants to develop and maintain boating facilities and protect water quality. In addition, in cooperation with SOLV (a non-profit organization founded in 1969 by Governor Tom McCall and other community leaders to enhance Oregon's livability) the board co-sponsors Oregon's Adopt-A-River program. This program brings together volunteers, businesses, and government agencies to clean up Oregon's streams. Individuals or organizations commit to a rule of "two, two, two," promising to cleaning up a 2-mile stretch of stream, twice a year, for two years. More than 700 miles of river have been adopted by over three thousand volunteers representing individuals, families, businesses, school groups, sports clubs, and civic groups (SOLV 2004).

CHAPTER 10

Groundwater Safeguards

ACCORDING TO THE OREGON DEPARTMENT OF Environmental Quality, 70 percent of Oregonians, including over 90 percent of rural residents, rely on groundwater as their primary or secondary drinking water source (Oregon Legislative Fiscal Office n.d.[a]). And that statistic does not even touch on the use of groundwater for agriculture or industry. Obviously, groundwater is of the utmost importance to the health of Oregon's individuals, communities, and economies. Beyond any safeguards provided in the permitting process for groundwater use, the state uses two major approaches for protecting that resource. The first has to do with the surprisingly technical field of well management. The second deals with protecting groundwater quality through preventive programs applied over a broader geographic area.

Well Management

Just how complicated can digging a hole in the ground get? It turns out it can be pretty complicated, especially if the goal is to protect the waters underlying Oregon. Wells puncture the skin of the earth. Like needle-sticks, they offer an avenue both for contamination and fluid loss. Poorly constructed wells can allow dirty surface water to drain into aquifers. They can also lead to "commingling," where wells punched through upper water-bearing zones drain water down the well hole to much lower levels—often to the chagrin of nearby well-owners. To maintain the integrity of the groundwater supply, wells have to be built correctly.

That is why the legislature has declared, "constructing new wells or altering existing wells is ... a business or activity affecting the public welfare, health and safety" (ORS 537.765). The Oregon Water Resources Department oversees the laws and regulations on well construction (ORS 537.747–537.780). (The common term, "well drilling," is avoided as too limiting, given that wells may also be driven, jetted, bored, or hand-dug.)

The department is aided in this effort by the Ground Water Advisory Committee. GWAC advises the agency on legislation and public policy for the development and protection of ground water, licensing of well constructors, and certain aspects of the budget. At least three of its nine members must be active in the well-construction industry (ORS 536.090).

The department has adopted ten divisions of rules regulating water wells, monitoring wells, geotechnical borings, and so-called "other holes" (OAR Chapter 690, Divisions 200–240). These regulations address constructing new wells or altering, abandoning, or maintaining existing wells. Wells are regulated primarily through activity notification and reporting, and construction standards.

▣ Water Wells

In Oregon, a well is "any artificial opening or artificially altered natural opening, however made, by which groundwater is sought or through which groundwater flows under natural pressure or is artificially withdrawn" (ORS 537.515). Oil, gas, and some high-temperature geothermal wells are not included in this definition and are not subject to regulation by the Water Resources Department.

WELL CONSTRUCTOR LICENSING

The most common way to have well work done is to pay someone to do it. In Oregon, anyone who advertises well-construction services, enters into contracts to construct wells, or operates well-drilling machinery has to have a well constructor's license. To obtain a license, a person: has to be eighteen; pass a written test covering laws, basic groundwater knowledge, and technology; have at least one year of experience operating well-drilling machinery; and pay the examination and license fees set in statute. The examination fee is $20. License fees are $50 for one year or $200 for five years. The department may (subject to contested case hearing) revoke, suspend, or refuse to renew licenses if the licensee violates any rules. Lastly, to assure compliance with all well-construction regulations, before constructing a well, the licensee must obtain a $10,000 bond or letter of credit. The Water Resources Department or any person injured by a constructor has a right of action on the bond or letter of credit (ORS 537.753).

Some landowners choose to do their own well work. Unlike most do-it-yourself projects where damage from goof-ups is limited to an owner's household, wallet, or ego, self-drilled wells can affect neighbors and risk the resource itself. That is why do-it-your-selfers must first get a $25 permit from the Water Resources Department, obtain a bond or letter of credit for $5,000, and give ten days' written notice before sealing their well (packing clay, concrete or other material around the well hole to keep it intact and protected) (ORS 537.753).

START CARDS AND WELL LOGS

With the exception of the landowner permit described above, there is no such thing as a state well-drilling permit—though some counties have adopted local siting requirements. Most wells can go in practically anywhere at anytime. All that is required are two notices to the Water Resources Department—one before work begins and one when done. The former, called a "start card," must be submitted in advance of any well construction, alteration, or abandonment. If submitted by a licensed well constructor, the start card must be accompanied by $125. If for a new well, the notice must include the proposed well's expected depth, diameter, and purpose. The start card must be in the hands of the department no later than the day the work begins (OAR 690-205-0200). This allows the department to conduct sample inspections to see if wells are being built "to standard." The department usually inspects 25 to 30 percent of all new wells. Each of the agency's five regions has a well inspector. In 2004, about 13 percent of the new water supply wells were found deficient and required repairs (Oregon Water Resources Department 2005[a]).

The completion notice is a well log (technically, a "water supply well report") that must be submitted within thirty days of completion for all water wells—even dry holes (Figure 30). The log must show all rock formations encountered and all materials used in constructing the well (OAR 690-205-0210). Well logs have been required since 1955, though their submittal was not routine until the late '50s. All told, the department has received about three hundred thousand well logs (Figure 31). According to well log records, since 1990 about sixty-four hundred water wells have been drilled each year (Oregon Water Resources Department n.d.[h];Oregon Water Resources Department 2005b).

It is important that the consumer understand that neither a start card nor a well log is a water right application, permit, or any other kind of authorization to use the water. If the well is for a small domestic or other exempt groundwater use (Chapter 2), this will not be a problem, since no use authorization is required. But for bigger community, commercial, or agricultural systems, water use permits are essential. Often people are lulled

09/03/24

(START CARD) # 80474

STATE OF OREGON
WATER SUPPLY WELL REPORT
(as required by ORS 537.765)
Instructions for completing this report are on the last page of this form.

(1) OWNER: Well Number 2861
Name
Address Meadowood Rd SE,
City Jefferson State OR Zip 97352

(2) TYPE OF WORK
[X] New Well [ ] Deepening [ ] Alteration (repair/recondition) [ ] Abandonment

(3) DRILL METHOD:
[X] Rotary Air [ ] Rotary Mud [ ] Cable [ ] Auger
[ ] Other

(4) PROPOSED USE:
[X] Domestic [ ] Community [ ] Industrial [ ] Irrigation
[ ] Thermal [ ] Injection [ ] Livestock [ ] Other

(5) BORE HOLE CONSTRUCTION:
Special Construction approval [ ] Yes [X] No Depth of Completed Well 225 ft.
Explosives used [ ] Yes [X] No Type ____ Amount ____

HOLE			SEAL			
Diameter	From	To	Material	From	To	Sacks or pounds
10	0	19	Bentonite	0	19	8 sacks
6	0	225	Bore			

How was seal placed: Method [ ] A [ ] B [X] C [ ] D [ ] E
[ ] Other
Backfill placed from ____ ft. to ____ ft. Material ____
Gravel placed from ____ ft. to ____ ft. Size of gravel ____

(6) CASING/LINER:
	Diameter	From	To	Gauge	Steel	Plastic	Welded	Threaded
Casing:	6	1	19	250	[X]	[ ]	[ ]	[ ]

Liner: NONE
Final location of shoe(s) NONE

(7) PERFORATIONS/SCREENS:
[ ] Perforations Method ____
[ ] Screens Type ____ Material ____
From	To	Slot size	Number	Diameter	Tele/pipe size	Casing	Liner
NONE							

(8) WELL TESTS: Minimum testing time is 1 hour
[ ] Pump [ ] Bailer [X] Air [ ] Flowing Artesian

Yield gal/min	Drawdown	Drill stem at	Time
75 gpm		225	1 hr.

Temperature of water 56 Depth Artesian Flow Found ____
Was a water analysis done? [ ] Yes By whom ____
Did any strata contain water not suitable for intended use? [ ] Too little
[ ] Salty [ ] Muddy [ ] Odor [ ] Colored [ ] Other ____
Depth of strata: ____

(9) LOCATION OF WELL by legal description:
County MARION Latitude ____ Longitude ____
Township 9 N or (S) Range 3 E or (W) WM.
Section 24 1/4 ____ 1/4 ____
Tax Lot 4400 Lot ____ Block ____ Subdivision ____
Street Address of Well (or nearest address) ____

(10) STATIC WATER LEVEL:
112 ft. below land surface. Date 8-18-95
Artesian pressure ____ lb. per square inch. Date ____

(11) WATER BEARING ZONES:
Depth at which water was first found 207

From	To	Estimated Flow Rate	SWL
207	208	75 gpm	112

(12) WELL LOG:
Ground Elevation ____

Material	From	To	SWL
Topsoil	0	1	
Brown Clay	1	3	
Brown Sandstone	3	7	
Blue Sandstone	7	225	112

Date started 8-17-95 Completed 8-18-95
(unbonded) Water Well Constructor Certification:
I certify that the work I performed on the construction, alteration, or abandonment of this well is in compliance with Oregon water supply well construction standards. Materials used and information reported above are true to the best of my knowledge and belief.
Signed ____ WWC Number 1279 Date 8-22-95

(bonded) Water Well Constructor Certification:
I accept responsibility for the construction, alteration, or abandonment work performed on this well during the construction dates reported above. All work performed during this time is in compliance with Oregon water supply well construction standards. This report is true to the best of my knowledge and belief.
Signed ____ WWC Number 514 Date 8-22-95

ORIGINAL & FIRST COPY-WATER RESOURCES DEPARTMENT SECOND COPY CONSTRUCTOR THIRD COPY-CUSTOMER

Figure 30: Well log example. A water supply well report ("well log") is an important information source for property-owners, developers, local governments and groundwater specialists. In this example, one can learn that a new well (Sec. 2) was drilled 225 feet deep (Sec. 5), and hit water at 207 feet, which then rose to 112 feet where it stabilized (Sec. 10). The yield is 75 gallons per minute, as estimated by an air test (Sec. 8) from "blue sandstone" (Sec. 12). Prospective buyers might be especially interested in logs from surrounding property, with an eye toward seeing how many deepenings there have been (Sec. 2), the general behavior (e.g., declines) of static water levels through time (Sec. 10), and the method of well yield tests (actual pumping over a period of hours with drawdown recorded is preferred).

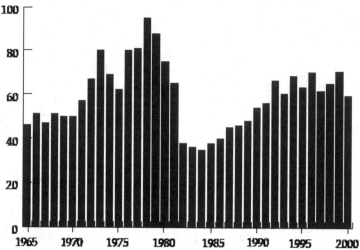

Figure 31: Well logs received (in thousands). Since 1955, the state has required a report be submitted for every water well drilled in Oregon. This graph shows the number of "well logs" received for water wells (not monitoring or geotechnical wells) in the last quarter century. On the average, about 5,000 water wells are drilled each year statewide. (Sources: 1965-1981: Oregon Water Resources Department 1993b; 1982-1990: Oregon Water Resources Department 1996c; 1991-2000, Oregon Water Resources Department n.d.[h])

into a false sense of bureaucratic security when given their official-looking customer copies of the well paperwork. Careful reading and contact with the Water Resources Department are recommended for the discriminating well owner.

WELL CONSTRUCTION STANDARDS

Well construction standards are the backbone of Oregon's well regulations. As the building code of wells, they are extremely detailed and technical. However, in general terms, their primary purpose is assuring that every well is built to fit local conditions, thus preventing adverse groundwater movement or contamination (OAR Chapter 690, Division 210).

The standards summarized below have been grouped by topic and selected from among the many delineated in OAR Chapter 690, Divisions 210 and 215. In addition to these, the Oregon Water Resources Commission may impose special area standards to address site-specific needs (OAR Chapter 690, Division 240).

QUALITY STANDARDS. The business of securing a clean supply of water is subject to the same rules of success as other businesses: location, location, location. Unless the Water Resources Department gives written permission, no wells may be constructed within:
- 50 feet of any septic tank, closed sewage or storm drainage system;
- 50 feet of a confined animal feeding or holding operation or animal waste holding facility;
- 100 feet of a septic drain line or sewage or sludge disposal area; or
- 500 feet of a hazardous waste storage, disposal, or treatment facility

If the first line of defense in keeping groundwater clean is good siting, the second is good design. Casing and sealing wells is one of the most important protections against both contamination and commingling of water. Though there are exceptions, most wells must be cased and sealed to at least a depth of 18 feet. Casing is material (e.g., steel or plastic pipe or tubing) that supports the sides of the well hole and prevents caving. The casing must extend at least 12 inches above the ground surface, which should be graded to drain surface water away from the well. Sealing refers to water-proofing the space between the well hole sides and the casing to prevent the inflow of surface water, shallow groundwater, and contaminants; or to prevent outflow of artesian groundwater (where groundwater has enough pressure to migrate upward and bubble out of wells or springs). Sealing material may be cement grout, concrete, or bentonite (OAR Chapter 690, Division 210).

All water used in well construction must be potable and all wells unattended during construction must be covered. Every new or altered water well and all pumping equipment, sand, gravel, and above-water-table casing must be disinfected (OAR Chapter 690, Division 210). Agricultural pesticides and fertilizers may not enter a well—back-siphon prevention devices are required for irrigation systems that apply fertilizers or chemicals through their lines (OAR Chapter 690, Division 215).

QUANTITY STANDARDS. Every well must be tested for yield (the rate at which the well gives up its water) and drawdown (the distance the water level drops in the well during the test). These and other measurements are recorded on the well log. No well may allow commingling or leakage of groundwater from one aquifer to another within a well, whether by gravity flow or artesian pressure (OAR Chapter 690, Division 210). In addition, artesian wells must have valves and casing maintained so that water flow is stopped when water is not being put to beneficial use (OAR Chapter 690, Division 215).

WELL ABANDONMENT

If making holes in the ground is no easy matter, getting rid of them is no simpler—at least not legally. There are state standards for this activity, as well. Oddly, given the frequency of their use, none of the standards reference bulldozers. It is probable that as a consequence of Oregon's rapid development, hundreds—if not thousands—of old wells are simply bulldozed into oblivion each year. At least apparent oblivion, in that a covered-up well is still open to anything that puddles, spills, or flows on top of it. Probably not very many developers or contractors know that to legally get rid of a well, it must be completely filled to stop any vertical of movement of water in the well bore. And it cannot just be filled with dirt, rubble, or junk. Generally, the filling material has to be concrete or cement grout. Any material that may interfere with proper sealing must be removed. Also, wells must be filled carefully, from the bottom up. If the developer encounters an old-fashioned dug well, the method of abandonment has to first be approved by the Water Resources Department before work is started. The proper abandonment of wells is just as important to water quality protection as their initial construction, and warrants very careful handling (OAR Chapter 690, Division 220).

Well drilling "rigs" are common sites in rural areas, but are also evident in cities, especially as knowledgeable developers or city officials require old wells to be properly abandoned, as shown above. (Photo by author)

▣ *Other Holes*

In terms of protecting the groundwater resource, a hole by any other name is still a hole—it just may not be a water well. The statutes give the Water Resources Department the authority to "... enforce standards for the construction, maintenance, abandonment or use of any hole through which ground water may be contaminated" (ORS 537.780(1)(c)(A)). This has given rise to a collection of standards for what has to be one of the most dubiously named categories in government: "Other Holes." The department recognizes three subcategories: monitoring wells, geotechnical holes, and residual other holes.

A monitoring well is any cased excavation or opening for determining the physical, chemical, biological, or radiological properties of groundwater. A geotechnical hole is one constructed to collect or evaluate subsurface data or information, monitor movement of landslide features, or to stabilize or dewater landslides. Furthermore, the definition of "well" does not include a hole drilled for the purpose of exploring for, or producing, petroleum, minerals, or geothermal resources (ORS 537.515). In 2004, of all the well reports received by the department, 45 percent were for water supply wells, 40 percent for geotechnical holes, and 15 percent for monitoring wells, while 75 percent of the water supply wells were for domestic use (Oregon Water Resources Department 2005a).

Basically, all these other wells and holes must be constructed, maintained, and used in ways that prevent contamination of groundwater (OAR 690-240-0030). Monitoring well regulations generally parallel those for water supply wells, including constructor licensing, start card and well log submittal, and construction standards. If a geotechnical hole is deeper than 18 feet, within 50 feet of a water supply or monitoring well, used to make a determination of water quality, or constructed in a contaminated area, a report to the Water Resources Department must be made by a responsible professional. Construction of holes other than water supply wells and monitoring wells does not require a license and licensing fee, bond, examination, well report, start card, or start card fee (ORS 537.515; OAR Chapter 690, Division 240).

▣ *Low-Temperature Geothermal Wells*

Two state agencies manage use of geothermal groundwater resources. The Oregon Department of Geology and Mineral Industries regulates production of groundwater that is 250 degrees Fahrenheit or hotter, or is from wells at least 2,000 feet deep. The Oregon Water Resources Department manages the use of shallower, cooler, "low-temperature" geothermal resources. The department's Division 230 rules establish standards for the development of low-temperature resources, such as those that heat many homes and businesses around Klamath Falls, Lakeview, and Vale. The rules acknowledge that low-temperature geothermal waters are pumped primarily to provide heat, not water, and that special management is therefore necessary. A number of management tools are available to protect the thermal characteristics of the groundwater resource (i.e., instead of just keeping them from being pumped dry, keeping them from being pumped cool, too), including declaring critical groundwater areas, prohibiting further use of wells, or ordering well repairs. The Water Resources Department can also restrict additional appropriations from the groundwater reservoir in order to limit thermal interference between wells. The well-construction regulations for low-temperature geothermal wells are the same as for water wells.

There are two types of wells associated with low-temperature geothermal uses: production and injection wells. A production well is one from which the hot groundwater is pumped. An injection well is one through which the used water, drained of much of its heat, is returned to an aquifer. The goal of the state is to maintain the thermal benefit provided by a given groundwater reservoir. An important protective mechanism is the control of injection wells. ORS 537.783 governs injection and requires a discharge permit be obtained from the Oregon Department of Environmental Quality before injection begins. All injection wells must have a plan approved by the Water Resources Department. Production and injection wells must be separated by a certain distance which varies depending on the amount of injection (OAR Chapter 690, Division 230).

▣ Well Controls

The Water Resources Department may impose conditions upon the use of any existing well necessary to prevent waste, undue interference with other wells, or contamination (OAR Chapter 690, Division 200). If problems are severe, the department may also order a halt to the use of the well and/or its permanent abandonment (ORS 537.775). Ultimately, it is the landowner who is responsible for maintaining and using on-site wells so they do not pose a health hazard (OAR Chapter 690, Division 215). In addition, owners of permitted or certificated wells are responsible for pump testing their wells (and reporting the results to WRD) every ten years. Pump tests are intended to provide aquifer and well information to help characterize the state's groundwater resources and help address well problems (OAR Chapter 690, Division 217).

◩ Well Information

Keeping track of three hundred thousand wells drilled across Oregon's hills and dales over the last half century is no easy matter. Therefore, the state started a well identification program in the 1990s. All wells drilled, deepened, or altered after 1996 are required to have an identification number. The number is on a stainless steel label that is usually attached to the well by the well constructor. Well identification numbers and location information must also be included in deed records during property transfers (ORS 537.788). Well constructors are required to place an identification tag on the wells they drill. If wells on lands being sold do not yet have such tags, the landowner must obtain them from the Water Resources Department (ORS 537.789) (Oregon Water Resources Department 2006f).

Pollution Prevention and Monitoring Programs

According to Oregon's Department of Environmental Quality, most wells are in shallow (less than 200 feet deep), unconfined aquifers common to Oregon's valleys. At the same time, most of Oregon's existing developed areas and those likely to grow in the next twenty years are located over these aquifers, which are generally the most easily polluted. Thus, much of Oregon's groundwater supply is subject to contamination—which is very difficult, or sometimes impossible, to clean up. Groundwater contamination may be caused by failing septic systems, fertilizer overuse, leaking underground tanks, buried wastes, unlined or improperly lined landfills, and runoff from livestock feedlots.

Overall, little is known about the quality of Oregon's groundwater resources. DEQ has evaluated groundwater quality for less than 7 percent of the state. Recent studies of forty-five areas found thirty-five with some impairment or reason for concern. Nitrate is the most commonly detected pollutant, followed by pesticides, volatile organic compounds, and bacteria. Additional information from over fourteen thousand residential real estate transactions that occurred between 1989 and 1998 involving water wells show a range of nitrate contamination. In some counties, no wells exceeded the federal drinking-water standard; in others, as many as 18 percent of the wells did. Approximately 12 percent of the groundwater-supplied public water systems in Oregon have had significant levels of contaminants detected (Oregon Department of Environmental Quality 2003c).

Potential point-source threats to groundwater quality in Oregon include nearly three thousand hazardous substance release sites; about twenty thousand leaking underground storage tanks; thirty-three dry cleaner sites with solvent releases; twelve Superfund sites with hazardous substance releases; forty thousand underground injection systems (like sumps, infiltration galleries, big septic systems, or any other system that places fluid below the ground surface); over eleven hundred permitted wastewater disposal facilities; two hundred thirty facilities that apply treated effluent or biosolids to land; four hundred eighty solid waste landfills; one hazardous waste landfill; and five hundred permitted confined animal feeding operations (Oregon Department of Environmental Quality 2003c).

DEQ oversees major programs dealing with groundwater quality; while a detailed description of them is beyond the scope of this book (indeed, each could be the subject of a book), these programs are the foundation of the state's groundwater quality protection efforts and thus warrant at least a brief mention. DEQ conducts groundwater quality investigations statewide; maintains Oregon's groundwater quality database; issues wastewater permits for Water Pollution Control Facilities and National Pollutant Discharge Elimination System programs, including groundwater protection requirements; certifies drinking-water protection plans for public water supply systems; administers an on-site sewage system program (about a third of Oregon's population is served by septic tanks), an underground storage tank program, solid waste and hazardous waste management programs, the Oregon dry cleaner program, the federal underground injection control program, and the federal Resource Conservation and Recovery Act program; and shares implementation of the drinking-water source assessment program with the Oregon Human Services Department (Oregon Department of Environmental Quality 2003d).

▣ Oregon Groundwater Protection Act

The Oregon Groundwater Protection Act of 1989 (ORS 468B.150–468B.190), and the groundwater quality rules that implement it (OAR Chapter 340, Division 40), focus on preventing groundwater contamination while conserving the resource for present and future beneficial uses. The

Act makes it "the goal of the people of the State of Oregon to prevent contamination of Oregon's ground water resource while striving to conserve and restore this resource and to maintain the high quality of Oregon's ground water resource for present and future uses." DEQ's objectives in implementing the Act target area-wide contamination resulting from non-point source pollution.

The Act established a statewide groundwater monitoring and assessment program and authorized DEQ to establish maximum measurable levels (MMLs) for groundwater contaminants. If DEQ finds an area with groundwater contamination believed to be caused by non-point source pollution, it must declare an area of groundwater concern. None have yet been declared. If monitoring detects nitrate at 70 percent or greater, or any other contaminant at 50 percent or greater, of the MMLs, DEQ must declare a groundwater management area (GWMA). DEQ appoints a local committee to recommend ways to remedy the groundwater problems, and designates a lead agency to develop a groundwater action plan to reduce existing, and prevent further, contamination (Oregon State University Extension Service 2004).

As of 2006, DEQ had designated three groundwater management areas: the Lower Umatilla basin, northern Malheur County, and the southern Willamette Valley. All were declared because nitrate contamination was found to exceed 70 percent of the MML. Children less than six months old who consume water with high levels of nitrate are at risk for "blue baby syndrome," where the nitrate interferes with the oxygen-carrying capacity of blood. Agricultural activities are believed to be major nitrate sources in all three of the GWMAs, with failing septic systems and residential application of fertilizer also contributing in the southern Willamette Valley (Oregon Department of Environmental Quality 2004b; Oregon Department of Environmental Quality 2004c).

◉ Well-Head Protection

The federal Safe Drinking Water Act was amended in 1986 to require states to develop a well-head protection plan (Oregon Department of Environmental Quality 1996d). As the agency in charge of Oregon's well-head protection program, DEQ has adopted rules to guide the voluntary development of local well-head protection plans (OAR 340-40-0140 through 0210). The goal of well-head protection is to prevent the contamination of public water supply wells by inventorying potential groundwater contamination sources and, where necessary, controlling activities in a well's vicinity that may contribute contaminants. Possible controls include public education, land use ordinances, or building codes.

Generally, contaminants fall into three source categories: microorganisms (e.g., bacteria, viruses, Giardia), inorganic chemicals (e.g., nitrate, arsenic, metals) and organic chemicals (e.g., solvents, fuels, pesticides). They can originate from a wide variety of activities or enterprises, including landfills, gas stations, farms, septic systems, garden chemicals, stormwater runoff, car repair shops, beauty shops, dry cleaners, or medical facilities. In Oregon, DEQ recommends designating an area large enough to cover the distance groundwater would travel in ten years. Accordingly, protected areas may extend several thousand feet from the well. Well-head protection plans are encouraged to address public education needs, delineate the protected area, inventory potential contaminants, identify ways to manage activities in the area, specify contingency options for addressing short- and long-term loss of drinking water sources, and establish standards for new wells. DEQ provides technical and some financial assistance for developing, and ultimately certifies, local well-head protection plans (Oregon Department of Environmental Quality and Oregon Health Division 1996).

▣ *Drinking Water Quality Tests During Real Estate Transactions*

Another important approach to preventing contamination and monitoring water quality conditions involves real estate transactions. In any transaction that includes a domestic supply well, the seller must have the well tested for nitrates and total coliform bacteria and submit the results to the Oregon Human Services Department Health Division. The division may require additional tests for specific contaminants in an area of groundwater concern or groundwater management area (ORS 448.271). It should be noted that there is no requirement that the tests meet any purity standards—the results just have to be reported. In addition, when any property that has a well on it is transferred, the property deed must include the well identification number and a standard form notifying landowners of the presence of any well and of monitoring, well maintenance, and abandonment obligations (ORS 537.788).

CHAPTER 11

Water Coordination and Planning

A CORPS OF ENGINEERS' DAM, STORING water in the name of the Bureau of Reclamation under a permit issued by the Water Resources Department, releases water—at a fish-friendly rate requested by the Fish and Wildlife Department—through a turbine generating power marketed by the Bonneville Power Administration, into a canyon managed as a scenic resource by the state Parks and Recreation Department and U.S. Forest Service, to downstream irrigation districts (operating as political subdivisions of the state), whose left-over irrigation water returns to the river and flows to a city intake for treatment to meet Health Division drinking-water standards, and is then routed through a county-approved golf course/subdivision to a sewage treatment plant operating in conformance with the Department of Environmental Quality's clean water standards for discharge back into the river, which empties into the Columbia (coursing water from seven states and one Canadian province), home to endangered salmon managed by the NOAA Fisheries Service, as advised by the Northwest Power and Conservation Council.

As water moves through the landscape it also moves through the hands of many different managers, who often have very different objectives. Harmonizing those objectives to avoid chaos and conflict requires careful coordination. And coordination is one of those things there is never enough of, except when talked about. It is often an effective cure for consciousness and always bureaucratic—in the best and worst senses of the term. Successful coordination requires good organizational circuitry, artful and timely communication, clear organizational goals, and alert staff. Successful coordination is also invisible. The quiet hum of a working machine more often provokes sleep than admiration. It is usually only when things go wrong that coordination becomes an issue.

There is some degree of coordination in all things, and rather a lot of it in water management. It comes to the fore in a number of areas in Oregon, as discussed below.

Reservoir Coordination

Major reservoirs are far more than inert bodies of slackwater. They are water machines designed and operated to meet a wide variety of often opposing objectives. How they are filled and emptied can have profound impacts on flood crests, fish migration, agricultural water supplies, and family vacations. Reservoir management requires careful coordination among a host of interests. Nowhere is this more true than in the mega-waterworks otherwise known as the Columbia River. However, it is no less true of smaller (that is, by comparison to the Columbia) systems in the state.

▣ *The Columbia River Coordinated System*

The Columbia is as much device as river. It is a power-generating, cargo-carrying, flood-arresting, desert-blooming combine—which by the way and after the fact is also expected to pass fish, unharmed, between its gears. Many of these objectives compete with one another, and achieving all of them may even be impossible. However, the overall goal of harmonizing

The Columbia River has been trained to perform a number of vital tasks, such as producing most of the region's power. Dams such as The Dalles, above, are carefully controlled through the Pacific Northwest Coordination Agreement to produce a range of benefits including flood control and navigation. (Oregon State Archives, Oregon Highway Division, OHD6732)

these needs is formalized in the Pacific Northwest Coordination Agreement (Bonneville Power Administration, U.S. Army Corps of Engineers, and U.S. Bureau of Reclamation 2001).

This agreement is basically an operations contract mandating an annual plan to meet all authorized purposes of the Columbia River dams. It also recognizes and seeks to accommodate the multiple uses of the river. While individual projects adhere to their own operating guidelines, their actions are coordinated through the agreement. Parties to the agreement include federal project operators (the Bonneville Power Administration, Corps of Engineers, and Bureau of Reclamation); private utilities (including Portland General Electric and PacifiCorp); municipal utilities (including the Eugene Water and Electric Board); and various Washington state public utilities. The agreement represents a complex, multi-layered approach to management in which the two basic actions of storage and release are timed to serve a difficult balance of major uses: flood control, power generation, and fish protection.

FLOOD CONTROL

In 1948, the second-largest city in Oregon disappeared. Vanport was a town created to house Portland's swelling wartime population, built, unfortunately, on the floodplain north of the city, some 15 feet below the Columbia's water level. On May 30th, when the river was in full spring flood, a railroad dike gave way and literally washed away the town of forty-eight thousand mostly poor, black inhabitants, killing fifteen (University Park Community Center n.d.). An Oregon version of Hurricane Katrina.

After the flood, the Corps developed the first storage plan for the Columbia basin. Today, nearly 40 million acre-feet of storage space is available for flood control purposes—over half from Canadian dams. There are two flood-prone seasons in the basin: winter (because of heavy rains) and spring (because of melting snow). Generally, reservoirs are drained from August through December at a pre-set rate based on historical flow patterns. They then stand ready to catch any floods from January through April. During this period operations are fine-tuned according to real-time forecasts of runoff provided by the federal Northwest River Forecast Center. From April through July, reservoirs are allowed to refill gradually, at a rate that carefully controls downstream flows—not so low as to injure fishlife or other uses, yet not so high as to cause local flooding.

POWER GENERATION

Streamflows and energy use are out-of-sync in much of the Columbia basin. Peak flows on the Columbia occur in spring and early summer, while energy demand peaks in fall and winter. The Columbia's reservoir system brings the two more into agreement by storing water during high flows and releasing it (through turbines) during low. This means that power generation generally works in concert with flood-control operations. The Columbia River's reservoirs are operated to satisfy three primary energy objectives: meeting the region's bottom-line energy demands, even if streamflows drop to a very low level; refilling all the reservoirs every year (an objective met on the average three out of every four years); then otherwise maximizing the production of energy.

FISH PROTECTION

Fish protection is where Columbia coordination is really being put to the test. Historically, up to sixteen million adult salmon and steelhead returned each year to their native tributaries in the Columbia Basin. Today, they number one million or fewer. The basin's hydroelectric dams are responsible for between five and eleven million of the loss from historic numbers (Northwest Power and Conservation Council 2000). Consequently, Snake River sockeye salmon were listed under the ESA in 1991, with over a dozen more listings of various stocks of salmon and steelhead following in the mid and late 1990s. In 2000, as provided by the ESA, the NOAA Fisheries Service and the U.S. Fish and Wildlife Service imposed a set of protective measures that restricted how the Columbia's dams would operate. The restrictions center on how fish get past the dams (both upstream and downstream), an issue that has come to focus on flow.

When the dams were first built, the focus was on how to get adult salmon returning from the sea up and over them. Accordingly, all the Columbia River dams that shoulder up against Oregon have fish ladders that route some flow to attract and convey fish.

Work is also underway to equip dams with fish by-pass facilities, which route juvenile fish away from turbines and around the dam. Sometimes fish are collected from a by-pass, placed on barges, and transported below a string of dams before release. In addition, a number of Oregon's fish hatcheries were established and are maintained with federal money provided as mitigation for the dams' effects on salmon and steelhead. The federal government, especially through the Bonneville Power Administration, has also funded habitat improvements and diversion-screening in many Columbia River tributaries.

As early as the 1980s, the Northwest Power and Conservation Council (see discussion under Interstate Coordination, below), in cooperation with tribes, utilities, and fish agencies, established a water budget to increase flows during the downstream spring migration of juvenile salmon. These "flushing" flows mimic natural spring freshets and speed young salmon through reservoir pools. The less time smolt (young salmon) spend in these expanses of slack water, the less they are exposed to predation or other injury. The water budget specifies flows at two Washington dams: Priest Rapids on the Columbia and Lower Granite on the Snake. Because these dams do not store much water, any flushing flows are expected to be drawn from dams further upstream, especially Grand Coulee (Washington's Columbia River), Dworshak (Idaho's Clearwater River), and Brownlee (Oregon and Idaho's Snake River).

Faster water is not the only fish objective for Columbia basin dams. What happens when fish-meets-dam is of critical importance to the survival of the Columbia basin's anadromous runs. While Woody Guthrie gave voice to the marvel of the Columbia's dams "turning our darkness to dawn," it is now terribly clear that they can also churn our smolts to smithereens. The NOAA Fisheries Service estimates that anywhere from 40 to 60 percent of Snake River system smolts die on their way downriver (National Marine Fisheries Service 2000). One solution may be to spill water, and smolts (juvenile salmon) along with it, over the dams rather than through turbines or by-pass structures.

In 1989, fisheries agencies, Indian tribes, and BPA signed a ten-year Long-term Spill Agreement calling for spilling water at four Corps dams: Lower Monumental and Ice Harbor (both in Washington on the Snake River) and John Day and The Dalles (on the Columbia River, between Oregon and Washington). The spill program has been controversial. When water goes over the dams instead of through turbines, power production suffers. And, reversing an old saying, power is money. Agencies that run the dams estimate that the system has lost, on average, about 1,000 megawatts of annual energy production due to operational requirements imposed under the Endangered Species Act (U.S. Army Corps of Engineers, Bonneville Power Administration, Bureau of Reclamation 2003). Between 1978 and 2003, that amounted to over $1.2 billion in forgone revenue—the estimated value of hydropower that was not sold because the water went to assist fish passage and improve fish survival (Northwest Power and Conservation Council n.d.).

Fish and environmental groups maintain that spilling water is the best method for getting young fish downstream. The agencies that run the dams, in addition to objecting to the lost revenue, counter that other methods (like trapping the young fish and barging or trucking them around the dams) are just as good, if not better (spilling water increases the amount of nitrogen in water, which can give fish the equivalent of the bends).

Although the ongoing and intense debate over salmon recovery has been termed the Northwest salmon crisis, it could just as easily and accurately be labeled a water crisis. While decisions about harvest levels and how or if we grow salmon in hatcheries are important, in the long-term salmon survival depends most directly on how we manage water. The biggest tools we have for doing that on the Columbia are the dams—and the stakes are high. Beyond power generation issues, there are navigation issues (Lewiston, Idaho is a seaport, thanks to the Columbia's system of reservoirs and locks, and roughly 17 million tons of goods are shipped on the river annually), cultural issues (the rights of Native Americans), agricultural issues (the ability to use Columbia basin water to continue or expand irrigation), fishing industry issues (the ability to have a commercial fishery), and tourism and recreation issues (the ability to have a salmon sport fishery) And, as in the Klamath, when water heats up, things can go a bit crazy.

To some, it seems strange that we are happily catching, killing, and eating a threatened species—in fact, we demand the continued ability to exercise our sport and commercial fishing prerogatives. In trying to sort through salmon recovery, we are forced to make nearly theological distinctions between wild and hatchery fish stocks (made difficult by decisions last century to gaily mix the genetic stuff of many salmon stocks). We are interested in preserving the seaport status of a town 465 river miles from the sea. To protect flows for fish, the State of Oregon restricts even relatively small additional diversions from the Columbia (flowing by the Dalles at an average volume of 134 million acre-feet), while irrigators in Washington propose a massive and tricky off-stream reservoir project to tap the Columbia costing anywhere from $1 to 2 billion.

On the other hand, many are beginning to question the benefit of having the dams at all, with a number of groups calling for the removal of the lower Snake River dams. And the agency primarily responsible for salmon protections, NOAA Fisheries, changes its tune as political winds shift, amending its earlier ESA findings so that dams are reborn as simply an unavoidable part of the landscape (like waterfalls, log-jams, or climate) and confines its vision of recovery to the best possible river operations with the dams in place. All this while a growing population of California sea lions and Caspian terns prey spiritedly on salmon with relative impunity because they're also protected, one under the Marine Mammal Protection Act, and the other under the Migratory Bird Treaty Act. A 1997 study estimates that West Coast sea lions' and seals' annual salmon consumption is about half of what is commercially harvested; a 2005 study indicated that, from 2000 to 2004, terns at the mouth of the Columbia ate an average 5.5 million juvenile salmonids a year (National Marine Fisheries Service 1997; U.S. Fish and Wildlife Service, U.S. Army Corps of Engineers, and NOAA Fisheries 2005).

Not surprisingly in this turbulent setting, lawsuits have abounded. In 2005, the U.S. District Court ruled on one of the major legal challenges. Judge James Redden held that NOAA's most recent ESA findings were invalid and remanded them to agency for revision. The court also instructed the dam agencies regarding operations while NOAA re-worked its finding, ordering them to increase spill in the late spring (when they had proposed eliminating it in favor of barging salmon). In his decision, Judge Redden pointedly called for greater cooperation, while calling attention to what many considered an unthinkable option—dam removal. "This remand ... requires NOAA and the Action Agencies to be aware of the possibility of breaching the four dams on the lower Snake River, if all else fails. The

Columbia River fish ladders, built to allow salmon to get around its giant dams, may attract as many tourists as fish. (Bonneville Power Administration)

Detroit Dam on the North Santiam River, like others in the Willamette system, must be drained and filled to meet a wide range of often-conflicting demands. The State of Oregon works with the Army Corps of Engineers to hammer out annual operating plans to achieve some sort of balance. (U.S. Army Corps of Engineers)

cooperation of the political branches (i.e., money) may mean such an action will be unnecessary ... 'Speeching' on the dams will not avoid breaching dams. Cooperation and assistance may" (Pacific Northwest Waterways Association 2006).

▣ *State Recommendations on Other Federal Projects and Programs*

The Oregon Water Resources Department leads other state agencies in reviewing the annual operating plans for the Army Corps of Engineers' Willamette, Rogue, and Willow Creek projects. The state joins the Corps in the difficult juggling act of reservoir regulation by recommending release schedules that protect Oregon's interests in fish, water quality, recreation, and irrigation water supplies. Reservoir filling and evacuation is carefully timed to produce the greatest benefit and fewest headaches for a basin's collective water interest (Figure 32). State participation in federal project management becomes especially important when water is in short supply during drought periods.

The state also reviews and seeks to influence federal proposals involving reauthorizing federal reservoirs (Chapter 7) or changing traditional reservoir operations to protect threatened or endangered species (e.g., the Klamath Project's effect on suckers and salmon). A number of state and federal agencies also cooperate in reservoir coordination for flood management purposes. The National Weather Service's River Forecast Center maintains a network of nearly nine hundred forecast points across the Northwest to inform flood prediction, instream flow estimates, and project operations (Northwest River Forecast Center 2005).

The Federal Emergency Management Agency administers the National Flood Insurance Program in cooperation with Oregon's Department of Land Conservation and Development, Building Codes Division, and Emergency Management Division. The NFIP is a federal program that enables property owners in participating communities to purchase insurance protection against flood-based property losses. The program provides an insurance alternative to disaster assistance to meet the ever-rising costs of repairing damage to buildings and their contents caused by floods (Federal Emergency Management Agency 2004).

Participation in the NFIP occurs through an agreement between local communities and the FEMA that states if a community adopts and enforces a floodplain management ordinance to reduce future flood risks to new construction in flood areas, the federal government will make flood insurance available within the community. The program relies on flood hazard mapping, flood insurance, and floodplain development standards applied at the local level. In Oregon, 255 cities and counties and two tribal nations participate in the NFIP. However, many flood studies and maps in Oregon were completed in the late 1970s and early '80s, and provide a snapshot of flood risk at that point in time. Oregon's metropolitan areas have had grown a lot since then, and increased development changes an area's hydrology as increased impervious surface area generates greater runoff volumes and velocities.

FEMA and the State of Oregon discourage development in floodways, allowing it only under very limited circumstances. The NFIP requires compliance with a no-net rise standard, which means there must be no increase in the flood elevation at the site or downstream resulting from the development. Some development, however, may occur at the edge of the floodplain. In Oregon, residential development must be elevated at least 1 foot above the base flood elevation.

DLCD is the coordinating agency for land use planning and the NFIP. Statewide planning Goal 7 requires local governments to address natural hazards, including flooding, in their comprehensive land use plans (the

Figure 32: Order of reservoir drawdown in the Willamette basin. Reservoir operations are harmonized in the Willamette basin to achieve both flood control and summer flow benefits. This schematic shows the order in which the U.S. Army Corps of Engineers releases water from its dams to meet flow objectives at Albany and Salem—flows critical to diluting waste discharges from Oregon's major cities which line the river's banks. (Source: U.S. Army Corps of Engineers 1996b)

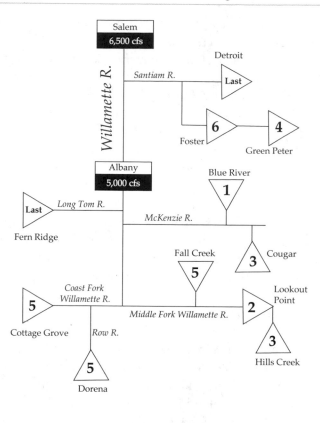

goals are described in more detail later in this chapter). Local governments are deemed to comply with Goal 7 for flood hazards if they adopt and implement local floodplain programs that meet the minimum NFIP requirements (Oregon Office of Emergency Management 2006).

Drought Management

Although not strictly limited to surface water supplies, drought is most often recognized and felt in a surface water environment. As events, droughts in Oregon are both relative and cyclical. A drought may mean something entirely different to a Lakeview rancher than to a Beaverton suburbanite. But however defined, droughts are sure things statistically. They come and they go, but they never stay away. Strictly speaking, despite the title of this section, no one manages a drought. What the state attempts to manage is people's response to drought.

The Emergency Management Division of the Oregon Department of Administrative Services facilitates the state's Drought Council. The council monitors weather, stream, and soil conditions across the state and makes recommendations to the governor as to which counties should be the subject of a drought declaration (actually, the statutes actually refer to declaring "a severe, continuing drought.") (ORS 536.740). The council is also responsible for developing a coordinated drought response and recovery plan for the state. In addition to the Emergency Management Division, it includes state and federal agencies, private organizations, and public interest groups. The council's Water Availability Committee, headed by the Oregon Water Resources Department, monitors precipitation, streamflow, and other indicators to assess current drought conditions and estimate the severity of any developing drought (Oregon Water Resources Department 2005c).

Declaring a drought activates a number of special state authorities. First, the governor may order state agencies and any political subdivisions (e.g., cities or water districts) in an area to implement water conservation or curtailment plans. These must be filed and approved by the Oregon Water Resources Commission. A conservation plan details efforts to reduce water usage for nonessential purposes, prevent waste, and promote water re-use. A curtailment plan specifies actions that will free up water necessary for human and livestock consumption, such as re-allocating supplies and limiting the times and manner in which water may be used. Second, the commission may issue temporary permits for emergency water use, allow immediate temporary transfers, issue temporary instream leases to convert all or a portion of a water right to an instream lease, authorize temporary substitution of a supplemental ground water right for a primary surface water right, grant use-preferences to human consumption or stock watering rights, or waive well-drilling and source-exchange notice requirements. Third, the commission, local governments, or public corporations may purchase options or enter into agreements to use existing permits or rights for any beneficial use for the duration of the drought (ORS 536.720–536.780).

Because they cut through a lot of regulations to allow more immediate water use during the drought emergency, drought declarations are popular. It is not uncommon for county governments or groups of local citizens to resist the council's recommendation to de-declare their drought. This resistance, coupled with sometimes lengthy natural recovery times in some basins, may keep counties in a state of a drought even as they weather torrential rains or heavy snows.

And just in case people want to take matters into their own hands, state law requires anyone (except airports) engaging in weather modification (whether they are trying to make it rain, hail, or snow, or stop the same) to first obtain a license from the Oregon's Department of Agriculture. The law also authorizes the formation of weather modification districts (for certain unnamed counties bordering the Columbia River and having populations less thantwenty-one thousand) (ORS 558.010–558.140).

Watershed Management

What happens in a watershed can have a great impact on the timing and quality of the water it releases. As water becomes increasingly important to Oregonians, so too does managing watershed activities such as logging, grazing, mining, road building, recreation, and urban development. There is a growing emphasis on watershed management, which might be described as the practice of selecting and harmonizing multiple land management objectives to preserve or enhance the benefits provided by the waters flowing from a basin. Arguably, any land or water management activity affects watershed outputs and therefore falls within the scope of watershed management. But most activities are already managed to some degree. Is watershed management simply the sum of all these separate management efforts? The answer is no, for what is missing is the critical need to link multiple management objectives to a central principle of hydrologic function. In other words, people playing by watershed rules, rather than the other way around.

The emerging popularity of watershed management seems to draw value from a change in emphasis in what is being managed. That is, a change from the grids stamped on maps by human beings (e.g., national forests, BLM checker-board ownerships, counties, congressional districts, property corners) to the irregular webbing of natural systems that arise from the landscape (e.g., watersheds, ecosystems, soil distribution, vegetation provinces). When the management target is the land (and thereby the water), rather than the grid, coordination among all agents becomes paramount. This section describes a few programs that impart an Oregon spin to what appears to be a national, if not global, watershed embrace.

▣ *Soil and Water Conservation: Districts and Commission*

Watershed management as a shared organizing principle may be newly elevated, but as a concept it is anything but new—nor is Oregon's experience with it. Soil and Water Conservation Districts (ORS Chapter 568) have been operating in the state since 1939. The legislature deputized SWCDs to carry out a state policy of seemingly heroic proportions:

> to conserve and develop natural resources, control and prevent soil erosion, control floods, conserve and develop water resources and water quality, prevent impairment of dams and reservoirs, assist in maintaining the navigability of rivers and harbors, preserve wildlife, conserve natural beauty, promote recreational development, protect the tax base, protect public lands and protect and promote the health, safety and general welfare of the people of this state. (ORS 568.225)

In pursuit of this mission, SWCDs can investigate soil erosion and flood and sediment damage, as well as anything to do with the conservation,

Controlling grazing, logging and urban development in streamside areas can bring help restore damaged watersheds. Under the Bureau of Land Management's watchful supervision, this stretch of Bear Creek in Central Oregon showed a remarkable recovery (more vegetation, greater bank stability, and more water in the channel) in just 8 years. (Both photos: U. S. Bureau of Land Management)

development, utilization, and disposal of water. In addition, districts can undertake demonstration projects and take necessary preventive and control measures in conjunction with cooperating landowners. Furthermore, they can build, maintain, and manage any flood-prevention or irrigation works and may offer financial aid to landowners for flood prevention or water management. SWCDs' water management responsibilities, however, are subject to Water Resources Commission policies and programs (ORS 568.552). Finally, the districts are given the authority that all red-blooded government subunits desire and that really makes them a going concern: the power of taxation. If given permission by the voters to organize as a taxing district, SWCDs may "assess, levy and collect an ad valorem tax each year on the real market value of all taxable property" in the district (ORS 568.806). At least four SWCDs have this authority: Polk, Marion, Yamhill, and East Multnomah.

There are forty-five SWCDs at work in every county in Oregon, and toether they cover nearly all of Oregon's land area. By statute, while the Oregon Department of Agriculture provides administrative oversight to the districts, each has a board of five or seven unpaid, locally elected members. SWCD funding comes primarily from grants (and, as noted above, in some instances property tax revenue), mostly from federal agencies such as the Natural Resource Conservation Service, the Oregon Watershed Enhancement Board, and the Bonneville Power Administration. While vested with considerable authority and a big mission, until recently many SWCDs have maintained a relatively low profile, often limiting their focus to farm-related activities, rather than the broad spectrum of watershed challenges recited in their statutory charter. Today, however, many SWCDs are enthusiastic participants in the wave of watershed improvement efforts that commenced in the early 1990s, including those promoted by the Oregon Watershed Enhancement Board (described later in this chapter).

The state Soil and Water Conservation Commission is seven-member body appointed by the director of the Department of Agriculture. Members must be SWCD directors, and serve staggered four-year terms. The commission chair also serves on the state Board of Agriculture. The primary mission of the commission is to coordinate the wide-ranging activities of the organizations involved in managing the interface between natural resources and agriculture, including the department, the Natural Resource Conservation Service, the Agricultural Stabilization and Conservation Service, Oregon State University's Extension Service, and SWCDs.

▣ *Watershed Councils*

There are about ninety watershed councils in Oregon (Figure 33). Watershed councils are locally organized, voluntary, non-regulatory groups established to maintain and restore watershed conditions in their area. The 1995 legislature unanimously enacted a bill providing guidance for establishing watershed councils. It emphasized that forming a council is a local government decision, with no state approval required. Watershed councils, which are required to represent the interests in the basin and be balanced in their makeup, catalyze partnerships between residents, local, state and federal agency staff and other groups.

A watershed council is "a voluntary local organization designated by a local government group convened by a county governing body to address the goal of sustaining natural resource and watershed protection and enhancement within a watershed" (ORS 541.351). Councils convene diverse interests in a non-regulatory forum to articulate a common vision for the ecological and economic sustainability and livability of their watershed. They often identify landowner participants for important projects, develop priorities for local projects, and establish goals and standards for future

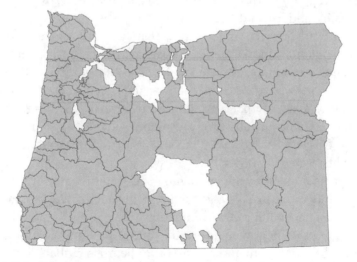

Figure 33: Watershed councils. Nearly every watershed in Oregon is represented by over 90 watershed councils or groups (shown in gray with watershed boundaries in black). If the map included areas covered by Oregon's county-based 45 soil and water conservation districts, which also have watershed protection responsibilities, all of Oregon would be represented. (Source: Information from Oregon Watershed Enhancement Board 2004)

watershed conditions. On-site projects are undertaken that improve a watershed's capacity to capture, store, and release water. Councils sponsor education projects to increase people's understanding of watershed processes and functions, and provide coordinated, broad-based review of land management plans to local, state, and federal decision makers (Oregon Watershed Enhancement Board n.d.[a]).

▣ *Oregon Watershed Enhancement Board*

The Oregon Watershed Enhancement Board was created in 1987. It actively encourages the collaboration of citizens, agencies, and local interests to restore salmon runs, improve water quality, and strengthen ecosystems critical to healthy watersheds and sustainable communities. OWEB administers a grant program funded from the Oregon Lottery that supports voluntary efforts by Oregonians to create and maintain healthy watersheds. (Measure 66, approved by citizen initiative in 1998, required that 15 percent of net proceeds from the State Lottery be deposited in a newly created parks and natural resources fund, with half going to creating, protecting, and operating state parks, ocean shore, and other natural recreation areas and half going to restoring and protecting native salmon runs, watersheds, water quality, and fish and wildlife habitat.)

OWEB funds projects that restore, maintain, and enhance the state's watersheds; support local watershed-based citizen groups; promote citizen understanding of watersheds; equip citizens with technical skills to restore urban and rural watersheds; and monitor the effectiveness of investments in watershed restoration. Typical projects involve, planting, reseeding, fencing, weed control, culvert replacement, wetland restoration, livestock watering, fish habitat, land purchases, conservation easements, instream water rights, water right transfers.

OWEB consists of seventeen members, eleven of whom may vote. The voting members are five at-large public members plus the representatives of the Fish and Wildlife, Environmental Quality, and Water Resources commission, the boards of Forestry and Agriculture, and a Native American tribe. The other six non-voting board members are from the OSU Extension Service, U.S. Forest Service, Bureau of Land Management, Natural Resources Conservation Service, Environmental Protection Agency, and NOAA Fisheries Service (Oregon Watershed Enhancement Board n.d.[b]).

▣ *Other*

A variety of other agencies play key roles in watershed management and warrant at least an abbreviated mention. The Oregon Department of Environmental Quality sets water quality standards for controlling run-off from urban, agricultural, and forested lands—termed "non-point" pollution because it comes from sources diffused through the landscape, rather than from a specific point, such as a pipe (ORS Chapter 468b). DEQ's non-point pollution control strategy emphasizes watershed protection and enhancement, voluntary stewardship, and partnerships among all watershed interests (Oregon Department of Environmental Quality 1995). In addition, DEQ prioritizes watersheds for funding and works with watershed councils through its water quality limited stream program (see Chapter 9).

The Department of State Lands oversees a number of wetlands protections programs, controls stream fill-and-removal activities on state waters (Chapter 9), and manages state-owned grazing lands. The Oregon Department of Forestry administers the Oregon Forest Practices Act (ORS 527.610-.992), which applies to both public and private lands. The Act restricts logging operations in riparian zones and in some circumstances controls the amount of cutting that can take place over a given area. Reclaiming mined lands is the responsibility of mine operators as overseen by the Oregon Department of Geology and Mineral Industries under the Reclamation of Mined Lands Act (ORS 517.750–517.951). The Department of Land Conservation and Development reviews local land use plans to assure compliance with statewide planning goals, including Goal 5, which requires the inventory and, at times, protection of riparian areas and wetlands. DLCD also reviews and offers assistance in developing local flood plain ordinances. Of course, the U.S. Forest Service and Bureau of Land Management, which in combination own over half the land area of Oregon, exercise a tremendous influence on state watersheds by virtue of practically anything they do. The major federal laws controlling land management by these agencies include the National Environmental Policy Act (42 U.S.C. 4321-4347), National Forest Management Act (16 U.S.C. 1600 et seq.), the Federal Land Policy and Management Act (43 U.S.C. 1701 et seq.), Multiple-Use Sustained-Yield Act (16 U.S.C. 528 et seq.), Wild and Scenic Rivers Act (16 USC 1271-1287), Wilderness Act (6 USC 1131 - 1136), and Taylor Grazing Act (43 U.S.C. 315, 316).

Lastly, and now probably only a legal curiosity, cities are specially authorized under ORS 448.295–448.320 to pass ordinances defining "offenses against the purity of the water supply" in watershed areas where their distribution system or water sources are located. They may also appoint "special policemen" to arrest "with or without warrant" any person committing any offense against the purity of the water supply source. While it is unlikely we will soon witness water-purity cops prowling the forests, ticketing reckless 'dozer operators and wayward tinklers, the idea of actively policing our lands to protect our waters has a certain charm.

Interstate Coordination

As noted in Chapter 1, wherever Oregon's boundary is a straight line, watersheds are split in two. The Klamath River heads in Oregon, but meets the Pacific in California. The Walla Walla, Washingtonian by reputation, is Oregonian by birth. The Owyhee assembles itself here, but is imported in large measure from Utah and Idaho. With this kind of watery give and take, formal agreements with neighboring states have long been recognized as important parts of the state's water management system.

▣ Water Compacts

Interstate agreements, or compacts, are considered big deals—the Constitution of the United States actually forbids them without consent of Congress (Article One, Section 10, Subsection 3). While Oregon has nothing to rival the water compact governing the use of the Colorado River (although there is a Columbia River *fish* compact and a Columbia River *Gorge* Compact, curiously, there is no Columbia River *water* Compact— see below), it does maintain a number of important water agreements with other states. Oregon is party to two interstate water compacts: the Klamath River and the Goose Lake Interstate compacts. The Goose Lake Compact, approved by the states of Oregon and California in 1963, is rather unremarkable. Codified in ORS 542.520, it prohibits export of water outside the basin without permission from both state legislatures. It also allows storage facilities built in one state to supply uses in the other.

The Klamath River Basin Interstate Compact (ORS 542.610–542.630), on the other hand, appears to do a lot. This compact between Oregon and California went into effect in September 1957 and governs the use of surface water. While its language may be convoluted, its effect is unmistakable:

in the Klamath basin, domestic use and irrigation rule. The compact sets up a system for post-1957 water appropriations which ranks uses in the following order of preference:

1. Domestic use
2. Irrigation
3. Recreation (including fish and wildlife uses)
4. Industrial
5. Hydroelectric power production
6. All other uses

But it does not stop there. Not only does the preference apply to issuing new permits, but to their exercise, as well. The compact suspends the prior appropriations doctrine for rights issued after the compact's effective date. Domestic and irrigation rights issued after that date get water before any other post-compact rights, priority date notwithstanding. These superior rights, however, are limited to the amount of water needed to irrigate 100,000 and 200,000 acres in California and Oregon, respectively. Once the acreage limit is reached, presumably any additional domestic and irrigation rights would be managed by priority date.

The compact also prohibits any new out-of-basin diversions, requires all return flows and waste waters be returned to the Klamath River at points upstream from Keno (to protect flows for downstream hydro projects), and directs both states to manage flows to minimize power rates for irrigation and drainage pumping. Water pollution is recognized as a significant concern that is to be addressed through cooperation, study, and recommendations for "reasonable" water quality standards. The compact is administered by a three-person commission representing the two states and the federal government. The federal representative is the chair, but cannot vote. Oregon's representative is the director of the Water Resources Department (State Water Resources Board 1971).

To some, the Klamath Compact probably seemed to offer a weighty, nearly invincible protection against practically any threat. However, it has failed to play any appreciable role in sorting through the Klamath basin crisis (Chapter 9). Irrigators in 2006, however, argued that its provisions amounted to a requirement in state and federal law that PacifiCorp maintain its very inexpensive rates for power from Klamath River dams (Klamath Basin Crisis 2005).

Although there is no Columbia River water compact, it is not for want (forty years' worth) of trying. One was first proposed in 1924. Although each of the governors of Washington (the state leading the charge), Idaho,

Unlike the Colorado River, the Columbia is not governed by any interstate compact. It is, however, subject to an international treaty, interagency agreements, and the recommendations of the Northwest Power Planning Council. (U.S. Army Corps of Engineers. Image File: Col0502.jpg. Date: 21JUN1990. Photographer: Bob Heims)

Montana, and Oregon were quick to appoint negotiators, no compact resulted. In 1949 President Truman endorsed the establishment of the Columbia Valley Administration, modeled on the Tennessee Valley Authority. Once again, the above-mentioned states convened a group to develop an agreement (this time including Wyoming, Nevada, and Utah representatives). The group decided to investigate not only the apportionment of water, but also hydroelectric power issues, stream pollution, and the protection of fish and wildlife.

A pact was signed in Portland in 1955 and referred to the respective state legislatures. The agreement did not divide water among the states, but instead subordinated downstream non-consumptive uses to upstream consumptive uses. Only Idaho, Nevada, and Utah ratified it. Oregon and Washington took issue with what they saw as the unfair advantage given to upstream states. The group revised the pact, adding a provision prohibiting the out-of-basin diversion of Columbia waters without the consent of all member states. In 1956, the amended agreement was once again submitted to the states' legislatures and but none ratified it. Questions arose as to the character of any compact commission: Oregon and Washington asserted it should have authority to build and run hydro projects; others that it have a purely advisory role. The negotiating group submitted revised pacts featuring an advisory form of commission for legislative approval in 1961 and 1963. Each time, five states ratified it. Oregon and Washington not only failed to ratify, but finally withdrew financial support, and the effort ended in 1966 (Dunbar 1983). Today, the Northwest Power and Conservation Council serves some of the purposes proposed for the compact commissions in the '60s.

The council was authorized by Congress in the Northwest Power Act of 1980 and approved by a vote of the legislatures of all four member states: Idaho, Montana, Oregon, and Washington. The governor of each state appoints two members to serve on the council for terms of office that vary by state. Funded by revenue from the Bonneville Power Administration's marketing of power from federal Columbia River dams, the council has three primary objectives: 1) formulate a twenty-year plan to assure adequate and reliable energy at the lowest economic and environmental cost to the Northwest, placing a priority on energy conservation and the use of renewable resources (e.g., wind, solar and geothermal power); 2) develop a program to protect and restore fish and wildlife populations injured by hydro development in the Columbia River basin; and, 3) make information readily available to citizens and involve them in decision making. The council is an advisory body whose plans and policies are ultimately implemented (or not, given the intractable nature of Columbia issues) by others, including the Bonneville Power Administration, the U.S. Army Corps of Engineers, the Bureau of Reclamation, and the Federal Energy Regulatory Commission.

The council's major venue for influencing water management is through its Fish and Wildlife Program. The Northwest Power Act requires the council to develop the program to protect, mitigate, and enhance the basin's fish and wildlife populations and habitat that have been harmed by hydroelectric development, and to update the program every five years. The council makes annual funding recommendations to the Bonneville Power Administration for projects to implement the program. On the average, BPA spends about $140 million per year on these projects.

The 2000 program sets forth a basinwide vision for fish and wildlife and establishes biological objectives and action strategies. It addresses all of the impacts on fish and wildlife—hydropower, habitat, hatcheries, and harvest. The program recommends actions to improve dam-passage survival that are biologically sound, economically feasible, and compatible with natural fish behavior. It requires that fish hatcheries funded through the program change from a predominant focus on furnishing fish for harvest to supplying fish to rebuild naturally spawning populations. The program also identifies key areas needing additional research and monitoring. Finally, it places a premium on rebuilding healthy, naturally producing fish and wildlife populations by protecting and restoring habitats and biological systems.

The primary implementing mechanism for the 2000 program was the development of subbasin fish and wildlife plans. In 2003, the council designed and BPA funded one of the largest locally led watershed planning efforts of its kind in the United States. It divided the Columbia basin into ecological provinces and then again into subbasins and had watershed councils, non-profit organizations, tribes, and government agencies develop plans to prioritize the most important actions to protect and restore fish and wildlife impacted by dam development. Subbasin plans are used to select which local projects to recommend to BPA for funding each year. In 2004, fifty-eight subbasin plans were submitted to the council, including twenty that include areas either entirely or partly in Oregon (some subbasins are shared between Oregon and its neighbors) (Northwest Power and Conservation Council 2005a; Northwest Power and Conservation Council 2005b).

◙ Out-of-Basin Water Diversion

Generally, the prospect of diverting water out of state has been viewed coolly in Oregon. As California's economy and population boomed after World War II, the Golden State began searching for additional water supplies. A number of preliminary plans were broached, some calling for continental-scale engineering to divert water from Canada or the Pacific Northwest. For Oregon, the diversion proposals generally involved transporting Columbia River water in amounts equal to two-thirds the flow of the Willamette for over 1,200 miles, with nearly a vertical mile of lift. This caught the attention of Northwesterners. At least one Oregon study concluded piping Columbia River water to California might not be so bad, especially if some could be dropped off in the drier parts of Oregon along the way. The downside— mellow by today's standards—was framed in terms of impacts to Oregon's ability to expand its irrigation base, continued water deficiencies in some areas, and loss of revenue to the hydropower system resulting from the waylaid water (State Water Resources Board 1969a). Environmental impacts were apparently not viewed as important, if they were considered at all.

Overall, however, the state's reaction was not that low-key. There was plenty of support for a congressional measure that became law as 33 USC Sec. 2265 and which states: "No Federal agency shall study or participate in the study of any regional or river basin plan or any plan for any Federal water and related land resource project which has as its objective the transfer of water from the Columbia River Basin to any other region or any other major river basin of the United States, unless such study is approved by the Governors of all affected States."

Nor was the Oregon Legislature willing to be silent on the subject. However, its aims were complicated by a 1982 U.S. Supreme Court decision, *Sporhase v. Nebraska*. This decision had the effect of declaring water an article of interstate commerce, meaning a state may not establish preference rights favoring its own inhabitants over consumers in other states—at least not without violating the commerce clause of the constitution. State laws governing the transfer of water to other states have to be carefully crafted to pass constitutional muster. Oregon's crafting may be found in ORS 537.801 through 537.870, which mandate a special application process and, in some cases, legislative approval.

A special evaluation process is required for out-of-basin applications for permits (surface water, groundwater, or storage), hydroelectric licenses, or transfers. However, the special process does not apply to transfers proposing less than 0.5 cfs or to historic out-of-basin diversions by a city in support of regional water service. In addition to any other information usually required, an out-of-basin transfer request must assess the amount of water available for future appropriation in the basin of origin; projected water needs in the basin; return-flow benefits that will be eliminated if the water leaves the basin; groundwater/surface water relationships that may be affected by the transfer; injury to existing appropriators or interference with planned uses; and water supply alternatives to the out-of-basin transfer.

The Water Resources Commission must conduct a comprehensive review of the application, publish a preliminary analysis in local papers, and conduct a public hearing at the applicant's expense. Before approving an application, the commission must reserve water for future instream and out-of-stream needs in the basin and subordinate the out-of-basin use to that reservation. Of particular importance for any large water transport scheme is the prohibition against any out-of-basin diversion of 50 cfs or more without the express consent of the legislature. This consent is waived for applications consistent with the Klamath River basin or Goose Lake interstate compacts; appropriations by a city in support of regional water service if the city has historically transported water out-of-basin; and domestic water supply districts that serve properties in other states.

It is important to note that the strictures apply inside the state as well as outside, and are therefore designed to withstand a court test. However, their primary purpose may be inferred from a bit of legislative saber rattling. The legislature provided that no person or agency of any state or of the United States can attempt to condemn any Oregon waters without first complying with the out-of-basin transfer statutes. In addition, should there be any violation or attempt to violate these statutes, the governor is instructed to initiate suits and other actions "necessary to protect and defend the sovereign rights and interests of the state in the premises." To top it off, the legislature threatens to loose the hounds of litigation: "Persons are given right of redress against such violator at private suit or action under any appropriate remedy at law or in equity."

It may be worth noting that the story of California and Oregon's (or other western states') water may not be over. In 2040, when Oregon is projected to have a population of about 5.4 million, California will have a population of 54.8 million (State of California, Department of Finance 2004; Oregon Office of Economic Analysis 2004). That 10-to-1 difference represents a mighty big thirst, especially at a time when climate change may be putting the squeeze on supply.

▣ *Other Agreements*

There are a number of other activities concerning other states or governments that affect water management in Oregon.

THE COLUMBIA RIVER TREATY

This treaty between the United States and Canada was signed in 1961 and put into effect in 1964. It provided for building four storage reservoirs—three in Canada (Mica, Keeleyside, and Duncan) and one in the United States (Libby)—that represent almost half the water stored in the Columbia River system. The treaty specifies protocols for river operations and power-sharing agreements. Operating plans are drawn up each year under the oversight of the Columbia River Treaty Operating Committee, which is made up of the Corps of Engineers, the Bonneville Power Administration, and B.C. Hydro. While the treaty has no expiration date, either country may after ten years' notice terminate the treaty after 2024 (Bonneville

Power Administration, U.S. Army Corps of Engineers, and U.S. Bureau of Reclamation 2001).

WALLA WALLA RIVER

While there is no compact with the State of Washington governing the use of the Walla Walla River, an accommodation is made in Oregon's water laws. ORS 537.835 makes the City of Walla Walla eligible to appropriate, impound, and divert waters from Mill Creek, an Oregon tributary of the Walla Walla River, for the beneficial use of both the State of Oregon and the city. In a neighborly gesture, the Oregon Legislature made it clear that should the city build any project, it must foot the entire bill and employ only Oregon residents for construction and maintenance. Furthermore, the Oregon Water Resources Commission is given authority "from time to time" to claim project water for Oregon's fire protection, wildlife, or streamflow maintenance needs.

OREGON HANFORD CLEANUP BOARD

From the early 1940s to the late 1980s, the federal government made plutonium for nuclear weapons at the Hanford Reservation in southeast Washington, just 35 miles north of the Oregon border. The process left behind great quantities of radioactive and chemical waste. During its forty-year operation, Hanford released—both intentionally and by accident—radioactive materials that entered the local water table and the nearby Columbia River. Most notably, from 1944 to 1971, eight reactors discharged radioactive cooling water directly into the Columbia, which in the early 1960s was believed to be the world's most radiologically contaminated river (Oregon Office of Energy 1996).

Sixty percent of the nation's volume of high-level radioactive waste is now stored in deteriorating tanks at Hanford (U.S. Department of Energy n.d.). It may be the largest nuclear waste dump in the Western Hemisphere (Hanford Watch 2004). The Hanford site is a residual nuclear nightmare straddling the Columbia River. It includes more than 50 million gallons of high-level liquid waste in 177 underground storage tanks; 2,300 tons of spent nuclear fuel; 12 tons of plutonium; about 25 million cubic feet of buried or stored solid waste; and about 270 billion gallons of groundwater contaminated above drinking-water standards, spread out over about 80 square miles; more than 1,700 waste sites, and about five hundred contaminated facilities (U.S. Department of Energy 2005).

An estimated 444 billion gallons of contaminated liquids were dumped into soils at Hanford. This incredible volume of waste created a poisonous dome of groundwater underneath the site. The water table rose tens of feet, changing both the direction and velocity of groundwater movement. In addition, 67 of the 177 underground waste storage tanks at Hanford are known or suspected to have leaked, releasing about one million gallons of highly radioactive and chemical waste to the soil. This leaked waste, containing an estimated one million curies of radioactivity, has reached the groundwater and may enter the Columbia (Oregon Hanford Cleanup Board 2005).

Naturally (or perhaps, unnaturally), a Columbia River contaminated with radioactive waste counts as a subject of interest for the State of Oregon. Therefore, the Oregon Legislature established the Oregon Hanford Cleanup Board in 1987 (originally the Oregon Hanford Waste Board) to be the focal point for discussions within Oregon regarding the disposal and cleanup of high-level radioactive waste in the Pacific Northwest (ORS 469.573). The board makes policy recommendations to the governor and the legislature, and may, after consultation with the governor, make policy recommendations on other Hanford issues such as defense wastes, disposal and treatment of chemical waste, and plutonium production (Oregon Hanford Cleanup Board 2004).

Water Supply Planning

If a plan is an ordered sequence of policy-guided actions undertaken to meet specific objectives, then Oregon does not have a recognizable water supply plan. There is no single, coherent document, statement, Web site or wish list that captures how much water we have, how we use it now, how much we will need in the future, and how we will get it. It would be as if Oregon, with increasingly crowded roads and a rapidly growing population, had no clue about how to fix the situation nor interest in providing for future road capacity, or left it up to cities, counties, or Nike to fend for themselves. But Oregon does not do that. In 2003, in its highway modernization program alone, 175 people worked with a budget of $253 billion to design and build highway improvements that:

> add capacity to accommodate current or projected traffic
> growth. This includes adding traffic lanes for passing and
> climbing, turning, accelerating and decelerating; building

new road alignments or facilities, including bypasses;
reconstructing roads with major alignment improvements
or major widening; and widening bridges to add travel lanes.
(Oregon Legislative Fiscal Office. n.d.[b])

It doesn't seem so far-fetched that the preceding statement might be applied to water, to have a state function or body whose job it was to at least design (if not build) water supply improvements that:

add capacity to accommodate current or projected water
supply needs. This includes adding facilities for storage, water
diversion, water treatment, and distribution; continually
implementing improvements in conservation and water-
saving technologies; realigning the water allocation system
to substantially increase instream investment through water
exchanges and transfers; and reconstructing measurement
networks with twenty-first-century technology to allow for
real-time sensing of water inflows and use.

We invest in transportation projects in accordance with a state plan. We have transportation planning because we understand—and as importantly, feel—that transportation is important to us personally and essential to us economically. Not so with water. If a measure of caring is the degree to which society organizes its thinking and allocates resources to the future, then on this score there is no contest: in Oregon, roads trump water any day.

Oregon's water is a state asset still in search of a good portfolio manager and investment plan—a search that began nearly a half century ago.

▣ *"An Integrated, Coordinated Program"*

In 1955, the legislature directed the state's water agency to take an integrated, coordinated approach for the use and control of Oregon's water resources (ORS 536.220; ORS 536.300). The Oregon Water Resources Board (a predecessor to the commission) was to "progressively formulate" a program and issue program "statements." The legislature identified at least two types of statements. One was the designation of preferred water uses (ORS 536.340(3)). The other was water use classification (ORS 536.340(1)). The former has never been issued. The latter evolved into what the state now calls basin plans (Chapter 3).

The legislature attempted to get the ball rolling by issuing thirteen policy "declarations" that were to be incorporated in the integrated program (ORS 536.310). Most of them seem not so much policies as platitudes (balanced management, multiple use, and state water sovereignty are good) or topical forget-me-nots (fish, wetlands, recreation are important water considerations, too). Woven through these declarations are lots of "ifs," "due regards," and other qualifiers that tend to take the wind out of any policy sails the statute does manage to unfurl. That isn't to say the declarations are useless or even off base. They include important assertions on protecting water rights, preventing waste, providing for instream flows, and siting storage projects. It is just that they represent a collection of oppositely charged policies that tend to cancel each other out. Instead of an energy source that sparks activity, these declarations drain the battery of progress.

Notwithstanding the disappointing starter-kit it provided for building the integrated program, the legislature had high expectations for the ultimate product. Although never called a state water plan, the integrated program was clearly meant to guide, shape, and, at times, restrict the actions of nearly every government in the state:

> No exercise of any … power, duty or privilege by any … state
> agency or public corporation* which would tend to derogate
> from or interfere with the state water resources policy shall
> be lawful. … No exercise by any state agency or public
> corporation of this state … of any power, duty or privilege,
> including … any order, rule, regulation, plan, program,
> policy, project or any other activity, which would in any way
> conflict with the state water resources policy as set forth in the
> statement shall be effective or enforceable until approved by
> the commission. (ORS 536.360–536.370)
> * *Any city, county or district organized for public purposes.*
> (ORS 536.007)

In fact, ORS 536.350 through 536.400 lay out an involved (but never exercised) process of notifications, approvals, hearings, agreements, and court remedies to assure compliance with the state water policy. It is quite the statutory drum roll, yet the intended act never showed. What the state ushered onto the stage over the next decade was a collection of basin classification statements. Rather than emerging from an integrated, coordinated policy, these basin plans themselves became Oregon's de facto and patchwork water policy.

Basin plans establish menus of future allowable water uses. They are permissive, not directive. As groupings mostly of stream classifications, basin plans list water use potentials, rather than direct water management results. Their purpose is to map all routes open for travel, not select a destination. The destination should have been selected in the integrated, coordinated policy that never came into being. After over thirty years of basin-by-basin classifications, the absence of an over-arching water policy became too obvious to ignore. Since 1983, the state has made repeated and concerted efforts to build a policy hull for its basin plan outriggers.

The first attempt was a legislatively mandated course correction called strategic water planning. This new planning process emphasized consideration and integration of all state water interests, more thorough interagency coordination, electronic database creation, and increased public participation in water planning. A Strategic Water Planning (later changed to Management) Group was charged with coordinating the water-related activities of the state, representing state interests in fisheries restoration, and asserting the state's hydroelectric position before the Federal Energy Regulatory Commission. However, because it was given no authority, SWMG never developed a real identity or sense of mission and was dissolved by the legislature in 1993.

The strategic planning process that gave life to SWMG was first applied in the John Day basin and led to an improved form of basin planning. The process became more issue oriented, broadened its scope to include more resource considerations (such as groundwater, watershed, and water quality concerns), experimented with different approaches (such as conditional uses), and emphasized action by explicitly adopting implementing measures. Despite these improvements, basin planning increasingly came to be seen as either a water management dinosaur (too unwieldy and unproductive a process in today's world) or as irrelevant (especially given newer endeavors such as land use planning and stricter approaches to determining water availability). The Water Resources Department's basin planning efforts ground to a halt in the mid-1990s.

Another policy re-orientation effort, and perhaps the most successful, was the adoption by the Water Resources Department of eight statewide water policies between 1990 and 1992. These policies (OAR Chapter 690, Division 410) were to serve as the building blocks of a policy superstructure to guide agency decisions and programs (Chapter 3). Many have become like New Year's resolutions: espousing laudable goals, but forgotten under the crush of current events. Few have enough volume to penetrate the noise of daily departmental operations and are seldom used either to justify or compel agency actions. Two notable exceptions are the allocation and conservation policies, both of which set in motion important programs that changed the shape of Oregon's water management.

One fairly recent mechanism for getting at future water needs is "reservations for future economic development." In passing the 1987 legislation establishing what to some was the bitter pill of instream water rights, the legislature also provided some sugar: the ability to reserve water for future need (ORS 537.356–537.358). A water reservation works the same way as reserving a table at a restaurant. People coming in the door without reservations may be seated, but not at your table—at least not until the appointed hour has passed.

Reservations are a two-step process. First, there must be a request to put aside a quantity of unappropriated water. (Only state agencies can request a water reservation—but they can act as a booking agent for other interests.) If the request appears to be in the public interest, the Water Resources Department may approve it. Second, future applicants may then request water permits, identifying the -reserved water as their source. The department evaluates these applications according to its processing rules. If a permit is issued, it will reach back to the priority date of the reservation and may be superior to many uses issued through the normal application process since the reservation (OAR Chapter 690, Division 79). In other words, the permittee may move through the crowded lounge, directly to their table.

However, whether reservations actually represent a workable vehicle for securing a desired water future remains to be seen. The process has been messy, with a number of reservations invalidated, disputed, or hung-up by opposing interpretations of the statutes, multiple rule amendments, and technical disputes. Some requests have been approved; those made by the Department of Agriculture (mostly for future multipurpose storage projects) in the Powder (roughly 25,000 acre-feet) and Hood River (over 80,000 acre-feet) basins were approved in 1996.

Interestingly, the most explicit statement of water management goals may be found not in any water resources document, but in Oregon's Benchmarks. Benchmarks are statewide measurable indicators that Oregon uses to assess its progress toward broad strategic goals. They act as goal-setting reference points and priority-setting aids for programs and budgets. Benchmarks measure results, rather than efforts. Instead of merely projecting current trends, the benchmarks set oftentimes ambitious resource protection goals (Oregon Progress Board 2005). At least seven of the state's ninety benchmarks set specific water resource-related targets (Table 41).

Oregon's blasé approach to water might be contrasted with that of a state where sky-rocketing water demand has created more of a water-consciousness: California. In 1957—about the same time the Oregon Legislature mandated an integrated state water program—the Department of Water Resources published the California Water Plan to coordinate current and future use of California's water. The Water Plan is California's strategic plan for managing and developing its water resources. It does not mandate actions nor authorize spending, but provides a framework for the coordinated control, conservation, development, management, and efficient use of water resources. The plan describes the role of state government and the growing role of California's regions in managing the state's water resources.

The plan includes "water portfolios" for the state as a whole and for each of its ten hydrologic regions. Portfolios are collections of diverse strategies that cover the entire hydrologic cycle and water quality conditions, encompassing more than eighty categories of water use, supply, and management. Rather than just assuming the state is on one path to the future, the plan presents three plausible but distinct scenarios for 2030. The scenarios describe future conditions that could develop if water resource use continues along current trends, becomes less intensive, or becomes more intensive. The water portfolios are evaluated against these future possibilities. The California Water Code requires the plan to be updated every five years to incorporate revised estimates of future water demands and the delivery capability of existing and planned facilities. The plan has been updated eight times between 1966 and 2005 (State of California 2005).

The closest Oregon has come to really getting interested in its water future (aside from the bold and progressive establishment of its 1909 water code) was the publication of the State Water Resources Board's "Oregon's Long-Range Requirements for Water" in 1969—a comprehensive analysis of water supply and demand that has not been repeated.

If Oregon's water planning has had meager results, its better-known land use planning system may serve to some degree as a substitute process.

Table 41. Oregon's water-related benchmarks

Benchmark	Description	Target 2005	Target 2010	Reported (year)	Comment (paraphrased from source)
69. Drinking Water	% of Oregonians served by public drinking water systems that meet health-based standards	95%	95%	95% (2005)	Oregon achieved the 2005 target level in 2003. However, 16% of community water systems still have a health based violation in 2003.
77. Wetlands					
a. Freshwater	Number of freshwater wetland acres gained or lost in any given year	0	0	75 (2004)	Oregon achieved the 2005 target from 2001-2004. While only a few new acres have been added, this is a positive trend for the state.
b. Estuarine	Number of estuarine wetland acres gained or lost in any given year		250	13 (2004)	Despite high hopes that restoration programs would add wetlands in river estuaries, little progress has been made since 2001. Data collection problems could be masking actual progress that is going unreported
78. Stream Water Quality					
a. Increasing	% of monitored stream sites with significantly increasing trends in water quality (over 10 years)	75%	75%	24% (2004)	The continuing drop in the % of sites with improving water quality trends reflects a tapering off of the benefit from water quality management plans implemented in the early 1990s.
b. Decreasing	% of monitored stream sites with decreasing trends in water quality (over 10 years)	0%	0%	10% (2004)	Despite long-term improvements in most streams, a small % of Oregon streams have worsening water quality, with about one stream in 10 showing a worsening trend.
c. Good or Excellent	% of monitored stream sites with water quality in good to excellent condition	40%	50%	49% (2004)	Oregon reached its 2005 target of 40% of monitored streams rated good or excellent in 2000 and has steadily increased since.

Benchmark	Description	Target 2005	Target 2010	Reported (year)	Comment (paraphrased from source)
79. Minimum Stream Flow Rights					
a. 9 mos./year	% of key streams meeting minimum flow rights 9 months per year	60%	65%	65% (2003)	After being above the 2005 target for most of the 1990s, Oregon may not meet that target in 2005 due to a string of recent years of low precipitation.
b. 12 mos./year	% of key streams meeting minimum flow rights 12 months per year	35%	40%	35% (2003)	Low precipitation in 2005 jeopardizes Oregon's ability to meet the 2005 target.
84a. Hazardous Substance Clean-up: Tank Sites	% of identified Oregon hazardous substance tank sites cleaned up or being cleaned up	80%	85%	92% (2004)	Since 1998 Oregon has made steady progress in cleaning up hazardous substance tank sites.
85. Freshwater Species					
a. Salmonids	% of monitored salmonids not at risk:	not set	not set	50% (2004)	The % of salmonids (salmon and steelhead) not at risk has remained at 50% since 1999 despite major recovery efforts by many sectors of society.
b. Other	% of monitored other fish & organisms not at risk	not set	not set	92% (2004)	The % of freshwater non-salmonid fish (e.g., Oregon & Borax Lake chub) considered at risk has held steady since 1999.
88. Species Protection - Rivers & Streams	% of at-risk species residing in rivers and streams protected in dedicated conservation areas	35%	38%	28% (2004)	The % of species residing in rivers and streams that are protected in dedicated conservation areas went up modestly in 2004. However, Oregon is unlikely to achieve the 2005 target.

Source: Oregon Progress Board 2005

◘ *Land Use Planning Can Be Water Planning*

Oregon's land use planning laws define the "land" in "land use plan" to include "water, both surface and subsurface" (ORS 197.015(5)). In addition, fifteen of the nineteen statewide goals specifically reference water resources, usually in terms of considering carrying capacity (i.e., the ability of water sources to support planned land uses). According to the goals, water is supposed to be taken into account when planning to preserve agricultural land, manage forestlands, and protect natural resources, or when providing for economic development, housing, transportation, and recreation. Clearly, the legislature had something watery in mind for local governments as they developed their state-mandated comprehensive land use plans. But beyond the lofty guidance in the goals, very little practical direction was provided local planners to help them deal with water as they wrote the plans. Therefore, local land use plans focused on critical issues of the 1970s (when Oregon's land use planning program began) such as farmland protection or establishing urban growth boundaries. Water was not a hot-button issue. Consequently, today most local plans are not well coordinated with local water supply or water management realities.

One exception may be the specific instructions given local communities in drafting public facilities plans. These plans describe the water, sewer, and transportation facilities needed to serve designated land uses in a timely, orderly, and efficient manner. Any community with twenty-five hundred or more people inside its urban growth boundary is required to adopt a public facilities plan (ORS 197.712(2)(e)). Water systems include the water source itself and the treatment, storage, pumping, and primary distribution system (OAR Chapter 660, Division 11).

However, Oregon's land use planning system does not explicitly tie water facility planning to the state's water planning and management system. Consequently, it is all too possible for jurisdictions to target water sources that may already be fully appropriated and effectively off-limits to development. Nonetheless, for urbanizing areas, public facilities plans may be the nearest thing there is to local water planning. These plans at least raise for public debate the suitability of future water sources, the costs of piping water to existing and future residents, and the role of existing water providers in meeting the demands of growth.

The role of existing water providers is especially important. Water control districts, domestic water associations, domestic water cooperatives, and irrigation districts (see Chapter 8) are explicitly listed as special districts in

ORS 197.015. Oregon's land use laws state that all plans, programs, rules, or regulations affecting land use adopted by special districts must comply with the statewide planning goals (ORS 197.250). Cities and counties (and metropolitan service districts, where present) must enter into cooperative agreements with special districts. These agreements must describe how the special district will be involved in local land use planning; establish responsibilities relating to new development; and define city and county roles with respect to district interests including water sources, capital facilities, and real property (ORS 195.020).

Explicitly making water providers key participants in land use planning would seem to set the stage for successful land and water coordination, especially in local urban areas. However, few special districts have fully participated in land use planning, nor have any substantive agreements been struck between districts and local governments. Thus, even this legally well-developed (and therefore rare) aspect of meshing water and land management fails to bridge the gap between the two.

In summary, land use plans were supposed to consider water resources. However, other pressing issues received more attention and water was largely overlooked. Today most plans are decoupled from water supply realities. Each plan has an implied water demand (the sum of the water needed by urban, agricultural, and environmental land uses) that in most cases has yet to be quantified. Until the land and water gap is bridged, the intricate and imposing systems designed to manage these resources may work at cross-purposes—causing confusion rather than creating the certainty originally promised under Oregon's planning system. That certainty disappeared by vote of the people in 2004 with the approval of Ballot Measure 37.

During the thirty previous years that Oregon's land use planning system had operated, a significant part of its citizenry never liked it. It had been challenged in the courts and at the ballot box multiple times. The notion of government restricting private property rankled, especially in rural areas, where people felt under the gun of regulations imposed by the urban elites of the Willamette Valley. The anger intensified as logging levels on federal forestlands plummeted and endangered species protections bit into the ability to support rural economies. At the same time, the conservative wave that swept the nation around the turn of the twenty-first century was felt in Oregon as well. Private property and individual rights were embraced more tightly than the older values of public interest and community commonwealth, the values that drove the passage of Senate Bill 100, which created Oregon's land use program in 1973.

The title of Ballot Measure 37 couldn't have been simpler: "Governments must pay owners, or forgo enforcement, when certain land use restrictions reduce property value." The proposition seemed like a no-brainer for many: if the Government was going to take away value from a private property owner, it was a matter of simple fairness that it pay that owner. And, if it couldn't pay, then it shouldn't be messing with the owner to begin with. Significantly, the measure was written not just prospectively, but included land use regulations restricting private property enacted prior to its passage.

Ballot Measure 37 passed with a margin of 60 percent in 2004. It amended Oregon statutes (not the Constitution) to provide that owners of private real property are entitled to compensation when a land use regulation—if enacted after the owner or a family member had possession of the property—restricts its use and reduces fair market value. The measure provided no funding mechanism for compensation, but allowed that instead of compensation, the enacting government may choose to "remove, modify or not apply" the regulation (Oregon Land Conservation and Development Department 2006).

The measure exempts certain land use regulations that control public nuisances (including nude dancing and porn shops) and risks to the public's health and safety (including "pollution control regulations"), or are required by federal law. Measure 37 broadly defines "land use regulation" to include "any statute regulating the use of land or any interest therein"; "statutes and administrative rules regulating farming and forest practices"; local government comprehensive plans, zoning ordinances, land division ordinances, and transportation ordinances; DLCD goals and rules; and many of Metro's (the Portland area's regional government) plans.

It is clear that Measure 37 altered the foundations of Oregon's planning system. It is anything but clear, however, how that alteration will be manifest in the landscape and in the very function of government. As of 2006 (when the Oregon Supreme Court overturned a lower court decision by reinstating the measure), there were many unsettled—and fundamental—questions. To name a few, it was not clear whether the latent development rights the measure freed up (if compensation were not provided) were transferable to anyone but the original owner or family; what proof of qualification and value reduction should accompany claims; or even which government body to file claims with (because many land use regulations are required by the state, claims in those cases need to be filed with both the local government and the state—and they may issue different decisions) (Sullivan 2005).

Measure 37 is also unsettled in terms of natural resource protections, including water. Although it exempts "pollution control regulations" for public health and safety, it is noteworthy that this differs significantly from the normal statutory construction, "public health, safety, *and welfare*." The measure does not specify what it means by "pollution control regulations" (Sullivan 2005). It does not incorporate by reference either state or federal statutes, nor does it offer a definition.

As of 2006, the measure had only begun to take its course, leaving many issues to the realm of informed speculation. Clearly, however, there are many potential planes of intersection between Measure 37 and water. Any regulations that impose conditions on land use to protect water supply seem ripe for claims. Protecting water supply for existing and future uses would seem to fall largely under the heading of "public welfare" (after all, Oregon's water code keys off protecting the public interest), but that doesn't buy anything in terms of cover from Measure 37.

For example, is keeping groundwater tables from dropping out from underneath landowners an exercise that protects public health and safety? Groundwater withdrawal doesn't create giant suck-holes, inhaling houses and industries (at least in most Oregon landscapes), or spread contagion. Therefore, it would seem a given that property owners feeling injured by a state or local government rule prohibiting them from sinking new wells could file a Measure 37 claim.

For a 2004 analysis of the measure's potential impacts, the Water Resources Department was advised by the Attorney General's office to focus on claims resulting from "new restrictions on a right to use water that had previously been granted by the Water Resources Department or Legislature" (Oregon Water Resources Department 2004b). For the most part, this principle would seem to frame Measure 37 impacts to water supply. Boiled down, the measure targets regulations that take away options originally available when property was first acquired. Since 1909, having the right to use water on your land was never guaranteed, but was subject to a separate state decision. Therefore, only new limits on already-granted rights would be subject to Measure 37. So, where might the state impose such new limits?

One of the first areas that comes to mind is groundwater. In a Critical Groundwater Area, for example, the Oregon Water Resources Department may restrict exercise of existing groundwater permits and rights (Chapter

4). In its 2004 analysis, the department estimated claims could total as high as $45 million. It is not unthinkable that some might argue that, because the groundwater uses listed in ORS 537.545 (domestic use, lawn and garden watering, small commercial purposes) do not require a permit, they might be said to occur by right on lands where such uses are allowed (i.e., areas that allow houses or commercial use). Although it never has, should the Water Resources Commission feel compelled to withdraw all groundwater in an area from further appropriation, including from heretofore exempt uses (perhaps because of long-term drought, impacts to surface water, etc.), Measure 37 claims could easily arise.

In addition, a growing number of local governments (such as Marion County) have developed groundwater protection ordinances that control land use intensities. In other words, some county governments are using their own authority to protect sensitive groundwater areas (often the state's designated groundwater limited areas) from overdraft due to development. Of course, these, too, fall under the scope of Measure 37. Some view such ordinances as having been effectively invalidated by the measure. If so, by some estimates this may lead to seventy-five hundred to ten thousand new domestic wells, perhaps doubling the level of domestic use statewide. Because each domestic well is entitled to pump up to 15,000 gallons each day (a containerized cargo truck's volume), the impacts could be significant (Jarvis 2005).

Theoretically, a number of other water supply issues may arise as Measure 37 is implemented. For example, if because of a drought or intense pressures on the resource, the Water Resources Department were to take a harder line on waste, enforce conditions on a water right for the first time, or set water efficiency standards for all existing and future agricultural rights, Measure 37 claims might result, either on the basis that an existing right was restricted or, more likely, that the action met the measures definition of being a statute or rule "regulating farming ... practices."

Really, anything is possible. Claimants don't have to be right, just believe they are right, to file a claim. Aside from some claim-filing fees, government bodies foot the bill for the claims process—and there is no provision for them to recover attorney fees and costs if a claimant brings a failed claim (Sullivan 2005). Through the many years that Measure 37 is litigated and amended (in early 2006, there were over thirty court actions pending), there will be many claims filed, both with and without merit. Governing bodies will have to deal with (and pay for) them all. In the universe of potential claims, some will likely square off against water laws, for there is no telling what may come out of the woodwork.

As of late 2005, over a thousand claims had been filed against local governments. Over 85 percent were for land divisions and dwellings. The measure was promoted primarily on the basis of the ability of a property owner to build a home. However, a significant number of claims were for much more than that, including the right to build a casino complex, a 300-acre motorcycle-ATV-paintball park, a rural commercial center with more than 1 million square feet, and a subdivision of over eight hundred homes on one-quarter-acre lots (Georgetown Environmental Law and Policy Institute n.d.).

Interestingly, water played a role in the lawsuit that brought the constitutionality of Measure 37 before the Oregon Supreme Court. A plaintiff alleged that he would suffer a number of injuries stemming from a neighbor's successful Measure 37 claim, including diminished water quantity and quality. Those supporting Measure 37 argued he did not have standing to sue, because he could not show actual, concrete impacts. The court, while concluding Measure 37 was indeed constitutional, also ruled that the plaintiff's assertions were sufficiently plausible to grant him standing. Thus, the stage seems well set for future court battles over water-related injury (Oregon Supreme Court 2006).

CHAPTER 12

Reflections

LIKE THE ALPENGLOW FROM MT. HOOD caught and given back by Lost Lake, reflections are fleeting. Rippling, approximate, and framed by a single point of view, how accurately they represent their source is dependent on wind, water, and illumination. This book has attempted to throw some light on water and better understand the winds of change. As even a quick look in any grocery check-out line or airport gift store will confirm, reflections make for good postcards. So, consider this last chapter a collection of postcards—some reflections on Oregon waters; a few notes scribbled on the way to a new century of water management.

The first part of any journey is understanding where you are and where you need to go. If we really want to get somewhere (as opposed to "anywhere," seemingly our current destination), we need to get serious.

While Oregon has enough water to count its blessings, it doesn't have near enough to ignore the basics. Given continued rapid growth and the likely impacts of climate change, Oregon needs to take water seriously. We really don't, for if we did, we'd know:

SYSTEM INPUTS: What's coming in each year, water-wise? We have some idea, but it's not very precise. We continue to drop gaging stations and our well monitoring network is pretty spread out. Compiling data is not a sexy budget line item. Thus, we're stuck using a water abacus.

SYSTEM OUTPUTS: How much water are we using each year? Again, we have some idea, thanks to the federal government's estimates of water use, and they're helpful. But there isn't a good way to verify them, since we don't measure most water use.

WHO OWNS WATER RIGHTS: The State's records of water right ownership start and stop with the issuance of the right. Who buys the land afterwards— that is, the true owner—is invisible because water rights records are not updated with real estate transactions. When it is time to track down the real owner (or owners, if the land was divided) for a transfer, or instream lease, or other reason, the project requires the wiles of a detective and the patience of a monk.

HOW MUCH WATER WE'LL NEED: No one is keeping track on a state basis of the water we'll need for a population that doubles every fifty years, a changing rural landscape, sustaining and expanding agricultural markets, or making sure fish and other species don't wink out on our watch.

Getting serious would likely mean intersecting some or all of the following waypoints on our journey.

⊞ *Measure the Treasure*

On any given day, we probably know our weight, the approximate level of the Dow Jones Industrial Average, the President's approval rating, the price of gas, and what's in our checking account. We track what is important to us. If water were, we would:

- Support, if not expand, the existing stream gaging system.
- Expand the state's well-monitoring network to better characterize Oregon's major aquifers.
- Better link land and water planning by tallying the water demands of planned land uses, comparing the aggregate demand to available water supplies, and flagging supply gaps.
- Tie water rights to county tax lot records to have up-to-date records of water right ownership.
- Measure water use either by phasing in a requirement for all water right holders to measure their use, or by expanding reporting programs, or by conducting a periodic water use census employing accepted statistical sampling methods.
- Every decade or so, assess Oregon's water supply and needs over the next fifty-year period and use it to frame water management decisions.

⊞ *Manage Water Like Oregon's Petroleum*

Though seldom stated this way, the State of Oregon has a fiduciary responsibility to the people to manage their water. If we think about Oregon's store of water as an investment portfolio, where managers deal in different currencies (surface water, groundwater, storage) and invest in municipal funds, irrigation funds, industrial and utility funds, and green instream funds, we would likely conclude the portfolio hasn't been managed very well when we look at our declining water tables, disappearing fish, dried-up creeks, and grim water recessions like the Klamath. There has, of course, been some return on investment—we would have neither big cities nor big agriculture without the benefit of past management efforts.

But Oregon's water management has been largely passive, as if the investment objective was not profit, but simply to not lose money. And management is surprisingly "retro." So much of it is caught up in water distribution (the ritual of protecting the rights of oldest users), adjudications (the archeology of figuring out century-old vested rights), and water right geneology (the craft of trying to determine, based on land transactions, who actually owns a right). There are few tools to consider, let alone provide for, future water users. As a state, Oregon has its back turned to its water future.

The lack of future focus is not the fault of state or federal water agencies, city public works departments, irrigation districts, watershed groups, environmental non-profits, or the thousands of people working Oregon's water beat today. On a day-to-day basis, they keep water flowing to hundreds of thousands. Oregon's water workforce is overwhelmingly professional and passionate. It's just that the system it operates within is the eight-track tape (rotary phone, 7.5 inch floppy, black and white TV, wind-up watch) of natural resource management schemes.

Comparing water to money may seem a bit of a stretch, but it shouldn't be. We know water has actual monetary value. Various studies show that having irrigation water available can add at least $2,000 to the value of an acre of land in Oregon, and that the value of an acre-foot of water can range from around $10 to over $700, with most transactions settling in at around $300 (Butsic and Netusil n.d.). While the value of water is all over the map (varying widely by season, type of use, priority date, and, in the case of irrigation, the quality of land it is applied to), there can be no doubt it is worth real money.

The U.S. Geological Survey estimates Oregon's annual water use amounts to about 8 million acre-feet (Chapter 7). If we assign the low-end value of $10 per acre-foot to that quantity, it suggests the State of Oregon's annual water portfolio may have a latent value of around $80 million. If this water, so priced, were viewed as an annual hydrologic harvest of public water, it would rival the value of timber harvested from state forests (which was worth about $100 million in 2000) (Oregon Department of Forestry n.d.).

Of course, people pay to harvest state-owned timber, just as they pay to graze animals, extract minerals, mine gravel, or place a dock or dam on state lands. All told, in the State's 2001-2003 budget period, these types of transactions brought in over $27 million to Oregon's Common School Fund (Oregon Department of State Lands. n.d.[e]). Needless to say, the amount brought in from charges to use the public's water (excepting hydroelectric use) was $0. Either we have it right in managing other state assets through a use fee and have it wrong with water; or, we are right with water and have wronged everyone we have ever charged for profiting from use of public resources.

But to stop charging for cutting state timber seems as far-fetched as to start charging for using its water. But, rightly or wrongly, in our society paying money for something is the truest measure of value, our most familiar way of ordering and expressing choice. Putting things in dollar-equivalents is one of the things we're really good at. We may read movie reviews, but what gets the headlines is the weekend box office; we may loathe an athlete, but nearly always respect the size of the free agent contract; we may want more meaning out of life, but net income trumps; time may fly, but it's always money.

Admittedly, a proposition to charge for a commodity that for a hundred years has been free would seem dead in the water. However, Oregon is a place where unlikely things can and do happen—our beaches are public, not because of an environmental crusade, but because they were deemed necessary public highways; I-5 and the rest of the interstate system was sold as a way to promote national defense; we don't have a sales tax or self-serve gasoline, but we do have death with dignity.

While a long-shot, treating water more like a valuable state asset and less like an article of religion would do a lot. If we paid even a little for the water we used (not just for its treatment and delivery) for drinking, lawns, crops, and factories, it would become painful to waste. Every minute the sprinkler was on beyond what was required, every dripping faucet, every jet from loose fittings would have us hemorrhaging money and we'd be driven to stop it. If water were truly money, we'd track it a lot better. The state would certainly have an incentive to have up-to-date records of who owns water rights. Snowpack surveys and gaging station reports would be anticipated like quarterly revenue projections. Monitoring wells would no longer be expensive holes for water wonks, but powerful indicators of future profit potential. Lastly, water fees could be a new source of revenue for schools, or local water supply investments. But seldom has the distance between vital and valued been so great. It's time to starting treating water like Oregon's petroleum.

Putting water on an equal footing with public timber or forage would entail: 1) finite terms of use; and, 2) paying something for the use. Doing either of these would require Oregonians to confront extremely difficult questions:

- What would the impact be on small operators and rural communities, and what would be the public's obligation to cushion hardship imposed by new costs?
- Should the changes be applied only to new permits, or should they apply to already-issued water rights, as well?
- What is the object of any use fee—to generate funds for water management? to support conservation? to support schools?
- Where would the money go? To Salem for distribution to schools or to increase water budgets of state agencies? Or would it stay in the locality, to be used for county water planning and water services, local watershed councils, and other local water management needs?
- What is water worth?

Whether we choose to answer these hard questions is a matter of public will, mutual respect, and strength in leadership—should any leaders step forward. However, if exploring the potential for water asset management opens a Pandora's box of factional bitterness, regional intolerance, political ego, bureaucratic edema, or economic turmoil, perhaps it is better off left alone.

It may be enough to ask out loud, at least once per century, if water is worth paying for. If the answer is no, and Oregonians choose to re-affirm the current contract between water users and the public, then in return for free use of its water, the public would seem entitled to an activated thrift by users and to their uncomplaining and creative cooperation in addressing water supply problems.

▣ *Pony Up*

Knowing that a) water demands will continue to intensify from population growth, at-risk species needs, and agricultural diversification; b) we don't have enough water to go around as it is; c) it's getting warmer; and, d) water is a finite resource might be cause for concern among fringe elements in society who view water as essential. Or cause for investment.

What kind of water supply investments might be needed? Our changing Oregon could require quite a few. Given present rates of growth, Newport, Lincoln City, Yachats, and Waldport will exhaust their current water supply in twenty years (*Oregonian* 2005). Plus, it is estimated that Oregon's drinking-water infrastructure needs $2.7 billion over the next twenty years (American Society of Civil Engineers n.d.). Looking at population trends and endangered species needs in the West, the Bureau of Reclamation identified where water supply crises and conflicts might erupt by 2025. The Klamath Basin was tagged—no surprise—but so also were the Columbia River and the Rogue, Lower Willamette, Umatilla, and Upper Deschutes basins. Noting that "crisis management is not an effective solution for addressing long-term, systemic water supply problems," the bureau is targeting investments in conservation, improved technologies, developing markets, and better collaboration (U.S. Bureau of Reclamation 2003).

It is possible for states to mean business in this regard. Of course, California has long invested in water through its State Water Project; by the end of 2001, about $5.2 billion had been spent on project facilities (California Department of Water Resources n.d.[b]). And in 2006, the governor proposed spending another $35 billion between 2006 and 2014 for flood control and water projects to be funded through state and local funds and general obligation bonds (California Budget Project 2006).

In 2006, the State of Washington passed a bill directing its water agency to "aggressively pursue the development of water supplies to benefit both instream and out-of-stream uses." It established a fund "to assess, plan, and develop new storage, improve or alter operations of existing storage facilities, implement conservation projects, or any other actions designed to provide access to new water supplies within the Columbia river basin" (Washington 2006). The Washington Legislature authorized the state to issue up to $200 million in general obligation bonds to undertake studies and pay for conservation projects over a ten-year period. And this is just part of a possibly even bigger picture, one where proponents envision huge new storage projects (with big off-stream components) costing as much as $8 billion (Energy NewsData 2006). Admittedly, this may be pie-in-the-sky for Northwesterners. But it also just might be thinking big—one of the symptoms of taking water seriously.

▣ *Be Civil*

Five hundred seventy-six miles. That is the longest distance shown between selected cities (Gold Beach to Ontario) on a recent Oregon Department of Transportation mileage table. But that's nothing compared to the light years separating urban and rural Oregon, especially on natural resource issues—including water. Because they are based more on direct experience, rural views on water tend to be traditional, utilitarian, informal, individual, pragmatic, and economic. Economic imperatives are in the foreground, environmental concerns in the background. Because water is less a raw material than a defining amenity to urbanites, their views are typically progressive, communitarian, legalistic, theoretical, and ecologic. Environmental imperatives occupy the foreground, economic concerns the background. Rural Oregon likes storage and the unfettered use of water for farming. Urban Oregon likes to see water free-flowing for nature and recreation. That these views clash is neither a surprise nor ought it be a problem—at least not in a functioning democracy. Yet it is a problem, given today's intolerant political climate and two great inequalities that cleave our water management system:

1. Rural Oregon, through its agricultural roots, has control of most of the state's water.
2. Oregon is an urban state.

"Control" not only means an 80 percent share of water withdrawals, but the senior priority dates to demand it—the accumulated advantage gleaned from a system built to promote development. "Urban" means that about 70 percent of Oregonians live in urban areas (Cai 2005) and only 2 percent live on farms (U.S Bureau of the Census 1996). It also means that over 40 percent of the state's voters live in the Portland metropolitan area alone (Oregon Secretary of State 2006). This haves vs. have-nots syndrome (both for water and votes) leads to growing stress which, left unrelieved, may sunder the system along a fault of intolerance and misunderstanding.

Even the briefest self-assessment by either side should prompt enough humility to temper any intolerance. Rural (agricultural) Oregon should admit that water is just as public as a state park, and that rural people have received the lion's share of the right to use it, and that using it has turned many ecosystems upside-down. It might also acknowledge that irrigated agriculture is not a virtue, but a business; and that, legalities aside, keeping flows in streams is not radical, but pumping them dry is. Urban Oregonians should own up to not knowing beans about their own dam and diversion systems that long ago extirpated local fish populations; and to seeking environmental balance most often where they don't live, even as water-craving subdivisions, factories, and golf courses multiply in their own backyards. They might also consider that irrigated agriculture is not a vice, but a cornerstone of local and state economies; or that somewhere among the 66 million acre-feet of water and 60 million acres in Oregon, there might be room for storage that offers both environmental and economic benefits.

The stress induced by the inequalities might be relieved by civil sharing. Rural Oregon should know that a death grip on water is not the answer, lest it provoke an urban electorate to pry it loose through the ballot box. Agricultural interests should be the first to articulate and live up to a water code of conduct and identify ways to remedy instream shortages or address municipal concerns. If urban Oregon wants more water, it should drop the attitude and work with the people who have it, by pursuing market-oriented water purchases, leases, or other arrangements; developing a more tolerant attitude toward storage; actively supporting incentives for conservation; encouraging investment in rural communities; and just generally behaving as if there were a world of real people beyond the urban growth boundary.

Conclusion

Today, Oregon's once-progressive system of public ownership and management of waters too often operates, not in support of the public's interests, but in isolation from them. The resulting inequities in water allocation have led to an unrewarding conflict pitting new users against old, urban against rural, and environmental against agricultural interests. None of these are pre-ordained enemies of the other. Conceivably, each has something to contribute to the other in securing the benefits of water for the people of the state. However, the object of wise water management has been obscured by the invective of battle. Not that opposition and competition over water are avoidable—they are not. In fact, they are necessary to sharply etch the choices we must make in shaping our water future. But an informed opposition generates debate. And debate is preferred over the often pointless battle we are now witnessing.

Leaders stepped forward in 1909 to imprint the dawning century with a new wave of water management. We badly need that kind of leadership today. As unlikely as that may be given the meat-grinder nature of our current politics, in many ways, Oregon has it too good not to succeed: we have a considerable store of water wonderfully expressed through world-class rivers; a well-established statewide planning system; respectable (and sometimes cutting-edge) water management tools which can result in a fair balance, fairly achieved; and a population, still of manageable size, which may yet act upon a sense of duty and privilege from living and working among the tumult and trance of our Oregon waters.

Oregon State Archives, Oregon Highway Division, OHD5816

Bibliography

Achterman, Gail L. 1992. Reflections on owls, salmon, and suckers: Current developments under the Endangered Species Act. *Rocky Mountain Mineral Law Institute* 38:5-1.

American Society of Civil Engineers. n.d. Report Card for America's Infrastructure Web site. Oregon. Accessed April 17, 2006 at http://www.asce.org/reportcard/2005/page.cfm?id=77

Baldwin, Pamela. 1990. *Water Rights and the Wild and Scenic Rivers Act*. March 30, 1990; 90-196A. Washington, D.C.: Congressional Research Service, the Library of Congress.

Bishop, Ellen Morris. 2003. *In Search of Ancient Oregon: A Geological and Natural History*. Timber Press, Portland, Oregon.

Bonneville Power Administration, U.S. Army Corps of Engineers, and U.S. Bureau of Reclamation. 2001. *The Columbia River System Inside Story*. Second Edition. Downloaded March 31, 2006 as http://www.bpa.gov/power/pg/columbia_river_inside_story.pdf

Borden, John. 1989. Oregon's minimum perennial streamflows. In *Instream Flow Protection in the West*, edited by Laurence J. MacDonnell, Teresa A. Rice, and Steven J. Shupe. Boulder: Natural Resources Law Center, University of Colorado School of Law.

Broad, T. M. and D. D. Nebert. 1990. Oregon: Water supply and use. In *National water summary 1987 — Hydrologic events and water supply and use*. U.S. Geological Survey Water-Supply Paper 2350. Compiled by Jerry E. Carr, Edith B. Chase, Richard W. Paulson, and David W. Moody. Denver: U.S. Government Printing Office.

Brown, Steve. 1996. Water Use Reporting Program Coordinator, Oregon Water Resources Department. Electronic mail to the author, 17 July.

Butsic, Van, and Noelwah R. Netusil. n.d. Estimating a price for water rights in the Umpqua basin, Oregon. In Proceedings of 2004 Universities Council on Water Resources Annual Conference, Allocating Water: Economics and the Environment. Downloaded April 12, 2006 as http://www.ucowr.siu.edu/proceedings/2004%20Proceedings/2004%20UCOWR%20Conference%20Proceedings/Tuesday/AM%20Technical%20Sessions/Session%202/Butsic&Netusil.pdf

Cai, Qian. 2005. 2004 Oregon Population Report, Population Growth in Oregon: 2000 to 2004; Population Estimates for Oregon, July 1, 2004. Population Research Center, College of Urban and Public Affairs, Portland State University.

California Budget Project. 2006. Governor releases proposed 2006-07 budget. Revised January 19, 2006. Accessed April 14, 2006 at http://www.cbp.org/2006/060110_govbudget.pdf

California Department of Fish and Game. 2004. September 2002 Klamath River Fish-Kill: Final Analysis of Contributing Factors and Impacts. July 2004. Northern California-North Coast Region, The Resources Agency, State of California. 183 p. Accessed March 28, 2006 from http://www.pcffa.org/KlamFishKillFactorsDFGReport.pdf.

California Department of Water Resources. 2005. State Water Project Annual Report of Operations 2001. Division of Operations and Maintenance. April 2005. 85 p. Accessed March 17, 2006 from http://wwwoco.water.ca.gov/annual/annual01.pdf

California Department of Water Resources. n.d. [a]. State Water Project--Overview Web site. Accessed March 17, 2006 from http://www.publicaffairs.water.ca.gov/swp/.

California Department of Water Resources. n.d.[b]. State Water Project Today Web site. Accessed April 14, 2006 at http://www.publicaffairs.water.ca.gov/swp/swptoday.cfm

Clark, Chapin D. 1983. *Survey of Oregon's Water Laws.* Oregon State University Water
    Resources Research Institute, WRRI 18. Corvallis: Oregon Law Institute.
Clewett, Ed, and Elaine Sundahl. 1990. A view from the south: Connections between
    southwest Oregon and northern California. In *Living with the Land: The Indians
    of Southwest Oregon.* The proceedings of the 1989 symposium on the prehistory of
    southwest Oregon. Edited by Nan Hannon and Richard K. Olmo. Medford: Southern
    Oregon Historical Society.
Climate Impacts Group, University of Washington. 2004. Overview of Climate Change
    Impacts in the U.S. Pacific Northwest. Downloaded April 6, 2006 as http://www.cses.
    washington.edu/db/pdf/cigoverview353.pdf
Climate Impacts Group, University of Washington. n.d. Hydrology and Water Resources,
    Key Findings Web site. Impacts of Climate Change on PNW Hydrology and Water
    Resources. Accessed April 5, 2006 as http://www.cses.washington.edu/cig/res/hwr/
    hwrkeyfindings.shtml#anchor4
Dart, John O., and Daniel M. Johnson. 1981. *Oregon: Wet, High, and Dry.* Portland,
    Oregon: Portland State University.
Drucker, Philip. 1963. *Indians of the Northwest Coast.* Garden City, New York: The
    Natural History Press.
Dunbar, Robert G. 1983. *Forging New Rights in Western Waters.* Lincoln and London:
    University of Nebraska Press.
Dziegielewski, Ben. 2000. Efficient and inefficient uses of water in North American
    households. Paper presented at the Xth IWRA World Water Congress in Melbourne
    Australia, March 12-17, 2000. Accessed March 15, 2006 from http://www.
    watermagazine.com/secure/jc/0240.htm.
Energy NewsData. 2006. NW Fishletter #210, February 22. Accessed April 14, 2006 at
    http://www.newsdata.com/fishletter/210/2story.html.
Eugene Water and Electric Board. n.d. Project Web site accessed March 16, 2006 at http://
    www.eweb.org/news/carmensmith/about.htm
Federal Emergency Management Agency. 2004. Flood Insurance, National Flood
    Insurance Program, Introduction to the NFIP Web site. Last updated: Friday, 22 Oct
    2004. Accessed April 1, 2006 at http://www.fema.gov/nfip/intnfip.shtm#1.
Federal Energy Regulatory Commission. 2005. Outstanding Licenses as of 07/08/2005;
    Downloaded February 20, 2006 from http://www.ferc.gov/industries/hydropower/gen-
    info/licenses.xls.
Foster v. Foster, 107 Or. 291, 213 P. 895; In re Willow Creek, 74 Or 622
Fox, Stan. 1996. Hydrologist, U.S. Natural Resources Conservation Service. Telephone
    conversation with author, 14 May.
French, Dwight. 2006. Oregon Water Resources Department, Administrator, Water Rights
    Section Manager. Phone conversation with author, February 9.
Gannett, Marshall W., and Kenneth E. Lite. 1996. Effects of climate, stream stage, and
    irrigation canal seepage on groundwater levels in the Middle Deschutes Basin, Oregon.
    Accessed 14 June 1996 through U.S. Geological Survey site: http://wwworegon.
    wr.usgs.gov/projs_dir/or161/aih.abs.mwg.html.
General Accounting Office. 1996. *Information on Allocation and Repayment of Costs
    of Constructing Federal Water Projects.* GAO/RCED-96-109. Washington, D.C.:
    United States General Accounting Office, Resources, Community, and Economic
    Development Division.

Georgetown Environmental Law and Policy Institute. n.d. Summary of Measure 37 Web site, downloaded March 31, 2006 from http://www.law.georgetown.edu/gelpi/takings/stateleg/Measure37.htm

Gleeson, G. W. 1972. *The Return of a River — The Willamette River, Oregon.* WRRI 13. Corvallis: Oregon State University Water Resources Research Institute. In Broad and Nebert 1987.

Glick, Richard M. 1995a. The public trust and other doctrines affecting water rights. Paper presented at Law Seminars International Fourth Annual Conference on Oregon Water Law, November 3, 1995, at Governor Hotel, Portland, Oregon.

Glick, Richard M. 1995b. Oregon: Water rights for growing cities. Accessed 14 February 1997 through Hieros Gamos site: http://www.hg.org/1365.txt.

Gonthier, Joseph B. 1985. A Description of Aquifer Units in Eastern Oregon. U.S. Geological Survey Water-Resources Investigations Report 84-4095. Portland, Oregon: U.S. Geological Survey.

Gould, George A. 1990. Water rights system. In *Water Rights of the Fifty States and Territories.* Edited by Kenneth R. Wright. Denver: American Water Works Association.

Guinness Book of World Records 1998. Guinness Media, Inc. New York, Toronto, London, Sydney, Auckland: Bantam Books.

Hanford Watch. 2004. Introduction to Hanford issues Web site, by Lynn Porter. Accessed April 2, 2006 at http://www.hanfordwatch.org/

Hayes, Beverly. 1996. Hydroelectric Program Manager, Oregon Water Resources Department. Electronic mail to the author 12 June1996.

Houston, Laurie L., Michio Watanabe, Jeffrey D. Kline, and Ralph J. Alig. 2003. Past and Future Water Use in Pacific Coast States. General Technical Report PNW-GTR-588. July 2003. United States. Department of Agriculture, Forest Service, Pacific Northwest Research Station. 44 p. Downloaded March 15 as http://www.fs.fed.us/pnw/pubs/gtr588.pdf

Hutson, Susan, Nancy L. Barber, Joan F. Kenny, Kristin S. Linsey, Deborah S. Lumia, and Molly A. Maupin. 2004. Estimated Use of Water in the United States in 2000. U.S. Geological Survey Circular 1268. U.S. Department of the Interior, U.S. Geological Survey. 52 p.

Idaho Power Company. n.d. Project Web sites accessed March 16, 2006 at: http://www.idahopower.com/riversrec/relicensing/hellscanyon/brownlee/brownleefacts.asp; http://www.idahopower.com/riversrec/relicensing/hellscanyon/oxbow/oxbowfacts.asp; http://www.idahopower.com/riversrec/relicensing/hellscanyon/hellscanyon/hellscanyonfacts.asp

Institute for Natural Resources, Oregon State University. 2004. Scientific Consensus Statement on the Likely Impacts of Climate Change on the Pacific Northwest. Downloaded April 6, 2006 as http://inr.oregonstate.edu/download/climate_change_consensus_statement_final.pdf

Institute for a Sustainable Environment, University of Oregon. 2005. The Economic Impacts of Climate Change in Oregon. A Preliminary Assessment. October 2005. Downloaded April 5, 2006 as http://cwch.uoregon.edu/publicationspress/Consensus_report.pdf

Jackson, Philip. 2003. Climate. In *Atlas of the Pacific Northwest*, 9th edition, edited by Philip L. Jackson and Jon A. Kimerling. Corvallis: Oregon State University Press.

Jarvis, Todd. 2005. Measure 37 and Oregon's Groundwater: A Paradigm Shift from Water Management to Water Planning? Powerpoint presentation downloaded April 5, 2006 as http://water.oregonstate.edu/news/Measure37.pdf

Johnson, Daniel M., Richard R. Petersen, D. Richard Lycan, James W. Sweet, Mark E. Neuhaus, and Andrew L. Schaedel. 1985. *Atlas of Oregon Lakes.* Corvallis: Oregon State University Press.

Johnson, Norman K. 1992. *Western State Water Right Permitting Procedures.* Midvale, Utah: Western States Water Council.

Kammerer, J. C. 1990. Largest Rivers in the United States: U.S. Geological Survey Open-File Report 87-242, 2 p.

Kaufman, Joseph Q. 1992. An analysis of developing instream water rights in Oregon. Willamette *Law Review* 28:285—332.

Kitzhaber, Annabel. 1984. *Water Resources in Oregon.* Salem: League of Women Voters of Oregon Education Fund.

Klamath Basin Crisis. 2005. KWUA electrical power legal brief: Rate was part of original deal asserted by federal law. Posted to KBC Web site 3 July 2005. Accessed April 1, 2006 at http://www.klamathbasincrisis.org/Poweranddamstoc/powrlegalbrief030705.htm

Klamath Basin Irrigation Dist. V. United States, 67 Fed. Cl. 504 (2005)

Kulongoski, Ted. 1995. *Oregon Attorney General's Administrative Law Manual and Uniform and Model Rules of Procedure under the Administrative Procedures Act.* Salem: Oregon Department of Justice.

LaLande, Jeff. 1990. The Indians of southwestern Oregon: an ethnohistorical review. In *Living with the Land: The Indians of Southwest Oregon.* The proceedings of the 1989 symposium on the prehistory of southwest Oregon. Edited by Nan Hannon and Richard K. Olmo. Medford: Southern Oregon Historical Society.

Lee, Janet. 1996. Executive Director, Oregon Water Resources Congress. Telephone conversation with the author, 1 April.

Leopold, Luna B. 1994. *A View of the River.* Cambridge, Massachusetts: Harvard University Press.

Lewis, John H. 1907. State Engineer. *The Need of Water Legislation in Oregon, Special Report No. 1.* Salem, Oregon.

Liberman, Kenneth. 1990. The native environment: Contemporary perspectives of southwestern Oregon's Native Americans. In *Living with the Land: The Indians of Southwest Oregon.* The proceedings of the 1989 symposium on the prehistory of southwest Oregon. Edited by Nan Hannon and Richard K. Olmo. Medford: Southern Oregon Historical Society.

Lissner, Fred. 1996. Manager, Groundwater / Hydrology Section, Oregon Water Resources Department. In conversations with the author 28 June (in person) and 17 July (by telephone).

Los Angeles Times. 2006. U.S. Acts to Help Wild Salmon in Klamath River. As stocks plummet, agencies demand remedies from the owner of hydroelectric dams. Options include fish ladders, demolition. By Eric Bailey, Times Staff Writer. March 30, 2006.

Lower Columbia River Estuary Partnership. n.d. Information downloaded February 24, 2006 from http://lcrep.org/index.htm

Loy, William. 2001. *Atlas of Oregon,* second edition. Eugene: University of Oregon.

Mack, Joanne M. 1990. Archaeology of the Upper Klamath River. In *Living with the Land: The Indians of Southwest Oregon.* The proceedings of the 1989 symposium on the prehistory of southwest Oregon. Edited by Nan Hannon and Richard K. Olmo. Medford: Southern Oregon Historical Society.

Mail Tribune (Medford). 2005. Savage Rapids removal passed. Dam would come down after pumps installed. By Mark Freeman. 20 November 2005.

McArthur, Lewis. L. 1992. *Oregon Geographic Names*. Portland: Oregon Historical Society Press.

McFarland, William D. 1983. *A Description of Aquifer Units in Western Oregon*; U.S. Geological Survey Open-File Report 82-165. Portland, Oregon: U.S. Geological Survey.

McKee, Bates. 1972. *Cascadia: The Geologic Evolution of the Pacific Northwest*. New York: McGraw-Hill Book Co.

Mead, Elwood. 1906. The Evolution of Property Rights in Water, Bulletin No. 157, Office of Experiment Stations. In *First Biennial Report of the State Engineer to the Governor of Oregon for the Years 1905 - 1906*.

Metropolitan Service District. 1991. *Role of the State in Water Management*. By Laurence R. Sprecher. Portland, Oregon.

Moffatt, Robert L., Roy E. Wellman, and Janice M. Gordon. 1990. *Monthly and Annual Streamflow, and Flow-Duration Values*. Vol. 1 of *Statistical Summaries of Streamflow Data in Oregon*. U.S. Geological Survey Open-File Report 90-118. Portland, Oregon: U.S. Geological Survey.

National Marine Fisheries Service. 1997. Investigation of Scientific Information on the Impacts of California Sea Lions and Pacific Harbor Seals on Salmonids and on the Coastal Ecosystems of Washington, Oregon, and California. NOAA Technical Memorandum NMFS-NWFSC-28. Accessed April 1, 2006 at http://www.nwfsc.noaa.gov/publications/techmemos/tm28/tm28.htm

National Marine Fisheries Service. 2000. Endangered Species Act—Section 7 Consultation, Biological Opinion, Reinitiation of Operation of the Federal Columbia River Power System (FCRPS), Including the Juvenile Fish Transportation System, and 19 Bureau of Reclamation Projects in the Columbia Basin (BOR). December 21, 2000. Downloaded April 1, 2006 as pdf document from http://seahorse.nmfs.noaa.gov/pls/pcts-pub/sxn7.pcts_upload.summary_list_biop?p_id=12342

Northwest Power and Conservation Council. 2000. Columbia River Basin Fish and Wildlife Program 2000. A Multi-Species Approach for Decision Making. Council document 2000-19. Executive summary Web site. Accessed April 1, 2006 at http://www.nwcouncil.org/library/2000/2000-19/summary.htm

Northwest Power and Conservation Council. 2005a. Northwest Power And Conservation Council Briefing Book. Council document 2005-1. Accessed April 2, 2006 at http://www.nwcouncil.org/library/2005/2005-1.pdf.)

Northwest Power and Conservation Council. 2005b. Overview of subbasin planning Web site. Accessed April 2, 2006 at http://www.nwcouncil.org/fw/subbasinplanning/admin/overview.htm

Northwest Power and Conservation Council. n.d. Fourth Annual Report to the Northwest Governors On Expenditures of the Bonneville Power Administration Web site. Accessed March 31, 2006 at http://www.nwcouncil.org/library/2005/2005-9.htm

Northwest River Forecast Center. 2005. Powerpoint slideshow presentation accessed April 1, 2006 at http://www.cses.washington.edu/cig/outreach/workshopfiles/seattle2005drought/king.ppt

Oregon Climate Service. n.d. 2000-2001 water year sets record low in parts of Oregon. Accessed 7 February, 2006 at http://www.ocs.oregonstate.edu/index.html.

Oregon Department of Agriculture. 2005. Oregon ranks #3 of all states in farms and ranches using irrigation; Oregon agriculture still depends on irrigation. January 26, 2005 press release.

Oregon Department of Energy. 2005. State of Oregon Biennial Energy Plan, 2005-2007. ODOE-003. January 2005. Downloaded on March 16, 2006 as http://oregon.gov/ENERGY/docs/EnergyPlan05.pdf

Oregon Department of Energy. n.d. Governor's Initiative on Global Warming Web site. Climate Change and Oregon. Accessed April 5, 2006 at http://www.oregon.gov/ENERGY/GBLWRM/climhme.shtml Oregon

Oregon Department of Environmental Quality. 1995. Non-point source pollution program. Accessed 5 March 1996 at http://www.state.or.us/wq.

Oregon Department of Environmental Quality. 1996a. Accessed 21 June 1996 at http://www.state.or.us/wq; file ORNQ0070

Oregon Department of Environmental Quality. 1996b. Total maximum daily load, dated January 1995. Accessed 5 March 1996 at http://www.state.or.us/wq; file ORNQ0019

Oregon Department of Environmental Quality. 1996c. Accessed 5 March 1996 at http://www.state.or.us.; file ORNQ0003

Oregon Department of Environmental Quality. 1996d. Ground water quality program. Acccessed March 5, 1996 at http://www.state.or.us;wq;file ORNQ0043

Oregon Department of Environmental Quality. 1997a. Permits Handbook. June 1996. Accessed 24 January 1997 at http://www.deg. state.or.us./general/ permitbk.htm.

Oregon Department of Environmental Quality. 1997b. Lower Columbia River Bi-State Water Quality Program: Final executive summary & steering committee priority recommendations. June 1996. Accessed 10 January 1997 at http://www.deq.state.or.us/wq/wqfact/bistrpt.htm.

Oregon Department of Environmental Quality. 1997c. Willamette River Basin water quality study. Accessed 10 January 1997 at http://www.deq.state.or.us/wq/wqfact/willstdy.htm.

Oregon Department of Environmental Quality. 2003a. 2003-2005 Budget Program Narrative, 2003-2005 Legislatively Adopted Budget (LAB). I. Oveerview of Water Quality. Downloaded , February 22, 2006 as http://www.deq.state.or.us/msd/budget/0305LAB_External/06_WQ/WQNARRTV.pdf

Oregon Department of Environmental Quality. 2003b. Fact Sheet. The 2002 303(d) List of Impaired Waters in Oregon. Last updated: 2 July 2003 by M. Fonseca. Downloaded on February 20, 2006 as http://www.deq.state.or.us/wq/wqfact/Final2002_303(d)list.pdf

Oregon Department of Environmental Quality. 2003c. 2003 Oregon Groundwater Conditions. 2p. Downloaded March 31, 2006 as http://www.deq.state.or.us/wq/wqfact/ORGWConditionsFSFeb03.pdf

Oregon Department of Environmental Quality. 2003d. Groundwater Quality in Oregon. Downloaded March 31, 2006 as http://www.deq.state.or.us/pubs/legislativepubs/2005Reports/groundwater.pdf

Oregon Department of Environmental Quality. 2004a. Introduction to drinking water protection in Oregon Web site, last updated: Wednesday June 09 2004. Accessed March 31, 2006 from http://www.deq.state.or.us/wq/dwp/DWPIntro.htm

Oregon Department of Environmental Quality. 2004b. Southern Willamette Valley Groundwater Management Area Declared. DEQ 04-WR-003. Last Updated April 2004 by Audrey Eldridge. Accessed March 31, 2006 from http://www.deq.state.or.us/wq/wqfact/SWVGWMAFactSheet.pdf

Oregon Department of Environmental Quality. 2004c. Declaration of a Groundwater Management Area in the Southern Willamette Valley. Open memo from Director

Stephanie Hallock, accessed March 31, 2006 from http://www.deq.state.or.us/wq/groundwa/SWVGroundwater/GWMA_SWV_Declaration.pdf

Oregon Department of Environmental Quality. 2005a. Query of map-based database on Total Maximum Daily Load (TMDL) Program Web site. Last updated September 8 2005. Downloaded Feb. 22, 2006 from http://www.deq.state.or.us/wq/TMDLs/tmdls.htm

Oregon Department of Environmental Quality. 2005b. Water Quality Program Web site last updated Friday December 09 2005. Downloaded February 22, 2006 fromhttp://www.deq.state.or.us/wq/

Oregon Department of Environmental Quality. 2005c. Nonpoint Source Pollution 319 Grants Web site, last updated September 6, 2005. Downloaded February 23, 2006 from http://www.deq.state.or.us/wq/nonpoint/sec319.htm

Oregon Department of Environmental Quality. 2006a. Enforcement Database Web site. Last updated February 23, 2006. Query results downloaded February 23, 2006 from http://www.deq.state.or.us/programs/enforcement/enfquery.asp

Oregon Department of Environmental Quality. 2006b. Protecting and Restoring the Willamette River Fact Sheet. 03-WR-008. Accessed June 30, 2006 at http://www.deq.state.or.us/wq/factsheets/fswillametteprotect.pdf.

Oregon Department of Environmental Quality. 2006c. Request For Proposals for Fiscal Year 2006, Oregon 319 NPS Program. Downloaded on February 24, 2006 as http://www.deq.state.or.us/wq/nonpoint/docs/NPS319GrantAppl.pdf

Oregon Department of Environmental Quality. n.d. [a}. Water quality monitoring factsheet ; downloaded Feb. 20, 2006 fromhttp://www.deq.state.or.us/wq/wqfact/mon.htm

Oregon Department of Environmental Quality. n.d.[b] Oregon TMDLs Approved by USEPA. Last updated October 11 2005. Downloaded February 22, 2006 from http://www.deq.state.or.us/wq/TMDLs/ApprovedTMDLs.htm

Oregon Department of Environmental Quality and Oregon Health Division. 1996. Oregon Wellhead Protection Guidance Manual Web site. Accessed March 31, 2006 from http://www.deq.state.or.us/wq/WhpGuide/Frontpage.htm

Oregon Department of Fish and Wildlife. 1984. Determining minimum flow requirements for fish. Duplicated.

Oregon Department of Fish and Wildlife. 1995. *Biennial Report on the Status of Wild Fish in Oregon.* Edited by Kathryn Kostow.

Oregon Department of Fish and Wildlife. 1997. Oregon Department of Fish and Wildlife Sensitive Species. December 1997. Downloaded February 24, 2006 as www.dfw.state.or.us/wildlife/pdf/sensitive_species.pdf

Oregon Department of Fish and Wildlife. 2004. Oregon List of Threatened and Endangered Fish and Wildlife Species Web sitelast updated September 17, 2004. Accessed February 24, 2006 at http://www.dfw.state.or.us/threatened_endangered/t_e.html

Oregon Department of Fish and Wildlife. 2006. March 10, 2006 email from Bernie Kepshire, Fish Screening State Coordinator, to author.

Oregon Department of Fish and Wildlife. n.d.[a] Fish Passage Requirements: Legislative Background Web site. Accessed February 21, 2006 from http://www.dfw.state.or.us/fish/passage/background.asp

Oregon Department of Fish and Wildlife. n.d.[b] Fish screening backgrounder. Draft document. Duplicated.

Oregon Department of Fish and Wildlife. n.d.[c]. Sales and fees of selected hunting and fishing licenses by calendar year. Downloaded March 30, 2006 as http://www.dfw.state. or.us/ODFWhtml/economic_information/salesbycalendar_1985-2003.pdf

Oregon Department of Fish and Wildlife. n.d.[d]. Economic Activity Associated with Oregon Wildlife and Fishery Resources. Downloaded March 30, 2006 as http://www. dfw.state.or.us/agency/budget/2005-2007_economic_activity.pdf

Oregon Department of Forestry. n.d. State Forests Management Web site. Accessed April 14, 2006 at http://egov.oregon.gov/ODF/STATE_FORESTS/state_forest_management. shtml

Oregon Department of Human Services. 2003. Protecting public health by assuring safe drinking water. In *Pipeline, Oregon Drinking Water News.* Vol. 18 Issue 3. Oregon Drinking Water Program. Downloaded February 16, 2006 as http://oregon.gov/DHS/ ph/dwp/docs/pipeline/pipesu03.pdf

Oregon Department of Human Services. 2006. Active Oregon Water Systems table. Provided in February 23, 3006 email to author from Paul Cymbala, Office of Public Health Systems, Drinking Water Program.

Oregon Department of Human Services. n.d. SDWIS Online database, Drinking Water Program Data Access Page accessed February 2, 2006 at http://170.104.158.16/ inventorylist.php3

Oregon Department of Justice. 1994. Forfeiture of rate and duty. DOJ File No. 690-001- NR002-94, Department of Justice, General Counsel Division. By Stephen E.A. Sanders, Assistant Attorney General. Duplicated.

Oregon Department of State Lands. 1999. Just the Facts #6: About Compensatory Mitigation for Wetland Impacts. Accessed March 29, 2006 from http://www.oregon. gov/DSL/WETLAND/docs/fact6.pdf

Oregon Department of State Lands. n.d.[a]. Navigability: Introduction Web site. Accessed March 29, 2006 at http://www.oregon.gov/DSL/NAV/about_us.shtml

Oregon Department of State Lands. n.d.[b].Waterway Leasing Q&A Web site. Accessed March 29, 2006 at http://www.oregon.gov/DSL/LW/leaseqa.shtml

Oregon Department of State Lands. n.d.[c]. Mitigation Banks Status Report and Contact Information Web site. Accessed March 29, 2006 at http://www.oregon.gov/DSL/ PERMITS/mitbank_status.shtml

Oregon Department of State Lands. n.d.[d]. DSL Statewide Programmatic General Permit Program Web site. Accessed February 3, 2006 at http://www.oregon.gov/DSL/ PERMITS/spgp.shtml

Oregon Department of State Lands. n.d.[e] Fund Diagram, 2001- 2003 Actuals. Downloaded April 12, 2006 as http://www.oregon.gov/DSL/DO/docs/csf_final_0103. pdf

Oregon Division of State Lands. 1995a. Proposed asset management plan: A plan to guide the care and management of land, waterways and minerals entrusted to the Oregon State Land Board. Salem.

Oregon Division of State Lands. 1995b. Unique among resource agencies. Salem.

Oregon Division of State Lands. 1996a. Who owns the waterways? Revised 28 February. Duplicated.

Oregon Division of State Lands. 1996b. State ownership of navigable waterways. Memorandum to Interested Persons, from Paul R. Cleary, Director, 24 May 1996. Duplicated.

Oregon Habitat Joint Venture. 2004. Wetlands in Oregon Web site, updated September 22, 2004. Accessed March 29, 2006 at http://www.ohjv.org/oregons_wetlands.html

Oregon Hanford Cleanup Board. 2004. Bylaws of Oregon Hanford Cleanup Board. Accessed April 2, 2006 at http://www.oregon.gov/ENERGY/NUCSAF/HCB/docs/BYLAWS-revised10-04.pdf

Oregon Hanford Cleanup Board. 2005. Position Paper on Capping Waste Sites located on the Hanford Nuclear Site. July 2005. Downloaded April 2, 2006 as http://www.oregon.gov/ENERGY/NUCSAF/HCB/docs/Capping.pdf

Oregon Land Conservation and Development Department. 2006. History of the Program Web site. Accessed April 3, 2006 at http://www.oregon.gov/LCD/history.shtml

Oregon Legislative Counsel Committee. 1995. *1995 Oregon Revised Statutes.*

Oregon Legislative Fiscal Office. n.d.[a] Natural Resources Agencies Budget Analysis. 2003-2005. Downloaded March 29, 2006 as http://www.leg.state.or.us/comm/lfo/03_05_leg_adopt_budget/1%20Natural%20Resources.pdf

Oregon Legislative Fiscal Office. n.d.[b]. Transportation Budget Analysis on Oregon Bluebook Web site. Accessed April 2, 2006 at http://www.leg.state.or.us/comm/lfo/03_05_leg_adopt_budget/1%20Transportation.pdf

Oregon Office of Economic Analysis. 2004. Demographic Forecast Web site. Downloaded April 2, 2006 from http://www.oea.das.state.or.us/DAS/OEA/demographic.shtml

Oregon Office of Emergency Management. 2006. Oregon's State Natural Hazard Mitigation Plan. Flood Chapter – State Flood Damage Reduction Plan. Downloaded April 1, 2006 as http://csc.uoregon.edu/pdr_website/projects/state/snhmp_web/SNHMP_pdf/OR-SNHMP_flood_chapter.pdf

Oregon Office of Energy. 1996. Accessed 11 June 1996 at http://www.cbs.state.or.us/external/ooe/nucsafe/hw-board.htm.

Oregon Parks and Recreation Department. 2003. Oregon Statewide Comprehensive Outdoor Recreation Plan, 2003-2007. Downloaded March 29, 2006 as http://www.prd.state.or.us/images/pdf/scorp_00_complete.pdf

Oregon Parks and Recreation Department. n.d. The Oregon Scenic Waterways Program: A Landowner's Guide. 23 p. Downloaded March 29, 2006 as http://www.oregon.gov/OPRD/RULES/docs/sww_log.pdf

Oregon Progress Board. 2005. Achieving the Oregon Shines Vision: The 2005 Benchmark Performance Report. Is Oregon Making Progress? Report to the Oregon Legislature and the People of Oregon – 2nd Printing. April 2005. Downloaded April 6, 2006 as http://www.oregon.gov/DAS/OPB/docs/2005report/05BPR.pdf

Oregon Public Utility Commission. 2006. Commission Sets Rates for Klamath Falls Irrigators. April 12, 2006 press release. Accessed June 27, 2006 at http://oregon.gov/PUC/news/2006/2006005.shtml.

Oregon Public Utility Commission, n.d. 2004 Oregon Utility Statistics. Accessed March 27, 2006 from http://www.puc.state.or.us/PUC/commission/statbook.pdf

Oregon Secretary of State. 2006. Voter Registration website. 2006 Monthly Voter Registration Statistics, January 2006. Accessed April 14, 2006 at http://www.sos.state.or.us/elections/votreg/jan06.pdf)

Oregon State University Extension Service. 2004. Groundwater Stewardship in Oregon Web site, last updated on May 25, 2004. Accessed March 31, 2006 at http://groundwater.orst.edu/regs/laws.html#gwquality

Oregon Supreme Court. 2006. Opinion *MacPherson vs. Department of Administrative Services, et al.* CC No. 05C10444; SC S52875.

Oregon Water Resources Department. 1981. Determination of flow available for Board consideration. Inter-office memorandum to Jim Sexson, Director, from Darrell Learn, 2 July 1981. Duplicated.

Oregon Water Resources Department. 1984. State of Oregon Land Use Classification by Basin. Duplicated.

Oregon Water Resources Department. 1987. Directory of Water Users' Organizations.

Oregon Water Resources Department. 1993a. The water availability program: A progress report, 1993. By Richard M. Cooper. Duplicated.

Oregon Water Resources Department. 1993b. Biennial report, State of Oregon, Water Resources Department: Report for the period January 1991 to December 1992. Salem.

Oregon Water Resources Department. 1994. Status report on reauthorization of the Corps of Engineers' Willamette Basin Reservoirs, 3 November 1994 Memorandum to Water Resources Commission from Director. Attachment 2, Preliminary statement of work, Willamette Basin Reauthorization Study. Duplicated.

Oregon Water Resources Department. 1995a. 1995 — 1999 strategic water resource management plan, submitted to the Oregon Legislative Assembly. Salem.

Oregon Water Resources Department. 1995b. Federal reserved water right negotiations with the Confederated Tribes of the Warm Springs Reservation: Negotiation Summary, 21 June 1995. By A. Reed Marbut. Duplicated.

Oregon Water Resources Department. 1995c. Informational report on water use reporting waivers, Work Session Item 2, 1 June 1995, Water Resources Commission Work Session. By Bill Ferber. Duplicated.

Oregon Water Resources Department. 1996a. Untitled table displaying instream water right and minimum perennial streamflow statistics. Hand dated 21 May. Duplicated.

Oregon Water Resources Department. 1996b. Dam and reservoir inventory. Dated January 23, 1996. Duplicated.

Oregon Water Resources Department. 1996c. Water Rights Information System, an on-line database accessed repeatedly in May through OWRD INTERNET site: http://www.wrd.state.or.us/dbaccess.html.

Oregon Water Resources Department. 2001. Internal Management Directives for Establishing Enforcement Priorities.

Oregon Water Resources Department. 2003. Summary of Water Legislation, 2003 Legislative Session. 23 p.

Oregon Water Resources Department. 2004a. Water Rights in Oregon, An Introduction to Oregon's Water Laws and Water Rights System. 52 p.

Oregon Water Resources Department. 2004b. July 19, 2004 memorandum to Christina Shearer from Mike Auman on Initiative #36, Estimated fiscal impact. Downloaded as Measure 37 Water Resources Dept. testimony 07-19-04 as http://old.orcities.org/currentissues/M37/M37WRDTestimony.pdf

Oregon Water Resources Department. 2004c. Memorandum from Barry Norris, Technical Services Division Administrator to Water Resources Commission. Informational Report on 2003 Field Regulation and Enforcement Activities. Agenda Item E, October 22, 2004 Water Resources Commission Meeting. 6 pages plus attachments

Oregon Water Resources Department. 2005a. Memorandum from Barry Norris, Technical Services Division Administrator to Water Resources Commission. Informational Report on 2004 Field Regulation and Enforcement Activities. Agenda Item J, July 28, 2005 Water Resources Commission Meeting. 6 pages plus attachments

Oregon Water Resources Department. 2005b. Water well constructor examination
   information packet. Accessed March 30, 2006 from http://www1.wrd.state.or.us/pdfs/
   WW_Studyguide.pdf

Oregon Water Resources Department. 2005c. Water Availability Committee of Oregon
   Web site. Accessed April 1, 2006 at http://egov.oregon.gov/OWRD/WR/drought_waco.
   shtml

Oregon Water Resources Department. 2006a. Oregon Water Resources Department Fee
   Schedule. Last modified: January 6, 2006

Oregon Water Resources Department. 2006b. Email from Bob Rice, Conservation and
   Instream Lease Coordinator, to author on February 23, 2006.

Oregon Water Resources Department. 2006c. Report to Governor's Office of Legal
   Counsel, Activities of the Water Resources Department Under Executive Order 96-30.
   January 19, 2006. Downloaded from www.leg.state.or.us/cis/2005gov_to_gov/2005_
   annual_report_wrd.pdf, March 7, 2006

Oregon Water Resources Department. 2006d. Email from Kathy Boles, Water Rights
   Information System manager, to author on March 23, 2006.

Oregon Water Resources Department. 2006e. Instream water right Microsoft Excel
   spreadsheet posted by Kathy Boles Water Rights Information System manager, to
   Oregon Water Resources Department file pick-up website and downloaded by author
   on June 22, 2006. Calculations of counts and totals performed by author.

Oregon Water Resources Department. 2006f. Well Identification Program Web site
   accessed March 31, 2006 at http://www.wrd.state.or.us/OWRD/GW/well_id.shtml, as
   updated January 2006

Oregon Water Resources Department. n.d.[a] Estimated stream miles in Oregon.
   Duplicated.

Oregon Water Resources Department. n.d.[b] Oregon Administrative Rules Web site.
   OAR 690-080, Programs for and Withdrawal from Control and use of State´s Water
   Resources , History Nov 1991; downloaded May 3, 2006 as http://www1.wrd.state.
   or.us/pdfs/law/690-080.1991-Nov-7.pdf

Oregon Water Resources Department. n.d.[c] Achieving more benefit from Oregon's water
   resources. Accessed April 2, 2006 at http://www1.wrd.state.or.us/pdfs/Economic_
   Environmental_Plan.pdf

Oregon Water Resources Department. n.d.[d] Aquifer Storage and Recovery Web site
   accessed March 6, 2006 at http://oregon.gov/OWRD/mgmt_asr.shtml

Oregon Water Resources Department. n.d.[e]. Summary of Water Legislation, 1999. 167 p.

Oregon Water Resources Department. n.d.[f] Hydroelectric Information database.
   Accessed February 20, 2006 from http://www.wrd.state.or.us/OWRD/SW/hydro_info.
   shtml

Oregon Water Resources Department. n.d.[g]. Klamath Adjudication Claims Web site
   accessed March 29, 2006 at http://oregon.gov/OWRD/ADJ/klamath_claim_summary.
   shtml

Oregon Water Resources Department. n.d.[h]. Online Well Log Database. Queried March
   30, 2006 from http://apps2.wrd.state.or.us/apps/gw/well_log/Default.aspx

Oregon Water Resources Department. n.d.[i]. Formal Enforcement Referrals (Final
   Orders). 1991-2004. 1 page.

Oregon Watershed Enhancement Board. 2004. Oregon Watershed Councils map.
   Downloaded May 5, 2006 as http://oregon.gov/OWEB/WSHEDS/images/Watershed_
   Councils_Map.pdf.

Oregon Watershed Enhancement Board. n.d.[a] An Overview of Oregon Watershed
    Councils Web site accessed April 1, 2006 at http://oregon.gov/OWEB/WSHEDS/
    wsheds_councils_overview.shtml
Oregon Watershed Enhancement Board. n.d.[b] About Us Web site accessed April 1, 2006
    at http://oregon.gov/OWEB/about_us.shtml
Oregon Secretary of State. 2006. Voter Registration Web site. 2006 Monthly Voter
    Registration Statistics, January 2006. Accessed April 14, 2006 at http://www.sos.state.
    or.us/elections/votreg/jan06.pdf
Oregonian. 12 February 1909.
Oregonian. 16 April 1996.
Oregonian. 23 August 2001. A Tale of Two Lakes—Detroit Lake
Oregonian. 4 April 2005. Water, water everywhere, but it's becoming scarce. By Michael
    Milstein.
Pacific Northwest River Basins Commission. 1969. Appendix II: The Region. In *Columbia-
    North Pacific Region Comprehensive Framework Study.* Vancouver, Washington.
Pacific Northwest Waterways Association. 2006. Columbia Snake River System
    BiOp Lawsuit 2004; BiOp invalid; New BiOp due in October 2006 Web site.
    Updated February 22, 2006. Accessed April 1, 2006 from http://www.pnwa.net/
    Issues%20Articles/Articles/Biop%20Lawsuit%20one-pager.pdf
Pagel, Martha O. 2002. The Intersection of Federal ESA Regulations and State Water
    Law. Downloaded March 27, 2006 from http://www.schwabe.com/showarticles.
    asp?Show=22
Palmer, Dr. Richard N., and Margaret Hahn. 2002. The Impacts of Climate Change on
    Portland's Water Supply: An Investigation of Potential Hydrologic and Management
    Impacts on the Bull Run System. Department of Civil and Environmental Engineering,
    University of Washington. Produced for January 2002. Downloaded April 5, 2006 as
    http://www.tag.washington.edu/papers/papers/PortlandClimateReportFinal.pdf
Palmer, Richard N., Erin Clancy, Nathan T. VanRheenen, and Matthew W. Wiley. 2004.
    The Impacts of Climate Change on the Tualatin River Basin Water Supply: An
    Investigation into Projected Hydrologic and Management Impacts. Department of
    Civil and Environmental Engineering. University of Washington. Downloaded April 5,
    2006 as http://www.cses.washington.edu/db/pdf/Palmer_etal_Tualatin240
Paulson, Richard W., Edith B. Chase, Robert S. Roberts, and David W. Moody, compilers.
    1991. Oregon: Floods and Droughts. 1991. In *National Water Summary 1988—89:
    Hydrologic Events and Floods and Droughts.* U.S. Geological Survey Water Supply Paper
    2375. Denver: U.S. Government Printing Office.
Portland General Electric. n.d.[a] Restoring runs above Pelton Web site accessed April 7,
    2006 at http://www.portlandgeneral.com/community_and_env/hydropower_and_fish/
    deschutes/restoring_runs.asp
Portland General Electric. n.d. [b] Project Web sites accessed March 16, 2006 at: http://
    www.portlandgeneral.com/about_pge/news/peltonroundbutte/factsheet.asp; and
    http://www.portlandgeneral.com/about_pge/news/sullivan_relicensing.asp
Portland Water Bureau. 1992. Water Supply 2050, Portland Metropolitan Region.
Portland Water Bureau. 2006. Supply/Service Area Web site accessed March 17, 2006
    from http://www.portlandonline.com/water/index.cfm?c=29460
Public Law 106–502
Rohse, Mitch. 1987. *Land Use Planning in Oregon: A No-nonsense Handbook in Plain
    English.* Corvallis: Oregon State University Press.

Root, Ann L. 1989. The Wild and Scenic River Act: Problems of implementation in Oregon. Master's thesis, Oregon State University.

Schonchin, Lynn J. 1990. Visions and values. In *Living with the Land: The Indians of Southwest Oregon.* The proceedings of the 1989 symposium on the prehistory of southwest Oregon. Edited by Nan Hannon and Richard K. Olmo. Medford: Southern Oregon Historical Society.

Service, Robert F. 2004. As the West goes dry. *Science*, Volume 303, pages 1124-27, 20 February 2004.

Sherrod, D. R., M. W. Gannett, and K. E. Lite, Jr. 2002. Hydrogeology of the upper Deschutes Basin , central Oregon—A young basin adjacent to the Cascade volcanic arc. In Moore , G. E., ed., Field Guide to Geologic Processes in Cascadia. Oregon Department of Geology and Mineral Industries Special Paper 36, p. 109-144. Accessed 7 February 2006, through U.S. Geological Survey Web site, http://or.water.usgs.gov/projs_dir/deschutes_gw/pubs.html

Sherton, Corinne. 1981. Preserving instream flows in Oregon's rivers and streams. *Environmental Law* 11:379.

Shupe, Steven J. 1989. Keeping the water flowing: Streamflow protection programs, strategies and issues in the West. In *Instream Flow Protection in the West.*, edited by Laurence J. MacDonnell, Teresa A. Rice, and Steven J. Shupe. Boulder: Natural Resources Law Center, University of Colorado School of Law.

Solley, Wayne B., Robert R. Pierce, and Howard A. Perlman. 1993. *Estimated Use of Water in the United States in 1990.* U.S. Geological Survey Circular 1081. Reston, Va.

SOLV. 2004. Oregon Adopt a River Web site accessed March 30, 2006 at http://www.solv.org/programs/oregon_adopt_a_river.asp

State of California. 2004. Department of Finance Population Projections by Race/Ethnicity for California and Its Counties 2000–2050. Downloaded April 2, 2006 from http://www.dof.ca.gov/html/Demograp/DRU_Publications/Projections/P1.htm

State of California. 2005. California Water Plan: Update 2005, A Framework for Action. Bulletin 160-05. December 2005. Volume1, Chapter 1. Introduction. The Resources Agency, Department of Water Resources. Accessed April 3, 2006 at http://www.waterplan.water.ca.gov/docs/cwpu2005/vol1/v1ch01.pdf

State Engineer. 1910. *Third biennial report of the State Engineer to the Governor of Oregon for the period beginning December 1, 1908, ending November 30, 1910.* Salem.

State Water Resources Board. 1958. *Umpqua River basin report.* Salem.

State Water Resources Board. 1959. *Rogue River basin report.* Salem.

State Water Resources Board. 1960. *Grande Ronde River basin report.* Salem.

State Water Resources Board. 1961a. *Deschutes River basin report.* Salem.

State Water Resources Board. 1961b. *North Coast basin report.* Salem.

State Water Resources Board. 1962. *John Day River basin report.* Salem.

State Water Resources Board. 1963a. *South Coast basin report.* Salem.

State Water Resources Board. 1963b. *Umatilla River basin report.* Salem.

State Water Resources Board. 1965a. *Hood River basin report.* Salem.

State Water Resources Board. 1965b. *Mid-Coast basin report.* Salem.

State Water Resources Board. 1967a. *Malheur Lake basin report.* Salem.

State Water Resources Board. 1967b. *Powder River basin report.* Salem.

State Water Resources Board. 1967c. *Willamette River basin report.* Salem.

State Water Resources Board. 1969a. *Oregon's Long-range Requirements for Water.* Salem.

State Water Resources Board. 1969b. *Malheur-Owyhee basins report.* Salem.

State Water Resources Board. 1971. *Klamath River basin report.* Salem.

Strahan v Cox, 127 F 3rd 155 (1st Cir 1997

StreamNet. n.d. Protected Areas (1988) database. Results from multiple queries February 2, 2006 of http://map.streamnet.org/protectedquery/default.htm

Sullivan, Edward J. 2005. Oregon's Measure 37 – crisis and opportunity for planning. Accessed April 4, 2006 at http://www.friends.org/issues/documents/M37/M37-Article-Ed-Sullivan.pdf

Task Force on Drinking Water Construction Funding and Regionalization. 1991. *Safety on tap, a strategy for providing safe, dependable drinking water in the 1990s.* Portland: Oregon Health Division.

Taylor, George H. 1997. Long-term Precipitation Cycles in Portland. Oregon Climate Service, October 1995. Accessed 10 January 1997 at http://www.ocs.orst.edu/reports/PDX_precip.html

Taylor, George H. 1998. Impacts of the El Niño/Southern Oscillation on the Pacific Northwest. Accessed April 10, 2006 at http://www.ocs.oregonstate.edu/index.html

Taylor, George H. 1999. Long-Term Wet-Dry Cycles in Oregon. Accessed April 10, 2006 at http://www.ocs.oregonstate.edu/index.html

Todd, David Keith. 1970. *The water encyclopedia; a compendium of useful information on water resources.* Port Washington, N.Y. : Water Information Center 559 p.

Trelease, Frank J., III. 1990. In-stream water rights. In *Water Rights of the Fifty States and Territories.* Edited by Kenneth R. Wright. Denver: American Water Works Association.

Tulare Lake Basin Water Storage District v. United States, 49 Fed. Cl. 313 (2001)

U.S. v. Adair, 478 F. Supp. 336 (D. Or. 1979)

U.S. Army Corps of Engineers. 1989. *Willamette River Basin Reservoir System Operation.* Portland District,Reservoir Regulation and Water Quality Section.

U.S. Army Corps of Engineers. 1992. Appendix E in *Authorized and Operating Purposes of Corps of Engineers Reservoirs.* Washington, D.C.

U.S. Army Corps of Engineers. 1996a. Water in the State of Oregon. Accessed 31 May 1996 at http://nppwm1.npp.usace.army.mil/ hydrology.html.

U.S. Army Corps of Engineers. 1996b. *Willamette basin review, Oregon: Addendum to the 1991 reconnaissance report.*

U.S. Army Corps of Engineers. 2000. Biological Assessment of the Effects of the Willamette River Basin Flood Control Project on Listed Species Under the Endangered Species Act. Final. Executive Summary. Prepared by USACE, Portland District Office and R2 Resource Consultants. Assisted by S. P. Cramer and Associates. 3 p

U.S. Army Corps of Engineers. n.d. Rogue River Basin Project website. Accessed June 26, 2006 at http://www.nwp.usace.army.mil/op/R/home.asp

U.S. Army Corps of Engineers, Bonneville Power Administration, and Bureau of Reclamation. 2003. Federal Columbia River Power System. Downloaded March 31, 2006 as http://www.bpa.gov/power/pg/fcrps_brochure_17x11.pdf

U. S. Board of Geographic Names. 1996. Accessed Geographic Names Information System 20 April 1996 at http://www-nmd.usgs.gov/www/gnis/index.html.

U.S. Bureau of Land Management. 2001. Western States Water Laws, BLM, Oregon Web site. Water rights fact sheet. Accessed April 1, 2006 at http://www.blm.gov/nstc/WaterLaws/oregon.html

U.S. Bureau of Reclamation [Water and Power Resources Service]. 1981. *Project Data.* Denver: U.S. Government Printing Office.

U.S. Bureau of Reclamation. 2003. Water 2025: Preventing Crises and Conflict in the West. 27 p. Downloaded April 14, 2006 as http://www.doi.gov/water2025/Water%202025-08-05.pdf

U.S. Bureau of Reclamation. 2005. Reclamation: Managing Water in the West. Overview 2005. Pacific Northwest Region. 67 p. Downloaded March 10, 2006 as http://www.usbr.gov/pn/about/pdf/overview.pdf

U.S. Bureau of Reclamation. n.d. AgriMet Irrigation Guide Web site. AgriMet, the Pacific Northwest Cooperative Agricultural Weather Network. Accessed March 14, 2006 from http://www.usbr.gov/pn/agrimet/irrigation.html#Efficiency

U.S. Bureau of the Census. 1992. Oregon, Part 37: State and county data. In 1992 Census of agriculture, AC 92-A-37, Volume 1: Geographic area series. U.S. Department of Commerce, Economics and Statistics Administration. Washington, D.C.

U.S. Bureau of the Census. 1996. *1990 Census of Population and Housing.* Accessed 18 May 1996 at http://govinfo.kerr.orst.edu/cgi-bin/buildit?3s-state.ors

U.S. Department of Agriculture. 2004a. Farm and Ranch Irrigation Survey (2003), Volume 3, Special Studies, Part 1, AC-02-SS-1, 2002 Census of Agriculture, National Agricultural Statistics Service. 216 p.

U.S. Department of Agriculture. 2004b. 2002 Census of Agriculture. Oregon. State and County Data. Volume 1, Geographic Area Series. Part 37. AC-02-A-37. 481 p.

U.S. Department of Energy. 1996. Accessed 3 July 1996 through Environmental Impact Statement for the Pantex Facility, Amarillo, Texas site at http://www.pantex.com/ds/pxeisc1.htm

U.S. Department of Energy. 2005. Hanford Site Overview Web site accessed April 2, 2006 at http://www.hanford.gov/?page=58&parent=6

U.S. Department of Energy. n.d. Office of River Protection. About us Web site accessed April 2, 2006 at http://www.hanford.gov/orp/?page=1&parent=0

U.S. Environmental Protection Agency. 2001. Drinking Water Infrastructure Needs Survey. Second Report to Congress. United States Environmental Protection Agency Office of Water. EPA 816-R-01-004. February 2001. Appendix B. Downloaded March 15, 2006 from http://www.epa.gov/safewater/needssurvey/needssurvey.html

U.S. Fish and Wildlife Service, U.S. Army Corps of Engineers, and NOAA Fisheries. 2005. Caspian Tern Management to Reduce Predation of Juvenile Salmonids in the Columbia River Estuary. Final Environmental Impact Statement. Accessed April 1, 2006 at http://www.fws.gov/pacific/migratorybirds/CATE%20FEIS%20Website%20Folder/Caspian_Tern_Final_EIS.pdf

U.S. Fish & Wildlife Service. n.d. Fisheries Restoration and Irrigation Mitigation Program. FY 2002-2004. 68 p. Downloaded March 27, 2006 as http://library.fws.gov/Pubs1/fishrestoration0204.pdf

U.S. Geological Survey. 2005. Assessment of the Klamath Project pilot water bank: a review from a hydrologic perspective. Prepared by U.S. Geological Survey, Oregon Water Science Center for U.S. Bureau of Reclamation, Klamath Basin Area Office. Downloaded April 7, 2006 as http://www.klamathbasincrisis.org/pdf-files/Final_USGS_Assessment_of_Water_Bank0505.pdf

U.S. Geological Survey. n.d.[a] Spreadsheet entitled "Estimated aquifer use." Emailed by Bruce Fisher, U.S. Geological Survey, geographer, to author on March 13, 2006.

U.S. Geological Survey. n.d.[b] County-Level Data files for Oregon, Estimated Use of Water in the United States for 2000. Downloaded March 2006 from http://water.usgs.gov/watuse/data/2000/index.html

University of Colorado. n.d. Center for Science and Technology Policy Research. Klamath Basin Project Web site accessed March 28, 2006 from http://sciencepolicy.colorado.edu/klamathbasin/

University Park Community Center. n.d. The History of Vanport Web site accessed March 31, 2006 at http://www.universitypark.org/vanport/

Wahl, Richard W. 1989. *Markets for Federal Water: Subsidies, Property Rights, and the Bureau of Reclamation.* Washington, D.C.: Resources for the Future.

Washington. 2006. HB 2068. Certification of Enrollment, Engrossed Second Substitute House Bill 2860, 59th Legislature, 2006 Regular Session. Downloaded April 14, 2006 as http://www.leg.wa.gov/pub/billinfo/2005-06/Pdf/Bills/House%20Passed%20Legislature/2860-S2.PL.pdf

Water Resources Committee. 1955. *Report of the Water Resources Committee, submitted to Forty-eighth Legislative Assembly.* Salem, Oregon.

Western States Water Council. 1986. *Western state ground water management.*

Willamette Partnership. 2005. Willamette Partnership Web site accessed February 24, 2005 at http://willamettepartnership.org/

Willamette Restoration Initiative. 2001. *Restoring the River of Life, The Willamette Restoration Strategy, Recommendations for the Willamette Basin Supplement to the Oregon Plan for Salmon and Watersheds.* February 2001.

Williamson, Kenneth J., David A. Bella, Robert L. Beschta, Gordon Grant, Peter C. Klingeman, Hiram W. Li, Peter O. Nelson, Gonzalo Castillo, Tom Lorz, Mark Meleason, Paula Minear, Stephanie Moret, Ilgi Nam, Maryanne Reiter, and Lisa Wieland. 1995. *Technical background report.* Vol. 2 of *Gravel disturbance impacts on salmon habitat and stream health: A report for the Oregon Division of State Lands.* Corvallis: Oregon Water Resources Research Institute, Oregon State University.

Subject Guide to Water Management Provisions of Oregon Revised Statutes

This guide's organization generally reflects that of the book, with an emphasis on water supply provisions in Oregon's laws. The subject categories and major subheadings are listed below. To better illustrate context and significance, the former are frequently quoted or paraphrased. Unless otherwise noted, "commission" and "department" refer to the Water Resources Commission and Department. Readers should be prepared for some "drift" in statutory references through time, as laws are amended or repealed. The Oregon Legislative Counsel's biennial publication of the Oregon Revised Statutes should be consulted as the authoritative source for current law. For additional information, readers are referred to Appendix B where administrative rules implementing major water laws are displayed.

Subject Guide Organization:

i. Public ownership, interest, welfare, protection
- A. Public ownership of waters
- B. Definitions of waters of this state
- C. Public interest, safety, welfare
    - 1. General
    - 2. Permitting
- D. Public use of waters
- E. Policies and planning
    - 1. Policies
    - 2. Planning
- F. Conservation /Waste
- G. Areas of water use restriction or special management
- H. Water quality protection
- I. Fish and wildlife protection

ii. Water Rights, General
- A. Bounds of rights
- B. Exemptions from need to obtain water right
- C. Water right application/evaluation process
- D. Measuring and reporting

iii. Water Rights, Types and Characteristics
- A. Irrigation
- B. Municipal
- C. Hydroelectric uses
    - 1. Permits
    - 2. Licenses, preliminary permits
    - 3. Standards
- D. Instream appropriations
- E. Storage sources
- F. Groundwater sources, wells

iv. Other Authorizations
  A. Transfers
  B. Registrations
  C. Miscellaneous authorizations
v. Water Use Enforcement
  A. Watermaster duties
  B. Illegal water use or practices
vi. Water Providers, Management Organizations, Authorities and Licensed Agents
  A. Water provision
  B. Water management
vii. Land use & land management
  A. Land use planning
  B. Real estate transactions
  C. Land management
    1. Removal-fill, wetlands, navigable waterways
    2. Scenic waterways
    3. Other stream-related protections
    4. Watershed protection

↭

i. Public ownership, interest, welfare, protection
  A. Public ownership of waters
    "All water within the state from all sources of water supply belongs to the public." [ORS 537.110]
    "...all of the waters within this state belong to the public for use by the people for beneficial purposes without waste." [ORS 536.310(1)]
    "...the right to reasonable control of all water within this state from all sources of water supply belongs to the public..." [ORS 537.525]
  B. Definitions of Waters of This State
    For water supply purposes: " 'Waters of this state' means any surface or ground waters located within or without this state and over which this state has sole or concurrent jurisdiction." [ORS 536.007(12)]
    For other purposes (as indicated): Fill and removal activities [ORS 196.800]
    Commercial fishing [ORS 506.006 et seq.]
    Water pollution controls [ORS 468B.005]
    Definition of groundwater [ORS 537.515(5)]
    Definition of "land" for land use planning purposes as including "water, both surface and subsurface" [ORS 197.015(6)]
  C. Public Interest. Safety, Welfare
    1. General
    "It is in the interest of the public welfare that a coordinated, integrated state water resources policy be formulated and means provided for its enforcement" [ORS 536.220(2)(a)]
    It is in the public interest that integration and coordination of uses of water and augmentation of existing supplies for all beneficial purposes be achieved for the maximum economic development thereof for the benefit of the state as a whole [ORS 536.310 (2)]
    Representatives carrying out water resource compacts and agreements to insure harmony with the public interest. [ORS 536.420 (2)]

Commission authorized to withdraw water from further appropriation when in the public interest to conserve water resources for maximum beneficial use and control [ORS 536.410]

Commission authorized to designate critical ground water area if ground water may become polluted to an extent contrary to the public welfare, health and safety [ORS 537.730(1)(f)]

State to insure that public interest in efficient use of water and heat resources is protected when addressing interference between geothermal and other wells [ORS 537.095]

Constructing new wells or altering existing wells declared activities affecting the public welfare, health and safety. [ORS 537.765]

Competitive exploitation of water resources for single-purposes discouraged when other feasible uses are in the general public interest. [ORS 536.310(5)]

Potential for future shortage of water risks public health, safety and welfare and therefore is a matter of statewide concern; element of policy on water storage facilities. [ORS 536.238(1)(d)]

Severe, continuing drought poses jeopardy to health, safety and welfare of people of Oregon; need for state authority [ORS 536.710 - 536.720]

Damage to life and property from failing dams to be prevented [ORS 540.350]

Controls over release of stored water when public safety endangered [ORS 541.510- 541.545]

Civil penalties imposed considering / reduced consistent with, the public health and safety and protection of the public interest. [ORS 536.915 - 536.920]

Denying stay of commission order based on determination of substantial public harm. [ORS 536.075]

   2. Permitting

Requirement that surface water permits, if issued, will "not impair or be detrimental to the public interest." [ORS 537.153]

Requirement that groundwater permits, if issued, "will ensure the preservation of the public welfare, safety and health." [ORS 537.621(2)]

Rejection or modification of applications for water use permits if found to impair the public interest (surface water) or risk preservation of public welfare, safety and health (groundwater) [ORS 537.170(6); ORS 537.625(1)]

Proposed water use presumed to be in public interest or ensure public welfare under certain conditions. [ORS 537.153; ORS 537.621]

Public interest "considerations" for permit issuance. [ORS 537.170(8); ORS 537.625(3)]

Water Resources Department may attach terms, limitations or conditions to make the use consistent with the public interest and protect public welfare. [ORS 537.153; .190; .211; 289; 292; 343, .625, .628, .629]

Multipurpose storage projects and municipal uses may take precedence over agency-requested instream rights if determined to be in the public interest   ORS 537.352

D. Public use of waters

Finding that public uses are beneficial uses. [ORS 537.334]

Instream public uses described, listed. [ORS 537.332]

The department as final authority in determining instream flows necessary to protect public uses. [ORS 537.343]

Certain agencies authorized to request instream water rights. [ORS 537.336]

Agencies authorized to reserve water for future economic development. [ORS 537.356–537.358]

Public agency water use registration process for roads. [ORS 537.040]

Annual water use reports required from government entities. [ORS 537.099]

Department to diligently enforce laws on water right cancellation to make water available for appropriation by the public. [ORS 536.340(1)]

E. Policies and Planning

    1. Policies

Policy considerations for formulating integrated state water resources program. [ORS 536.310]

Water storage policy. [ORS 536.238]

Emergency water shortage policy. [ORS 536.710]

Policy on hydroelectric development. [ORS 543.015]

Policy on groundwater appropriation. [ORS 537.525]

Groundwater quality protection. [ORS 468B.150 - 468B.190]

Geothermal development policy. [ORS 522.015]

Policy on interference between geothermal and other wells. [ORS 537.095]

Findings on diverting water from basin of origin. [ORS 537.801]

Findings on water rights of Indian tribes. [ORS 539.300]

Furnishing water for certain purposes declared public utility. ORS 541.010]

Policy/direction that department diligently enforce laws concerning cancellation and excessive unused water claims. [ORS 536.340(1)]

Finding that public uses are beneficial uses. [ORS 537.334]

Policy declaring establishment of minimum streamflows a high priority. [ORS 536.235]

Maintenance of minimum perennial stream flows sufficient to support aquatic life, to minimize pollution and to maintain recreation values to be fostered and encouraged. [ORS 536.310(7)]

Policy to restore native stocks of salmon and trout to historic levels. [ORS 496.435]

Wildlife policy; water. [ORS 496.012 (2), (4), (5)]

Watershed protection policy. [ORS 541.353]

Policy establishing soil and water districts for preventing soil erosion, controlling floods, promoting conservation. [ORS 568.225]

Policy promoting "the protection, conservation and best use of wetland resources, their functions and values." [ORS 196.672]

Scenic Waterway policy. [ORS 390.815; ORS 390.835]

Willamette River Greenway policy. [ORS 390.314]

Outdoor recreation policy; water. [ORS 390.010(3)(i)-(k)]

    2. Planning

Direction to adopt an integrated, coordinated approach for the use and control of Oregon's water resources. [ORS 536.220; ORS 536.300]

Direction to "progressively formulate" a water resources program and issue program "statements" consisting of preferred water use designations and classifications. [ORS 536.300(2); 536.340]

Authority to classify or re-classify waters. [ORS 536.340]

Variance process for unclassified uses. [ORS 536.295]

Prohibition against actions by agencies or public corporation that would conflict with state water resources policy. [ORS 536.360–536.370]

Process of notifications, approvals, hearings, agreements and court remedies to assure compliance with the state water policy. [ORS 536.350 — 536.400]

F. Conservation / Waste

"...all of the waters within this state belong to the public for use by the people for beneficial purposes without waste." [ORS 536.310(1)]

"the waters within this state belong to the public for use by the people for beneficial purposes without waste." [ORS 536.310(1)]

Willful waste of water illegal. [ORS 540.720]

Beneficial use as "the basis, the measure and the limit of all rights to the use of water in this state." [ORS 540.610(1)]

Assessment required to determine amount of water necessary for use proposed in permit applications. [ORS 537.153(3); ORS 537.621(3)]

Use of Conserved Water Program. [ORS 537.455–ORS 537.500]

Water-efficient plumbing fixtures mandated. [ORS 447.145]

Water management plans required for irrigation district water right re-mapping. [ORS 540.572–540.578]

Water company reservoir and ditches to be maintained so as to prevent waste. [ORS 541.060]

G. Areas of Water Use Restriction or Special Management

Streams withdrawn by Legislature from further appropriation. [ORS Chapter 538]

Authorization for commission to withdraw waters from further appropriation. [ORS 536.410]

Authorization for commission to classify and re-classify waters. [ORS 536.340]

Scenic waterways: prohibition of dams, reservoirs, impoundments, placer mines and certain water diversion facilities. [ORS 390.835]

Scenic waterways: limitations on certain upstream uses. [ORS 390.835 (5)–(13)]

Commission authorized to designate critical groundwater areas. [ORS 537.730]

Department of Environmental Quality authorized to declare "areas of groundwater concern" and "groundwater management areas" to control contamination. [ORS 468B.175; 468B.180]

Commission authorized to declare Serious Water Management Problem Areas. [ORS 540.435]

Direction to consider water availability in permitting decisions. [ORS 537.150(4), 537.153(2), 537.620(4)(b), 537.621(2), 537.170(8)(d), and 537.625(3)(d)]

Process for / restrictions on, transferring water out of basin of origin. [ORS 537.801 - 537.870]

Water use rotation agreements. [ORS 540.150]

Dams and hydroelectric facilities prohibited on portions of Umpqua River. [ORS 541.875]

Goose Lake Interstate Compact. [ORS 542.520]

Klamath River Basin Interstate Compact. [ORS 542.610–542.630]

H. Water Quality Protection

Surface water quality. [ORS 468B.040 - 468B.110]

Oil and other spills. [ORS 468B.300 - 468B.500]

Confined Animal Feeding Operation, Controls over. [ORS 468B.200]

Agricultural water quality management plans. [ORS 568.900 — 568.933]

Water quality regulations on forestlands. [ORS 468B.100-.110; ORS 527.765]

Drinking Water Quality Act. [ORS Chapter 448]

The Oregon Groundwater Protection Act of 1989. [ORS 468B.150 – 468B.190]

Hanford Waste Board, Purposes of. [ORS 469.566 - .587]

I. Fish and Wildlife Protection

Policy declaring establishment of minimum streamflows a high priority. [ORS 536.235]

Maintenance of minimum perennial stream flows sufficient to support aquatic life, to minimize pollution and to maintain recreation values to be fostered and encouraged. [ORS 536.310(7)]

Policy to restore native stocks of salmon and trout to historic levels. [ORS 496.435]

Oregon Department of Fish and Wildlife authorized to require dam owners to install and maintain fish passage facilities at all artificial in-channel obstructions in fish-bearing streams. [ORS 509.610 - 509.645]

Requirement that any person diverting water from a fish-bearing stream at a rate of 30 or more cubic feet per second must install, operate and maintain a fish screening or by-pass device. [ORS 498.311; ORS 509.615]

Provides any person diverting water at less than 30 cubic feet per second from a fish-bearing stream may be required to install screens or by-passes. [ORS 498.306]

Oregon Endangered Species Program. [ORS 496.171 – ORS 496.192]

Fish, wildlife and recreation declared highest and best water uses in state Scenic Waterways Act. [ORS 390.835]

Fishery resource declared an important economic and recreational asset; consideration of protection measures when planning or building impoundment structures mandated. [ORS 536.310 (4)]

Salmon Trout Enhancement Program projects and certain fish passage structures exempt from water right requirements. [ORS 537.142]

ii. Water Rights, General

A. Bounds of Rights

Amount of water authorized for use under a water right limited to beneficial use without waste. [ORS 536.310(1); ORS 540.610(1)]

Beneficial uses described, listed. [ORS 536.300]

"...no person shall use, store, or divert any waters until after the department issues a permit to appropriate the waters." [ORS 537.130(2)]

"...the use of ground water for any purpose, without a permit...or registration...is an unlawful appropriation of ground water." [ORS 537.535(2)]

Water rights appurtenant to premises. [ORS 540.510]

Forfeiture of right after 5 consecutive years of non-use. [ORS 540.610]

B. Exemptions from need to obtain water right

Groundwater exemptions described, listed. [ORS 537.545]

Groundwater exempt uses recognized as rights to appropriate ground water equal to those established by water right certificates. [ORS 537.545(2)]

Salmon Trout Enhancement Program projects; certain fish passage structures. [ORS 537.142]

Miscellaneous surface water uses exempt from permit requirements: emergency fire-fighting; livestock diversions with enclosed delivery systems and automatic shut-off devices; land management practices to prevent soil erosion or improve water quality; forest managers' use of water for mixing pesticides or controlling slash burning; diverting water to tanks from authorized reservoirs; and using rain water collected from an impervious surface. [ORS 537.141]

C. Water Right Application/Evaluation Process

Requirement that surface water permits, if issued, will "not impair or be detrimental to the public interest." [ORS 537.153]

Requirement that groundwater permits, if issued, "will ensure the preservation of the public welfare, safety and health." [ORS 537.621(2)]

Rejection or modification of applications for water use permits if found to impair the public interest (surface water) or risk preservation of public welfare, safety and health (groundwater). [ORS 537.170(6); ORS 537.625(1)]

"...the Water Resources Department shall approve all applications made in proper form which contemplate the application of water to a beneficial use, unless the proposed use conflicts with existing rights." [ORS 537.160(1)]

Application must be to the department and submitted prior to construction of water works. [ORS 537.130(1)]

Application fees. [ORS 536.050]

Application submittal requirements. [ORS 537.140; .537.615]

Establishment of priority date; initial review and public notice of application. [ORS 537.150; .537.620]

Direction to consider water availability in permitting decisions. [ORS 537.150(4), 537.153(2), 537.620(4)(b), 537.621(2), 537.170(8)(d), and 537.625(3)(d)]

Proposed final order: contents, timeline; public interest determinations; protests and requests for standing: contents, timelines; final order, scheduling of contested cases. [ORS 537.153; ORS 537.621]

Proposed water use presumed to be in public interest or ensure public welfare under certain conditions. [ORS 537.153; ORS 537.621]

Water Resources Department may attach terms, limitations or conditions to make the use consistent with the public interest and protect public welfare. [ORS 537.153; .190; .211; 289; 292; 343, .625, .628, .629]

Contested case hearing: timeline, participants, process. [ORS 537.170; ORS 537.622]

Exceptions to contested case orders. [ORS 537.173;.ORS 537.626]

Final decision timelines, extensions, provisions for writ of mandamus. [ORS 537.175; ORS 537.627]

Permit issuance. [ORS 537.211; 537.625]

Time for beginning/completing diversion allowed under permit; extensions. [ORS 537.230]

Permit cancellation. [ORS 537.410 - 537.450]

Water right certificate issuance. [ORS 537.250; 537.700]

Water right changes (transfers). [ORS 540.505 - 540.587]

Abandonment, forfeiture of water rights. [ORS 540.610 - 540.670]

D. Measuring and Reporting

Department authorized to require measuring devices along ditches or on streams feeding reservoirs. [ORS 540.310 - 540.330]

Government agencies required to report annually the amount, period and purpose of water use. [ORS 537.099]

Commission authorized to require water right owners in serious water management problem area to measure and report annual water use. [ORS 540.435]

Commission authorized to require any person or public agency diverting water for uses not requiring a water right permit or certificate to furnish information with regard to such water and the use thereof. [ORS 537.141 (6)]

Commission authorized to require any person or public agency using ground water for any exempt use to furnish information with regard to such ground water and the use thereof. [ORS 537.545]

Authority of commission to require reservoir operators to install automatic stream level recording devices. [ORS 541.535]

iii. Water Rights, Types and Characteristics

A. Irrigation

Additional, incidental uses allowed. [ORS 540.520(8), (9)]

Rates and duties, specific findings required for lesser amounts. [ORS 537.621]

Irrigation season extensions. [ORS 537.385]

Supplemental rights. [ORS 540.505(3)]

B. Municipal

Time to begin construction of surface water diversion works extended. [ORS 537.230]

Ten-year limit to complete construction of works associated with new reservoir. [ORS 537.248]

Immunity from permit cancellation. [ORS 537.410(2)]

Partial perfection of right allowed. [ORS 537.260(4)]

Overcoming presumption of forfeiture. [ORS 540.610(1) - (2)]

Application of appurtenancy requirements. [ORS 540.510(3)]

Preference over established instream water rights allowed. [ORS 537.352]

Preference for hydroelectric permits. [ORS 543.270]

Requirement to link water use requests with present and future population. [ORS 537.140(1)(e)]

Hydroelectric permits for municipal corporations. [ORS 537.282 - 537.299]

C. Hydroelectric Uses

Approval process. [ORS 537. 282 - 537.299 (permits for municipal corporations); ORS Chapter 543 (private licenses)]

Fees. [ORS 536.015 - 536.017; ORS 543.280]

Preference for hydroelectric purposes when in competition with instream water right request. [ORS 537.360]

1. Permits

Permit process for municipal corporations (e.g., cities, towns, water districts). [ORS 537.282 - 537.299]

2. Licenses, Preliminary Permits

Licenses required for citizens or private corporations. [ORS 543.050(2)]

State preliminary permit required for applicants to Federal Energy Regulatory Commission. [ORS 543.210]

Term of preliminary permits and licenses. [ORS 543.260]

3. Standards

Findings required on individual and cumulative effects of proposed projects. [ORS 543.255]

Natural resource protections as minimum standards. [ORS 543.017(1)(a) - (d)]

Finding required on need for power. [ORS 543.017(1)(e)]

Commission allowed to waive or modify standards. [ORS 543.300(7)]

D. Instream Appropriations

Maintenance of minimum perennial stream flows sufficient to support aquatic life, to minimize pollution and to maintain recreation values to be fostered and encouraged. [ORS 536.310(7)]

"...establishment of minimum perennial streamflows is a high priority of the Water Resources Commission and the Water Resources Department." [ORS 536.235]

Finding declaring public uses are beneficial uses. [ORS 537.334]

Certain state agencies authorized to request instream water rights. [ORS 537. 336]

Pre-existing minimum perennial streamflows to be converted to instream rights. [ORS 537.346]

Instream rights granted same legal status as other rights. [ORS 537.350]

Prohibition against instream rights impairing rights of senior water users.   [ORS 537.334]

Department as final authority in determining instream flows necessary to protect public uses. [ORS 537.343]

Precedence of multipurpose storage projects and municipal uses over instream rights. [ORS 537.352]

Preference for hydroelectric purposes when in competition with instream water right request. [ORS 537.360]

Purchase, lease, or donation of existing water right for conversion to instream right. [ORS 537.348]

Use of leased water right for instream purpose, protection against forfeiture. [ORS 537.348 (2)]

E. Storage Sources

Excess release as natural flow and available for appropriation. [ORS 540.045(3)]

Information submittal requirements for applications to store water. [ORS 537.140(1)]

Plans and specifications required for dams 10 or more feet high and impounding 3,000,000 or more gallons. [ORS 540.350 - 540.400]

Requirement for counties, municipalities or districts to complete newly approved storage projects within 10 years. [ORS 537.248]

Information submittal requirements for applications for secondary permits. [ORS 537.400(1)]

Additional permit required to make up for losses from storage. [ORS 537.400(2)]

Registration and management requirements for pre-existing ponds. [ORS 537.405 – 537.409]

Prohibition of splash dams. [ORS 541.455]

Penalties for splash dams. [ORS 541.990]

F. Groundwater Sources, Wells

Definition of groundwater. [ORS 537.515(5)]

Policies governing issuance of permits. [ORS 537.525]

Information submittal requirements for groundwater applications. [ORS 537.615]

Aquifer storage and recovery: declaration as beneficial use; relationship to existing rights; approval process. [ORS 537.531 - 537.534]

Well construction, generally. [ORS 537.747–537.783]

Definition of well. [ORS 537.515(9)]

Constructing new wells or altering existing wells declared activities affecting the public welfare, health and safety. [ORS 537.765]

Department charged to "...enforce standards for the construction, maintenance, abandonment or use of any hole through which ground water may be contaminated." [ORS 537.780(1)(c)(A)]

Ground Water Advisory Committee. [ORS 536.090]

Action by party injured by improper well construction. [ORS 537.753]

Self-drilled wells, regulations. [ORS 537.753]

Geothermal wells. [ORS 537.783]

Department authorized to order halt to the use of the well and/or its permanent abandonment. [ORS 537.775]

Well identification, location; requirements related to real estate transactions. [ORS 537.788 - 537.789]

Oregon Groundwater Protection Act of 1989. [ORS 468B.150–468B.190]

Health Division authorized to require tests for specific contaminants in an area of groundwater concern or groundwater management area. [ORS 448.271]

iv. Other Authorizations

  A. Transfers

Changes in type of use, point of appropriation or place of use requiring approval through transfer order. [ORS 540.520(1)]

Uses eligible for transfer. [ORS 540.505(4)]

Transfers not required to move points of diversion to follow a naturally changing stream channel, if distance moved is 500 feet or less. [ORS 540.510(5)]

Information submittal requirements for transfer application. [ORS 540.520(2)]

New certificates to record transfer. [ORS 540.530]

Requirement that new diversion facilities resulting from transfers must have fish screens. [ORS 540.520(3)]

Temporary transfers. [ORS 540.523]

Transfer of surface water points of diversion to underground points of appropriation. [ORS 540.531]

Irrigation district transfer requests; re-mapping. [ORS 540.574–540.580]

Transfer not required to move point of diversion, if necessitated by government action; notification requirements. [ORS 540.510(6)]

Transferring water outside of basin of origin. [ORS 537.801 - 537.870]

Transferring out-of-stream uses to instream purposes. [ORS 537.348]

  B. Registrations and Reserved Water Rights

Claims of surface water use prior to 1909 to be registered by December 31, 1992. [ORS 539.005 – 539. 240]

Claims of groundwater use prior to 1955 to be registered by May 29, 1961.   [ORS 537.605]

Process for negotiating reserved water rights of federally recognized Indian tribes. [ORS 539.300 - 539.350]

Registration process for road watering by public agencies. [ORS 537.040]

Ponds built prior to 1995 and meeting certain size requirements. [ORS 537.405–537.409]

Reclaimed water. [ORS 537.131 - 537.132]

  C. Miscellaneous Authorizations

Drought options and agreements for use of existing rights. [ORS 536.720 – 536.780]

Weather modification permits. [ORS 558.010–558.140]

Supplemental water rights. [ORS 540.505(3)]

Water exchanges. [ORS 540.533–540.543]

Drought temporary water use permits. [ORS 536.750]

Reservations of water for economic development. [ORS 537.356 – 537.358]

v. Water Use Enforcement

A. Watermaster Duties

Duties described. [ORS 540.045; ORS 540.100; ORS 540.210]

District water not subject to watermaster control. [ORS 540.270]

Controlling water diversions; posting notice. [ORS 540.045(1)]

Power to enter private property. [ORS 536.037(1)(e); ORS 537.780(1)(e)]

Requirement to distribute water according to terms of rotation agreement.   [ORS 540.150]

Filing affidavits of non-use for apparently-abandoned water rights. [ORS 540.660]

B. Illegal Water Use or Practices

Illegal to use water without permit. [ORS 537.130(2); 537.535(2)]

Willful waste of water illegal. [ORS 540.720]

Illegal acts: Interference with headgate, use of water in violation of watermaster order, unauthorized use of water, interference with waterworks; water use as prima facie evidence of guilt. [ORS 540.710 – 540.730]

Filing criminal complaints; arrest by watermaster. [ORS 540. 060; ORS 540.990]

Criminal penalties. [ORS 540.990]

Civil penalties. [ORS 536.900 – 536.935]

Injunction against watermaster action. [ORS 540.750]

Failure to comply with measuring requirement. [ORS 540.320 - 540.330]

Failure to comply with requirement to install suitable outlet in dam. [ORS 540.340]

Failure to comply with order of commission regulating release of stored water. [ORS 541.545]

Penalties for splash dams. [ORS 541.990]

vi. Water Providers, Management Organizations, Authorities and Licensed Agents

A. Water Provision

Cities granted authority to own and operate water supply facilities. [ORS 225.020]

Formation of water authorities. [ORS 450.600 – 450.700]

Furnishing water for certain purposes declared public utility. [ORS 541.010]

Public Utility Commission to set certain water provider rates. [ORS Chapter 757]

Boundary commissions to approve boundary changes for domestic water suppliers, any additional function proposed by water and other districts, extra-territorial water line extensions, or establishment of new community water supply systems. [ORS Chapter 199]

Irrigation Districts: formation, powers. [ORS Chapter 545]

Irrigation district water right mapping process and requirements. [ORS 541.325 - 541.333]

Ditch Companies: formation, powers. [ORS Chapter 541]

Water Control Districts: formation, powers. [ORS Chapter 553]

Drainage Districts: formation, powers. [ORS Chapter 547]

Water Improvement Districts & Companies: formation, powers. [ORS Chapters 552 and 554]

Diking Districts: formation, powers. [ORS Chapter 551]

Soil and Water Conservation Districts: formation, powers. [ORS Chapter 568]

B. Water Management

Water Resources Commission: authority, duties. [ORS 536.022 - 536.027]

Water Resources Department: authority, duties. [ORS 536.032 - 536.039]

Watermaster: districts and duties. [ORS 540.010 - ORS 540.150]

Ground Water Advisory Committee: membership, purpose. [ORS 536.090]

Environmental Quality Commission: authority, duties. [ORS 468.010 - 468.015; ORS 468B.010]

Department of Environmental Quality: authority, duties. [ORS 468.030 - 468.050]

Hanford Waste Board: duties. [ORS 469.566 - .587]

Fish and Wildlife Commission and Department: authority, duties. [ORS 496.090 - 496.166]

Fish Screening Task Force. [ORS 496.085; 498.306]

Watershed councils: formation, duties. [ORS 541.384, 541.388]

Oregon Watershed Enhancement Board: members, purpose. [ORS 541.360]

Energy Facility Siting Council: authority, duties. [ORS 469.450]

Certified water rights examiners: qualifications, purpose. [ORS 537.797 - 537.798]

vii. Land use & land management

A. Land Use Planning

Definition of "land" for land use planning purposes as including "water, both surface and subsurface." [ORS 197.015(6)]

Department to address land use compatibility in permit decisions. [ORS 537.153(3)(b); ORS 537.621(3)(b)]

Communities with 2,500 or more people within urban growth boundary to adopt public facilities plan. [ORS 197.712(2)(e)]

Special districts required to comply with land use planning goals; water districts included as special districts. [ORS 197.250; 197.015]

Cities, counties and metropolitan service districts to enter into cooperative agreements with special districts. [ORS 195.020]

Municipalities required to link water use requests with present and future population. [ORS 537.140(1)(e)]

B. Real Estate Transactions

Sellers required to disclose existence of valid water use permit on subject property. [ORS 537.330]

Well identification, location; requirements when selling. [ORS 537.788 - 537.789]

Well identification, location; requirements related to real estate transactions. [ORS 537.788 - 537.789]

C. Land Management

1. Removal-Fill, Wetlands, Navigable Waterways

Removal-Fill Law: regulation of streambed activities. [ORS 196.795–196.990]

Wetland protection; wetland conservation plans. [ORS 196.600–196.692]

Removal-fill permit exemptions: emergencies; permitted dams; forest activities. [ORS 196.810; 196.905]

Permit required for any removal-fill activity on state scenic waterways or streams providing essential salmon habitat. [ORS 390.835; ORS 196. 810]

Wetland protection and conservation encouraged. [ORS 196.672]

State title to navigable waterways: determination, procedures. [ORS 274.400–274.520]

Leasing of submerged or submersible lands. [ORS 274.040–043; 705-895; 915]

2. Scenic Waterways

Scenic Waterway Program, generally. [ORS 390.805 - 390.925]

Scenic waterway policy. [ORS 390.815]

Designation criteria: "...outstanding scenic, fish, wildlife, geological, botanical, historic, archeologic, and outdoor recreation values of present and future benefit to the public." [ORS 390.815]

Additional designations. [ORS 390.855 - 390.865]

Fish, wildlife and recreation declared highest and best water uses in state Scenic Waterways Act. [ORS 390.835(1)]

Dams, reservoirs, impoundments, placer mines and certain water diversion facilities prohibited on state scenic waterways. [ORS 390.835(1)]

Designations include stream and all land and tributaries within one-quarter mile. [ORS 390.805]

Permit required for any removal-fill activity on state scenic waterways.   [ORS 390.835(2)]

Limits on additional water rights issuance. [ORS 390.835(5) - (13)]

    3. Other Stream-Related Protections

Oregon Forest Practices Act: restrictions on forest operations in riparian zones and other areas. [ORS 527.676]

Willamette River Greenway. [ORS 390.310 - 390.368]

Livestock diversions with enclosed delivery systems and automatic shut-off devices exempt from permit requirements. [ORS 537.141]

    4. Watershed Protection

Watershed enhancement program, generally. [ORS 541.351-541.401]

Oregon Watershed Enhancement Board. [ORS 541.360]

Watershed councils defined; duties listed. [ORS 541.384; ORS 541.388]

Soil and Water Conservation Districts charged with conserving and renewing natural resources; implement projects and prevention and control measures; manage projects to prevent floods or develop water resources. [ORS 568.225; ORS 568.550-.552]

Reclamation of Mined Lands Act. [ORS 517.750–517.951]

Municipalities authorized to pass ordinances defining "offenses against the purity of the water supply" and appoint police. [ORS 448.295–448.320]

Major Water Management Responsibilities and Regulations By Agency

This appendix lists major (but definitely not all) water management functions and agencies responsible for each. (The annotated list of water management statutes in Appendix A also may be useful in determining authorities and duties.) Given the complexity and reach of Oregon's water management workings, any list of this nature is bound to be incomplete. It represents a "states' eye view" of water management, and thus places the activities of state agencies in the foreground, and others in the background. Only the most important administrative rules of the most important activities (in the author's estimation) are displayed. Readers should be aware that rules are subject to significant change over time. For the most recent version of rules and their content, consultation with individual agencies or with the Secretary of State's Office (responsible for maintaining the state's body of rule) is recommended.

The appendix is organized by the following topic categories:

Water Use Permitting and Approval
Water Allocation
Water Supply Planning
Water Use Requests for Public Purposes
Water Use Monitoring & Enforcement
Water Quality Control
Water Information Gathering and Analysis
Water Storage Facility Management
Water Service Delivery Watershed and Streambank Management
Water-related Habitat Management
Other

Abbreviations (federal agencies italicized):

BLM	*Bureau of Land Management*
BOR	*Bureau of Reclamation*
DEQ	Department of Environmental Quality
DLCD	Department of Land Conservation and Development
DOGAMI	Department of Geology and Mineral Industries
DSL	Department of State Lands
EFSC	Energy Facility Siting Council
FERC	*Federal Energy Regulatory Commission*
NMFS	*National Marine Fisheries Service*
NPS	*National Park Service*
NRCS	*National Resource Conservation Service*
NWS	*National Weather Service*
NPCC	Northwest Power Conservation Council
ODA	Oregon Department of Agriculture
ODF	Oregon Department of Forestry
ODFW	Oregon Department of Fish and Wildlife
OEM	Office of Emergency Management
OPRD	Oregon Parks and Recreation Department
OWEB	Oregon Watershed Enhancement Board
OWRD	Oregon Water Resources Department
PUC	Public Utility Commission
SWCDs	Soil and Water Conservation Districts
USACE	*U.S. Army Corps of Engineers*
USFS	*U.S. Forest Service*
USFWS	*U.S. Fish and Wildlife Service*
USGS	*U.S. Geological Survey*

Activity	Agency	Rules[1] [Chapter: -Division(s)]
Water Use Permitting and Approval		
Surface water and groundwater permits & certificates	OWRD	690: -003, -033, -077, -082, -310, -315, -320, -330
Water right transfers	OWRD	690: -380, 385
Miscellaneous water use authorizations	OWRD	690: -340
Low temperature geothermal water-use approval	OWRD	690: -230
"High temperature" geothermal water-use approval	DOGAMI, DSL	632: -20; 141: -75
Hydroelectric licensing and siting	FERC, EFSC, OWRD, DSL	690: -051; -052; -053
Reservations for economic development: approval	OWRD	690: -079
Use of Conserved Water Program	OWRD	690: -018
Out-of-basin diversions	OWRD	690: -012
Water right cancellation	OWRD	690: -017
Water Allocation		
Surface water and groundwater classification and withdrawal	OWRD	690: -010, -080, -082, -500 through -520
Determination of hydraulic connectivity between surface water and groundwater	OWRD	690: -009
Adjudication of vested water rights and determination of reserved water rights	OWRD	690: -028, -029, -030
Water Supply Planning		
Statewide	OWRD, DLCD	690: -400 & -410
Regional and local	Counties, special districts, cities	
Water management (conservation) plan approval	OWRD	690: -086
Water Use Requests for Public Purposes		
Instream water right requests	DEQ, ODFW, OPRD	690: -077
Reservations for economic development: requests	Any state agency	690: -079
Water Use Monitoring & Enforcement		
Distribution	OWRD	690: -225 & -250
Waste prevention	OWRD	690: -250
Determination of groundwater use interference with surface water	OWRD	690: -009

Activity	Agency	Rules[1] [Chapter: -Division(s)]
Civil penalties for violation	OWRD	690: -225, -240, -260
Water use measurement/ metering	Various special districts and cities	
Conservation assistance	OWRD, OWEB, SWCDs, OSU Extension, BOR, NRCS	

Water Quality Control

General surface water and groundwater quality authority	DEQ	340: -40, -41, -45, -51
Civil penalties for violation	DEQ	340: -12
Confined animal feeding operations & agricultural management plans	ODA	603: -74 & -90
Forest practices	ODF	629: -635 through -660
Drinking water purity	Health Division	333: -61
Well construction and abandonment	OWRD	690: -200, -205, -210, -215, -220, -230, -240
Land use plans, Goal 6 provisions to prevent pollution	DLCD, counties, cities, special districts	

Water-related Recreation Management

State scenic waterway management	OPRD, DSL, OWRD	736: -40; 141: -100; 690: -310-260
Federal Wild and Scenic Rivers	USFS, BLM, NPS	
Boating	Marine Board	250: -21 & 30
Angling, fish management	ODFW	635: -07, -11, -500

Water Information Gathering and Analysis

Streamflow, groundwater levels	OWRD, USGS	
Streamflow forecasts	National River Forecast Center	
Surface water and groundwater quality sampling	DEQ, USGS	
Drinking water sampling	Health Division, utilities	
Precipitation measurement	OWRD, NRCS, NWS	
Precipitation mapping and analysis	Oregon Climate Service, NWS	

Activity	Agency	Rules[1] [Chapter: -Division(s)]
Water Storage Facility Management		
Dam safety	OWRD, BOR, USACE	690: -020
Operations	USACE, BOR, Irrigation Districts, municipal water providers, utilities	
Water Service Delivery		
Urban	cities, water authorities, special districts	
Agriculture	irrigation, water control, water improvement, drainage, diking districts	
Coordination/carrying capacity	DLCD, counties, cities, special districts	660: -11
Rates	PUC	
Watershed and Streambank Management		
State Scenic Waterway Program	OPRD, DSL, OWRD	736:-40
Federal Wild & Scenic River Program	USFS, BLM, NPS	
Fill and Removal	DSL	141: -85, -89, -102
Wetlands	DSL, USACE	141: -85, -86, -120
Navigable waterways	DSL	141: -81
Watershed management	GWEB, local watershed councils, USFS, BLM	695: -20
Water-related Habitat Management		
Fish passage & screening	ODFW	
State Endangered Species Act	ODFW	635: -100
Additional public interest standards for new water appropriations	OWRD	690: -033
Federal Endangered Species Act	NMFS , USFWS, NPCC	

Activity	Agency	Rules[1] [Chapter: -Division(s)]
Wetlands	DSL	141: -85
Land use plans, Goal 5	DLCD, counties, cities, special districts	660: -16 & -23

Other

Drought emergency	OEM, OWRD	690: -019
Flood	OEM, River Forecast Center, USACE	
Weather modification permits	ODA	
Water efficient plumbing	Building Code Agency	918: -750 & -770

[1] Controlling Oregon Administrative Rules for state agencies

Glossary

ABANDONMENT. The act of voluntarily giving up a water right. *See also* FORFEITURE and CANCELLATION.

ACRE-FOOT. A volumetric measure of water: the amount of water needed to cover an acre of land one foot deep.

ADJUDICATION. The determination made by a Circuit Court, based on information provided by the Water Resources Department, regarding a claim of pre-1909 surface water use or pre-1955 groundwater use. The determination is set out in a court decree recording the amount, type, and location of water uses existing before the adoption of Oregon's Water Code. Adjudication is the process of "grand fathering" old uses.

ALLOCATION. Generally, the amount of water legally authorized by the state for use by a party or use-sector. In certain contexts, it may be used synonymously with "appropriation."

APPLICATION. 1. A request made by a party for permission to use public water in a specific amount for a specific purpose. Applications may be for surface water, groundwater, or storage permits, as well as for transfers, limited licenses, and other water use authorizations. 2. Also describes the act of delivering water to a place of use.

APPROPRIATION. Specifically, the act of taking physical control of public water for application to a beneficial use. Generally, the term may refer to the state authorizing use of public water. *See also* OVER-APPROPRIATION.

APPURTENANT. The attachment of a water right to the place of its use; applies most often to uses which apply water to (usually surveyed) expanses of land, such as irrigated fields or nurseries. Water rights held by cities or irrigation districts are roughly appurtenant, but are not held to the same specificity.

AQUIFER. "A water-bearing body of naturally occurring earth materials that is sufficiently permeable to yield usable quantities of water to wells and/or springs." [OAR 690-08-001]

AQUIFER STORAGE AND RECOVERY (ASR). A form of underground water storage in which surplus (usually winter) water is pumped through injection wells into aquifers, where it is held by the local rock formation until pumped back up during times of need.

ASSIGNMENT. The act of transferring interest in an application, permit or license from one holder to another, not binding unless filed with the Water Resources Department. Assignments cannot change the terms of any application, permit, or license. Assignments are not required, nor are they especially common.

BASIN. An area, defined by a perimeter of higher-elevation landforms such as ridges or mountains, that captures all precipitation falling on it. The source area for rivers and streams. A basin may broken down into multiple sub-basins. A "closed" basin is one in which captured surface water has no outlet to the sea. Synonyms: watershed, drainage, catchment. The State of Oregon recognizes 18 major river basins for water management purposes.

BASIN PROGRAM. A division of state administrative rules which set down future allowable water uses in, and other water regulations specific to, a given river basin. Sometimes called "basin plan."

BENEFICIAL USE. A prerequisite of obtaining permission to use the public's water. "Beneficial use shall be the basis, the measure and the limit of all rights to the use of water in this state." [ORS 540.610] The term is not defined in law; however, ORS 536.300 offers examples of uses which by their listing are declared beneficial.

CANCELLATION. Most frequently, an order of the Water Resources Commission de-authorizing use of water under a water right certificate that has been forfeited for non-use (*see* FORFEITURE).

CERTIFICATE. Short for "water right certificate," the final form of water use approval. The certificate confirms that the earlier form of permission to use public water, the water use permit, has been exercised in accordance with its terms. "Certificated" means having been issued a certificate of water right—often used in counter-distinction to permits or rights established through the adjudication process.

CERTIFIED WATER RIGHT EXAMINER (CWRE). A person certified by the State Board of Engineering Examiners based on criteria established by the Water Resources Commission. CWREs are allowed to conduct surveys of water used under a permit. The person must first be a registered, professional surveyor or engineer, or a geologist specializing in certain areas of expertise. Some CWREs also act as general water right consultants and help applicants prepare water use requests.

CIRCUIT COURT. According to the Oregon Blue Book, circuit courts are the state trial courts of general jurisdiction. ORS 536.075(1) states that any party affected by an order of the Water Resources Commission or Department that was issued without a contested case hearing may appeal the order to the Circuit Court of Marion County or local circuit court. Examples of such orders can include certain watermaster directives to water users or commission withdrawals of water bodies to further appropriation. There are 94 circuit court judges representing 22 judicial districts in Oregon. As local elected officials, judges often reflect local values in water resource management, often to the dismay of state officials. Petitioning the circuit court (or the Court of Appeals) results in an automatic stay of any department order. The department or commission can overcome the stay by finding in writing that substantial public harm will result from the stay. (*Compare with* COURT OF APPEALS)

CIVIL PENALTY. A fine; a monetary penalty imposed by the state on persons found to have violated law, rules, regulations or lawful orders of a responsible official.

CLASSIFICATION. The authority granted the Water Resources Commission by the Legislature [ORS 536.340] to specify which types of water use will be considered beneficial uses on specific water bodies. Classifications, which are expressed through administrative rules in basin programs, may apply to surface water or groundwater.

COMMINGLING. The mixing of groundwater among subsurface water-bearing zones. A common occurrence in basalt aquifers, where wells are constructed to drain water from each layer they penetrate so that it collects in the well bore. Commingling can take water away from less-deep wells relying on a water-bearing zone being drained.

CONSERVATION. Generally, the practice of using the least amount of water necessary to produce a benefit. Specifically, within the context of Oregon's Use of Conserved Water Program [ORS 537.455]: "eliminating waste or otherwise improving efficiency in the use of water while satisfying beneficial uses by modifying the technology or method for diverting, transporting, applying or recovering the water, by changing management of water use, or by implementing other measures."

CONSERVED WATER, USE OF. An ability granted water right holders by the state to retain and use elsewhere water they have conserved. Under ORS 537.455 through ORS 537.500, the State of Oregon is unique in promoting conservation by allowing water right holders to keep a certain proportion of conserved water, defined as the difference between the amount stated on their right or whatever they are actually able to divert (whichever is smaller) and the bottom-line amount needed to serve the beneficial use under the right.

CONSUMPTIVE USE. Any water use in which the quantity of water returned to the original source after utilization is diminished by loss during transport (such as leaky canals), evaporation (such as from sprinkler heads on hot, windy days or from water transpired by growing plants) or incorporation into a product (such as a watermelon).

Although technically speaking nearly all uses consume water, the term is usually reserved for uses characterized by significant consumption such as irrigation and municipal uses.

CONTESTED CASE HEARING. An administrative proceeding undertaken to determine the facts of a case or to apply agency policies to a particular situation. A trial-like process where opposing sides present their arguments to the presiding officer, called a hearings referee or an administrative law judge. Just as in a trial, witnesses are sworn, testimony is taken, and there is cross-examination. The hearings officer's decision is expressed through a written order. Contested case hearings should be distinguished from rule-making hearings or other public hearings which operate under a different set of procedures.

COURT OF APPEALS. The state Court of Appeals is a ten-judge body having jurisdiction over most civil and criminal appeals. Its members are elected on a statewide, nonpartisan basis for six year terms. ORS 536.075(2) states that any party affected by an order of the commission or department issued after a contested case hearing may appeal the order to the Court of Appeals. Petitioning the Court of Appeals (or the circuit court) results in an automatic stay of any department order. The department or commission can overcome the stay by finding in writing that substantial public harm will result from the stay. Because of its statewide nature, the Court of Appeals is often perceived as being more likely to find in favor of state agency actions than local courts. Accordingly, state agencies often select legal strategies that result in decisions that fall outside the jurisdiction of local courts.

CRITICAL GROUNDWATER AREA. An area designated under a rule of the Water Resources Commission as needing corrective action to remedy serious groundwater problems, including excessive water table declines, imminent overdraw of the groundwater supply,

substantial interference among wells, polluted groundwater and more, as provided in ORS 537.730. Any controls placed on previously established legal water uses are set through a contested case hearing.

CUBIC FOOT PER SECOND (CFS). The flow of a water body expressed in equivalents of the volume of water contained in a cube measuring 1 foot by 1 foot by 1 foot that passes by a given point every second.

DAM SAFETY. A program under which the Water Resources Department must approve the siting and plans of most dams, dikes, or other such structures that, in the event of failure, could damage life or property. [ORS 540.350–.400]

DISCHARGE. 1. The flow of a stream at a certain point; the amount of water produced by a watershed. 2. The process of groundwater entering or generating surface water sources.

DISTRIBUTION. The process by which a state watermaster assures senior water rights are satisfied by requiring junior water right holders to reduce or cease water use.

DISTRICTS. Bodies organized under specific Oregon statutes to deliver certain services, including providing water. Water providers can include cities, water associations, irrigation districts, water improvement districts, and many others formed in accordance with ORS Chapters 225, 264, 541, 545, 547, 551, 552, and 553. "Special districts," which include a wide range of water districts, are defined and given specific land use planning duties by ORS Chapter 197.

DROUGHT. A climatic period characterized by precipitation significantly lower than normal. Under ORS 536.740, the Governor may declare a "severe, continuing drought," thus giving the Water Resources Commission certain authorities to allow emergency water uses.

DUTY. The total volume of water authorized for use during an irrigation season, typically expressed in acre-feet. *Compare to* RATE.

ENFORCEMENT. The act of obtaining compliance with water law. Watermasters are the state's enforcement officers.

EXCHANGE. A substitution of one authorized water source for another in equal or reduced amounts. [ORS 540.533–.543]

EXEMPT USE. Any water use explicitly listed in statute as not being subject to the water right or any other approval process. Most frequently the term refers to relatively small groundwater uses (especially domestic groundwater uses of less than 15,000 gallons per day) as listed in ORS 537.545. Other exemptions include Salmon Trout Enhancement Projects, emergency fire fighting, and certain erosion-prevention practices.

EXTENSIONS. The additional time provided by the Water Resources Department to permit holders to make beneficial use of water.

FORFEITURE. The loss of a water right because of five successive years of non-use.

GAGING STATION (*also Gauging*). A point at which water levels (stage height) in streams are measured and recorded, usually for extended periods. By manually and regularly measuring stream depth and water velocity, a relationship is calculated between streamflow and stage height. Streamflow rates can then be inferred from recorded water levels.

GROUNDWATER. Generally, all water present below land surface. Defined in ORS 537.515 as: "any water, except capillary moisture, beneath the land surface or beneath the bed of any stream, lake, reservoir or other body of surface water within the boundaries of this state, whatever may be the geological formation or structure in which such water stands, flows, percolates or otherwise moves." However, where groundwater is immediately supplied by and behaves as surface water (*see* HYDRAULIC CONNECTION), it is frequently managed as surface water.

GROUNDWATER LIMITED AREA. An area established by rule in a Water Resources Commission basin program where a restrictive classification limits future groundwater uses. The classification is usually in response to evidence of limited groundwater supplies, including declining water tables.

GROUNDWATER MANAGEMENT AREA. An area established by the Environmental Quality Commission under ORS 468b.150–468b.190 in response to groundwater contamination problems. Designation is triggered by specific contaminant levels and results in a plan to prevent additional, and reduce existing, pollution.

HYDRAULIC CONNECTION. A condition in which water can move freely between a surface source and an aquifer. The term usually is used when groundwater is in immediate proximity (either geographic or temporal) to surface water and is subject to the same influences as the surface water body. Groundwater that is hydraulically connected to surface water may be managed in accordance with surface water regulations.

INCHOATE RIGHT. An "in progress" water right that had diversion works under construction when Oregon enacted its Water Code in 1909 and granted "grandfathering" potential should it be completed.

INJURY. An statutorily undefined adverse impact to existing water right holders that must be guarded against in issuing new water rights or approving transfers.

INSTREAM USE. Any use that supports benefits derived from keeping water flowing in-channel. Most often the term describes the public uses defined in ORS 537.332: recreation, pollution abatement, navigation. and an array of environmental purposes, including fish and wildlife preservation. Instream uses are eligible for protection through instream water rights.

INSTREAM WATER RIGHT. A water right for instream uses held in trust for the people of the state by the Water Resources Department. *See* INSTREAM USE.

JUNIOR USER. A holder of a water right with a more recent priority date than other water rights authorizing use from the same source. Junior users must stop using water when it is needed by holders of rights with older priority dates.

LICENSE. 1. Hydroelectric: A water use authorization given to non-municipal hydroelectric operators. Municipal hydroelectric projects operate under hydroelectric permits. Licenses have expiration dates, permits do not. The maximum term of a state license is fifty years. The federal government through the Federal Energy Regulatory Commission also issues hydroelectric licenses. 2. Limited: A temporary authorization to use water for a period not to exceed five years. Limited licenses are subordinate to all other authorized uses. 3. Well constructor's: A one- or five-year certification issued by the Water Resources Department to persons wishing to drill or otherwise construct wells and who have passed a written test.

MINIMUM PERENNIAL STREAMFLOW. An amount of water allocated by administrative rule to support fish, recreation, and water quality needs at a specific stream point or reach. As precursors to instream water rights, "minimum flows" were assigned a priority date and managed conjunctively with other water rights, except they could be waived. Oregon adopted over five hundred minimum flows, most of which have been converted to instream water rights, as required by ORS 537. 346.

NATURAL FLOW. 1. The streamflow generated by a watershed prior to any human use. 2. The flow of a stream minus any augmentation from water released by upstream reservoirs.

NON-CONSUMPTIVE USE. Any water use in which all the water utilized is returned undiminished to the original source, usually applied to hydroelectric projects and instream uses.

OTHER HOLES. Wells, borings, or excavations that allow sampling or measuring, or that expose, groundwater. Used primarily to distinguish standards (OAR Chapter 690, Division 240) for drinking water wells from those for sampling wells and other bore-holes.

OUT-OF-STREAM USE. A use which removes water from a stream, such as municipal or irrigation uses. Although often used synonymously with consumptive use, the two are not necessarily the same: ultimately some hydroelectric projects may return all water used back to a stream (i.e, be non-consumptive), but the points of uptake and return may be separated by considerable distance.

OVER-APPROPRIATION. Generally, the condition resulting from the state allowing more use than can be supplied by a particular water source. Specifically, "a condition of water allocation in which: (a) The quantity of surface water available during a specified period is not sufficient to meet the expected demands from all water rights at least 80% of the time during that period; or (b) The appropriation of groundwater resources by all water rights exceeds the average annual recharge to a groundwater source over the period of record or results in the further depletion of already over-appropriated surface waters." [OAR 690-400-010]

PERFECT. The act of completing the process of taking water and applying it to beneficial use in accordance with the terms of a permit. Perfection immediately precedes the issuance of a water right certificate.

PERMIT, WATER USE. A provisional authorization issued by the Oregon Water Resources Department allowing a party to divert or pump a specific amount of water for a specific use at a specific location. If water is put to beneficial use within the time period allowed by law and in accordance with any permit terms, the department finalizes the permission in the form of a water right certificate.

PLACE OF USE. The location specified in a permit or certificate where water is applied to beneficial use. For most water uses, field locations are surveyed and an official map recorded with the Oregon Water Resources Department.

POINT OF DIVERSION, APPROPRIATION. The location specified in a permit or certificate where water is taken up. For surface water rights, the term "point of diversion" is applied; for groundwater uses, "point of appropriation." For most water uses, diversion locations are surveyed and an official map recorded with the Oregon Water Resources Department.

PRIMARY RIGHT. Used mostly in counter-distinction to supplemental rights, the term refers to the original, older right underlying the lands also served by the back-up supplemental rights. (*See also* SUPPLEMENTAL RIGHT.)

PRIOR APPROPRIATON. A seniority system of allocation in which authorization to use water occurs through state permits (water rights) and in which the dates (priority dates) of water use requests control access to water in times of shortage. Older rights get first claim to water. (*Compare to* RIPARIAN RIGHTS.)

PUBLIC INTEREST. Oregonians' stake, collectively held by right of ownership in all state waters, that must be protected or advanced in all water management decisions.

RATE. A measure of the volume of water passing a certain point in a specific period, used to express streamflow or water uptake. In an irrigation right, the rate is the maximum instantaneous amount of water allowed for diversion. *Compare to* DUTY.

REACH. A segment, stretch, or other indeterminate length of stream.

RECHARGE. The return of water to an aquifer from rain, melting snow, other forms of surface seepage, or artificial injection. Annual recharge refers to the amount of water restored to the aquifer each year by natural means. Artificial recharge refers to the injection of water through wells either to restore declining groundwater levels or to store water for later use (*see* AQUIFER STORAGE AND RECOVERY).

RECLAIMED WATER. Water that has been diverted, used, collected and treated, usually by a municipality, and offered for another use. Usually this involves treated sewage water sold to entities needing an irrigation source (e.g., farms, golf courses).

REGISTRATION. A water use authorization characterized by either an automatic or simplified approval process. Registration procedures usually require the user to: provide notice of the use to the Water Resources Department, pay a nominal fee, and cease use if other water users are injured. Normally limited to small or historically established uses, registrations typically do not enjoy the legal status of a water use permit or right; holders do not receive a priority date and cannot make a call on water.

REGULATION. 1. The act of requiring junior users to diminish or cease their use to satisfy the demands of senior users. 2. The act of stopping illegal activities, including using water without a permit or in violation of the terms of a permit. 3. The effect on streamflow from draining or filling reservoirs.

RELICENSING. The process for determining whether, or under what conditions, to issue a new license for a hydroelectric project with an expiring license. The Federal Energy Regulatory Commission oversees a well-developed relicensing process for projects with federal hydro licenses.

RESERVATION. The assignment by the state of a priority date to a specific amount of unappropriated water from a particular source for a specific future use, as requested by another state agency for economic development purposes. The assignment reserves water as a source for future uses by placing a block of water off-limits to future appropriation for a certain amount of time. [ORS 537.356]

RESERVED RIGHTS. Rights to water held by federal or Native American tribal interests by virtue of establishment of federal reservations. When lands are reserved by the federal government (e.g., wildlife refuges, national forests, tribal lands), an amount of water sufficient to serve reservation purposes is also reserved. These rights, which may be reserved either explicitly or implicitly, often pre-date many rights issued by the state, but are seldom specifically identified or quantified—and therefore frequently the subject of negotiation and/or litigation.

RIPARIAN RIGHTS. An authorization to use water that is inherent, under some systems of appropriation, in ownership of streamside lands. Sometimes the right is limited to the amount of water that may be used without interfering with downstream landowners' uses. Common in the humid eastern U.S., rare in the arid West. *Compare to* PRIOR APPROPRIATION.

ROTATION. An agreement among water right holders whereby their rights to a common source are pooled for mutual and alternating use.

RULE. An administrative rule is "any agency directive, standard, regulation or statement of general applicability that implements, interprets or prescribes law or policy, or describes the procedure or practice requirements of any agency." [ORS 183.310] Agencies often adopt rules to interpret broad statutory policies or spell out specific procedures for their statutorily authorized programs. In the case of the Water Resources Department, by statute, every rule must be the subject of a public hearing [ORS 536.027].

SCENIC WATERWAYS. Streams or portions of streams protected by the State of Oregon because they "possess outstanding scenic, fish, wildlife, geological, botanical, historic, archeologic, and outdoor recreation values of present and future benefit to the public." [ORS Chapter 390] Dames are prohibited on scenic waterways and flows are protected from reduction from new upstream uses.

SECONDARY USE (Also "secondary permit."). The use of stored water or the authorization required to apply stored

water to a place of use. Oregon law maintains a distinction between the oft-coupled activities of impounding and then using stored water, by requiring two different permits.

SENIOR USER. A holder of a water right with an earlier priority date than other water rights authorizing use from the same source. Senior users are allowed to use water, in accordance with the terms of their water right or permit, before and to the exclusion of junior users.

SOURCE. The water body or type of water body tapped to support a beneficial use. The three major types of sources are surface water, groundwater and storage.

STORAGE. The act of impounding or accumulating water. Storage may be natural or artificial, and for the latter, either a primary or secondary objective. Although water is commonly stored to augment supplies for later use, a good deal of storage simply holds water back to prevent flood damage.

SUPPLEMENTAL WATER RIGHT. A right authorizing additional water to be applied to lands specified under a previous water right when the source identified in the previous right becomes exhausted during any given season of use.

SURFACE WATER. Water that collects on, issues from, or runs in channels on the land's surface, including ponds, lakes, springs, seeps, streams (both seasonal and permanent), and rivers.

TRANSFER. An authorization to change a water right's type of use, place of use, or point of uptake.

VESTED RIGHT. A right established by putting water to beneficial use prior to the establishment of state permitting systems. Such uses were "grandfathered-in" with the enactment of Oregon's 1909 Water Code, the 1927 eastern Oregon groundwater use law, and the 1955 Groundwater Act.

WASTE. A statutorily oft-referenced, but never defined, condition of water use: "the waters within this state belong to the public for use by the people for beneficial purposes without waste." [ORS 536.310(1)] Inferred to be the quantity of water in excess of the minimum needed to support a beneficial use authorized in a water right.

WATER AVAILABILITY. The potential of a water body to sustain additional use after considering existing water uses and water conditions. Expressed as the unappropriated flow in excess of the amount likely to be present 80% of the time, after accounting for out-of-stream and instream water rights.

WATERMASTER. A state officer appointed by the Oregon Water Resources Department to enforce the water laws of the state and perform other duties related to the measurement, monitoring, and management of state water resources.

WATERS OF THE STATE. For the purposes of water supply management, "any surface or ground waters located within or without this state and over which this state has sole or concurrent jurisdiction." [ORS 536.007] Definitions of this term for other purposes are found in ORS 196.800, 506.006, and 468b.005.

WELL. "Any artificial opening or artificially altered natural opening, however made, by which ground water is sought or through which ground water flows under natural pressure or is artificially withdrawn. " [ORS 537.515(9)]

WELL CONSTRUCTION. The act of creating direct access to groundwater, including by digging, drilling, driving, or jetting. Wells must be constructed to strict state standards.

WELL CONSTRUCTOR. A person who advertises well construction services, enters into contracts to construct wells, or operates well-drilling machinery and has obtained a state license to do so.

WELL LOG (*"water well report"*). A document that must be submitted within 30 days of completion for every water well that describes all rock formations encountered and all materials used in constructing the well. [OAR 690-205-080] Required since 1955, well logs also report initial water levels and well yields.

WILD AND SCENIC RIVER. Streams or portions of streams protected by the federal government under the 1968 Wild and Scenic Rivers Act because they possess "outstandingly remarkable scenic, recreational, geologic, fish and wildlife, historic, cultural or other similar values." The federal government cannot approve a dam or be involved in any project that adversely affects such a river.

WITHDRAW. 1. The act of closing waters to further appropriation, often used specifically to describe state orders of withdrawal issued under ORS 536.410. 2. The act of taking water out of a stream, lake, reservoir or aquifer.

Index